Contents

Computer Numerical Control of Machine Tools

G. E. Thyer

Second edition

 NEWNES

Newnes
An imprint of Butterworth-Heinemann Ltd
Linacre House, Jordan Hill, Oxford OX2 8DP

● PART OF REED INTERNATIONAL BOOKS

OXFORD   LONDON   GUILDFORD   BOSTON
MUNICH   NEW DELHI   SINGAPORE   SYDNEY
TOKYO   TORONTO   WELLINGTON

First published 1988
Second edition 1991

British Library Cataloguing in Publication Data
Thyer, G. E.
  Computer numerical control of machine tools.
  I. Title
  621.9

ISBN 0 7506 0119 1

Printed and bound in Great Britain by
Butler & Tanner Ltd, Frome.

Preface

Since this book was first published, computer numerical control of machine tools has continued to develop in a number of directions. These include the introduction of mill-turning centres and of turning centres with Y axis control, and new developments in FMS cells and the use of probes. This new edition deals with these and also provides more information on CNC grinding machines, turret punch presses and EDM spark erosion.

The chapter on computer-aided part programming has been extended and completely rewritten, and the updating amendments throughout the book include eighty-eight new illustrations which it is hoped will be found interesting and informative.

The main changes to the book are the following:

- Chapter 2 has been extended to provide information on the newer CNC machines and FMS cells.
- The sections in Chapter 3 dealing with tool holding and work holding have been greatly broadened and revised. The use of rotary spindle heads on turning centres, seen to be of great advantage, has added fuel to the discussion on the respective merits of lathe and milling machines.

- Additions have been made to Chapter 4 which it is hoped will provide a better understanding of the designation of axes.
- The introduction of in-process measurement on cylindrical grinding machines has required an extension of Chapter 5, which also deals with the principles of operation of touch trigger probes.
- Chapter 7 has been revised to meet the more general use of magnetic discs (floppy discs) as the storage medium; information on punched tapes is still included.
- Chapter 10 has been revised to contain more detail on the use of touch trigger probes.
- The biggest revision is to Chapter 13, which has been greatly extended to provide information on conversational programming and computer-aided machining, with examples.

The opportunity has also been taken to correct a few errors in the first edition.

I hope that the book will continue to be found useful in this new edition, and I should welcome advice on new topics that might usefully be included.

G. E. Thyer

Preface to the first edition

This book looks at computer numerical control from the viewpoint of the machine tool user rather than that of the computer operator or programmer. The evolution of numerical control is the latest result of a continuous development in component production since the first machine tools were made. The application of computers provides an extremely useful additional tool that allows the production engineer to use machine tools more economically for a very wide range of work. It must be remembered that although the computer is itself very powerful, the material removal techniques, cutting tools and machining operations used on NC machines are essentially no different from those used on manually operated or fixed cycle (mass production) machines. With the development of machining and turning centres, the computer has enabled multiple machining operations to be carried out simultaneously on small quantities of components.

The book is based on notes for lectures given to students ranging from sixth-form pupils to honours degree students. It is intended and hoped that students, at whatever level of study, will be able to obtain from the book the information on computerized numerical control that they need. BTEC students at levels NIII and NIV, and equally those following degree courses in universities or polytechnics, will find useful information in every chapter. Students on City & Guilds courses requiring part programming,

setting and operating information on CNC machine tools will probably be more interested in Chapters 7 to 12. The questions at the ends of chapters can be used for self-checking: the answers are to be found in the text of the chapter concerned. Three useful example programs are given in Chapters 11 and 12.

The book should also be useful for anyone in a training school involved with numerical control, and can be used in the retraining of technicians and skilled craftsmen or women who are being introduced to NC machines. Similarly, production managers needing to update their knowledge with the requirements for the efficient operation of CNC machines should find this a useful reference work.

Chapters 3 to 6 explain to the machine user the 'why' and 'how' of the use of certain constructional and instrumentation features developed for numerical control; they are not intended to be sufficiently detailed for the designer of machine systems. Various aspects of machining technology are introduced to remind readers of the necessary background knowledge, but do not cover all that is required. It is expected that readers will have studied or be concurrently studying manufacturing technology and undergoing practical training. In the author's opinion, it is essential for optimum utilization of CNC machine tools that everyone involved, particularly those engaged in part programming, should have a thorough understanding of material removal techniques and machining technology. When computer systems with stored technology become more economically viable and widely available, it will still be necessary for part programmers to use their judgement, based upon practical knowledge, in selecting methods of work holding, speeds, feeds etc. for some components.

There are a number of different systems of computer-aided programming and graphical numerical control, and the final chapter can only be regarded as a guide to how they are used. It is essential that the manual for the particular machine or system in use is studied to obtain the exact requirements, particularly in part programming. The only additional skill required in using a computer is keyboard skill; the knowledge of the various commands will be gained with practice, and in many cases prompts appear on the screen. The development of machines with numerical control capability has been the result of the work of different machine manufacturers, and consequently there is no one absolute or definitive machine tool or control system. The different manufacturers use

their own discretion, based on experience, in devising operating techniques, and thus it has been necessary to use the words 'generally', 'frequently', 'normally' etc.' a number of times in the book when explaining principles and techniques.

The section on punched tape in Chapter 7 is important; although the use of punched paper tape is decreasing, there are still control systems that use it. Moreover, from an educational viewpoint a punched tape is far more useful than a floppy disc, as it is possible for students to decode a section of the tape.

There has been a tremendous increase in the scope of CNC machine tools over a very short time, and I hope that the book will provide readers with a firm base on which they can build a fuller understanding of the subject. In such a wide coverage it is possible that there may be points of contention, and if so I will be pleased to be notified of any. In addition I would welcome advice on topics or chapters that would be more helpful if they were further developed.

G. E. Thyer

Acknowledgements

I am indebted to many persons, firms and organizations for help, advice and photographs used in the book. Sincere thanks are extended to the following who have been of particular help at various times.

The Principal and Governors of the South East London College for permission to write the book using my lecture notes
Mr C. Leach of Agie UK Ltd
Mr G. Davies of Anilam Electronics Corporation
Mr B. Hancock of Bridgeport Machines Ltd
Mr R. Ansell of Cincinnati Milacron UK
Mr P. Waller of Crawford Collets Ltd
Mr D. Collins of DBC Machine Tools Ltd
Mr L. Mustarde of Denford Machine Tools Ltd
Mr Banner of Dicksons (Engineering) Ltd
Mr D. P. Clark of Euchner UK Ltd
Mr D. Averill of Fanuc Europe, UK branch
Mr K. G. Pritchard of W. Frost Engineers (Coleshill) Ltd
Mr D. Stockton of George Kuikka Ltd
Mr W. D. Hogben of Graticules Ltd
Dr J. Liverton of Jones and Shipman plc
Mr H. Behrens and Mr N. Prescott of Heidenhain (GB) Ltd
Mr R. J. Lacey of Hightech Components Ltd
Mr M. Crabtree of J. C. Holt Ltd
Mr R. Ricketts of Marposs Ltd
Mr M. Powell of Matchmaker Machines Ltd
Mr M. Roberts of Mercury Ideographics for photographs of computer systems

Mr A. Stevens of Mills Marketing Services Ltd
Mr K. Rook of Monarch DS&G Ltd
Mr T. Newman of N. C. Engineering Ltd
Mr D. A. Goldsmith and Mr I. Robotham of Pathtrace Engineering Systems Ltd
Mr R. Rushbrooke of Pratt Burnerd International Ltd
Mr R. Barnes of Pullmax Ltd
Mr R. Jeffery of Sandvik Coromant UK Ltd
Mr M. Humphrey and Mr P. Williams of Renishaw Metrology Ltd
Stanmatic Precision Ltd, agents for Magnescales
Mr Rose of L. S. Starrett Ltd
Mr R. Eacott of Sumiden Hardmetal UK Ltd
Trimos-Sylvac Metrology Ltd
Mr M. Mathews of Unimatic Engineers Ltd

I would also like to thank a number of my colleagues, particularly:

Mr Tom Tebbutt of the Polytechnic of Central London and Dr Bera of the Polytechnic of the South Bank for reading the first draft of the book and making helpful suggestions.
Mr Stan Brion of SELTEC for the help and comments related to the electrical section in Chapter 3 and, finally, I thank my son Antony for the comments he made on Appendix C on computers.

Chapter One

Principles of machine tools

1.1 Material cutting techniques

The first cutting tool used by mankind is likely to have been a piece of broken bone or splintered flint. The deficiencies of these materials as cutting tools started a search to find better tool materials and more efficient cutting methods. This search is still going on: a new technique to cut materials is developed when the existing methods are not efficient or not quick enough.

The first and simplest cutting action would have been the splitting or dividing of materials using a chopping action with a thin wedge- or knife-shaped tool, as shown in Figure 1.1a. With this type of action and tool it is difficult to produce components to precise sizes or accurate shapes. The need arose for a cutting action where control could be maintained over the amount of material being removed. The shaving action shown in Figure 1.1b was developed, which makes possible the removal of thin sections of the material from selective areas requiring change of shape or size. With the shaving action the depth of cut can be accurately controlled.

The accuracy of the work produced by the cutting technique is entirely dependent on the relative movements of the work and the tool. It required the development of machines where the positional relationship and movements of the work and the tool could be controlled accurately for components to be produced to the

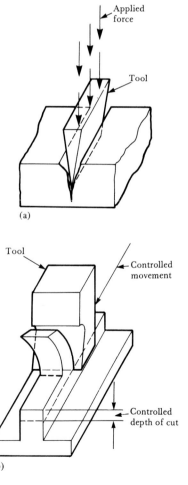

Figure 1.1(a) Dividing cutting action; (b) shaving cutting action

1

desired size and shape. The invention of the bow drill over six thousand years ago and a form of lathe some three to four thousand years later are evidence of the start of the development of machines for cutting material. The cutting tools used on these machines were made of copper or more probably bronze.

It was not until about 1772 when Wilkinson developed his boring machine that machine tools appeared which were capable of machining relatively large workpieces. Wilkinson's machine was powered by a water wheel and was used to bore the cylinders of the first steam engines. The cutting tools used were made of high-carbon steel suitably heat treated. In about 1792 Maudsley developed a screw cutting lathe with a slide rest, which was more efficient than any previous machine. Throughout the Industrial Revolution there was a great increase in the development of different types of machine tools. From the middle to the end of the nineteenth century came a whole range of new types of machine tools such as capstan lathes and turret lathes and cam-controlled machines with mechanical systems for automatic control. These machines were the first machines capable of the high rates of production required for the mass production of components.

The machine operating systems that have now been developed produce parts more cheaply and thus generally more quickly than ever before. This has been achieved with machine operating systems that have become more automatic in operation. The machine tools have control systems of varying complexity using mechancial, electromechanical, electrical or fluid power. *Numerical control* is the most recently evolved control system. On a numerically controlled machine tool, information on the shape of the component and on operating parameters such as the feed rate and the tools to be used is input to the machine in number form – hence the term 'numerical'.

Technological progress in many branches of engineering has resulted in the introduction of materials specially developed to withstand increased functional conditions such as high operating temperatures and high stresses. These enhanced mechanical and physical characteristics have created problems in the machining of these materials using the known machine tools, and alternative cutting techniques have had to be developed. This progress has continued and has resulted in the development of many different types of machine tools.

1.2 Material removal techniques on machine tools

Machine tools now use a variety of cutting tools manufactured from different materials, and have a range of material removal techniques such as:

(a) Cutting with a single-point metal tool
(b) Cutting with multipoint metal tools
(c) Cutting with abrasives
(d) Electrochemical machining
(e) Electric-discharge machining
(f) Ultrasonic impact machining
(g) Laser, plasma and electron beam machining.

Each of these techniques has particular applications for which it is the most suitable. The use of numerical control increases the efficiency of the functioning of most of the machine tools using these methods. The material removal techniques (d) to (g) can only be carried out on machine tools with special features or facilities.

The actual cutting process is exactly the same on a machine tool that uses numerical control as on machine tools that use a different control system but the same material removal technique. This is because numerical control is not a machine tool, but is a technique of *controlling the operation* of the machine. To improve the efficiency of operation of a machine, information obtained from the actual cutting process may be fed into the control unit where it can be used to control the rate of movement and other factors. However, for some machines the actual cutting process may not be directly linked with the numerical control system.

Numerical control ensures that the amount, direction and rate of movement of the tables on which the tool or work is mounted are accurate, repeatable and at the desired rate. In addition, numerical control is used for producing parts more quickly by controlling other machine operating features, such as:

(a) Selecting the cutting tool
(b) Turning the cutting fluid on and off
(c) Selecting and controlling the spindle speed.

It is extremely important that machine tools should not be considered merely as chip- or swarf-making machines. The criterion is the number of components produced, not the weight of chips resulting from the process.

Numerical control can be used for a wide range of other applications which require control of the operation of equipment, such as:

(a) Press tool work on turret presses
(b) Flame cutting on oxyacetylene machines
(c) Control of plotters
(d) Measurement and inspection equipment
(e) Control of robots.

1.3 Function of machine tools

The function of all machine tools regardless of the technique being used to remove material is to produce components of specified dimensional size, geometric form and desired surface texture which are not *economically* obtainable by other processes.

1.4 Size and form

The similarity of size of components produced is dependent on the *repeatability* and *constancy* of the work/tool positional relationship during the final cut. The accuracy or precision to which components are made is mainly dependent on the minimum *depth of cut* efficiently obtainable: if it is possible to remove a layer of material 0.001 mm thick, then corrections of that value can be made to the size of the workpiece. In order to produce cylinders on a lathe to a diameter tolerance of 0.01 mm it is necessary to reposition the tool and maintain its position relative to the axis of rotation of the work within 0.005 mm for the duration of the operation.

It will be found that the time required to position a tool manually is inversely proportional to the degree of accuracy required. The more accurate the work has to be, the longer is the time required to position the tool. It is in the speed of controlling, repeating and maintaining the work/tool positional relationship that numerical control has the greatest influence and can be used to the greatest advantage.

1.5 Kinematic principles of operation of machine tools

The geometric form of the workpiece is dependent on the relative movement of the work and/or tool during machining. There are three different kinematic principles which are used by machine tools to produce the required geometric forms: generation, copying and forming.

Generation
Under this principle, the work shape is produced as a result of the combined movements of the tool and/or work. The movements required are those which produce the parameters of the workpiece, which are generally *circles of revolution* and *straight lines*. Practically all components are composed of combinations of these two geometric shapes. It is convenient that only these shapes are used because, except when numerical techniques are used, it is difficult to control accurately the simultaneous movement of the tool or work in more than two directions.

The movements are controlled by the standard constructional features of the machine. Chapter 3 gives details of movement actuating mechanisms and slideways.

Generation of cylinders
A cylinder is composed of a circle and a straight line. There are four techniques of producing cylindrical forms, as shown in Figures 1.2a–d.

In Figure 1.2a the tool rotates and moves axially. This technique is suitable when the tool is small relative to the work; drilling is the most common example. Vertical boring machines also use this technique. The fact that the drill and boring tool produce internal cylinders does not change the principle involved.

3

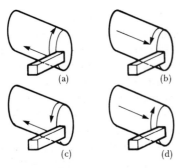

Figure 1.2 *Methods of generating cylinders*

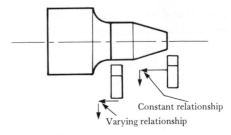

Figure 1.3 *Generation of forms requiring synchronized movements*

In Figure 1.2b the work rotates and moves axially. This technique is suitable when the work is small relative to the tool; cylindrical grinding is an example of this technique. Sliding head automatics also use this technique for small-diameter work.

In Figure 1.2c the work rotates and the tool moves axially. Turning on lathes is an example of this technique.

In Figure 1.2d the tool rotates and the work moves axially. This technique is suitable when the cylindrical form is required on a workpiece which would be difficult to rotate, such as the bearing housings on a lathe headstock. Horizontal boring machines can use this technique.

In these four methods it is comparatively easy to control the rotational movements using spindles rotating in bearings. The translational straight line movements are achieved using carriages guided by slideways. The roundness of the forms produced will be dependent on the precision of the rotation of the spindles in the bearings. The parallelism of the cylinders will be dependent on the accuracy of the alignment of the slideways with the axis of rotation of the main spindle.

To generate forms other than parallel cylinders it is necessary to synchronize axial and transverse movements, as shown in Figure 1.3. To produce the taper section of the component there has to be a constant uniform relationship between the axial and the radial movements of the tool. To produce the curved section the relationship between the axial and radial movements changes for different parts of the curve. At the start of the curve in a given time the axial movement has to be much greater than the radial movement, but at the end of the curve in the same time the radial movement has to be considerably faster than the axial movement. With the development of numerical control it is relatively easy to control the actual relationship of the tool and work movements, with the result that designers are able to specify complex shapes that can be machined comparatively easily.

Copying
Under this principle, the required work shapes are again produced as the result of the movements of the tool and/or work. However, the movements are now dependent on the passage of a stylus or tracer over a pattern which is the shape of the work.

Copying has been very effectively used for work which has a complex profile consisting of many changes in form or non-circular curves, such as multidiameter cylinders or cavities for moulds and dies. It is now more economical to use numerical control for this type of work. The shape of the work produced is dependent on the movements, which are controlled by the numerical information input to the control unit of the machine. There is no need to produce templates or patterns: when a firm has copying machine tools, numerically controlled machines can produce the templates required more economically than by traditional methods.

Forming
With this principle, the shape of the tool is the

reflected shape of the work. Examples of forming are shown in Figures 1.4a–d. Figure 1.4a shows an angle end mill. To produce this form on a numerically controlled machine, a standard end mill would probably be used with the spindle head being rotated under numerical control to the required angle; the flat section at the base of the angle would be produced as a separate operation. Figure 1.4b shows a form tool producing the end faces on a nut. Figure 1.4c shows a vee groove being formed; this could be an operation carried out on a numerically controlled machine. Figure 1.4d shows two form tools for producing a ball joint. The nut and the ball joint are likely to be required in extremely large quantities, and would justify the cost of producing the form tools required.

Machine tools have to be very rigid to use the forming principle for producing components, and generally the form has a maximum length of the order of 50 mm. Components can be produced more quickly when using the forming principle than when using the other principles, because the only movement required at feed rate is the movement of the tool to the depth of the form. However, the tools can be very expensive and can only be justified for large quantities.

There are very few applications of forming on numerically controlled machine tools. A typical operation is when using a ball-nosed end mill to produce a slot with a concave base. For lathe work it is easier to program a single-point tool to produce the various curves or angles required than to grind a special form tool.

1.6 Surface texture

The desired surface texture is mainly dependent on the tool geometry and the feed rate. In Figures

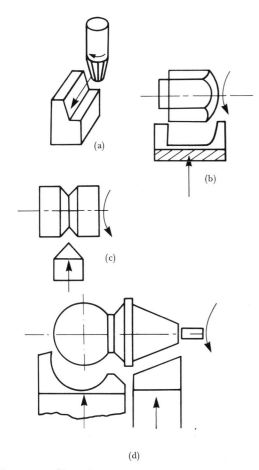

Figure 1.4 Forming operations

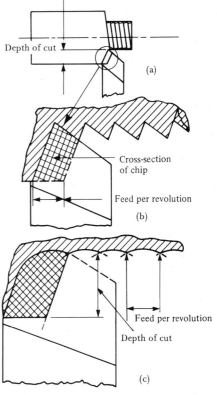

Figure 1.5 Surface texture produced with single-point tools

1.5a–c it can be seen that the shape of the tip of the tool is left on the workpiece during turning. To produce a smoother surface a larger radius can be ground on the tip of the tool. Figures 1.5a and b show the texture left on a turned surface· by a tool without a radius on the tip of the tool. Figure 1.5c shows the smoother texture left by a tool with a radius on the tip of the tool. The radius is usually of the order of 1–3 mm. Too big a radius can cause problems with the tip of the tool rubbing rather than cutting. Alternatively the value of the feed rate can be reduced, but this will increase the machining time.

When a smooth surface texture is a main requirement, it is necessary to employ grinding wheels or to carry out honing or lapping using abrasives as multipoint cutting tools. If the material being machined is either a copper or an aluminium alloy, a very smooth surface can be produced using a single-point diamond cutting tool on a special lathe.

Questions

1.1 Why is the shaving type of cutting action to be preferred to the dividing cutting action?

1.2 Name seven different material removal techniques.

1.3 To what does the term 'numerical' apply in the name 'numerical control'?

1.4 How does numerical control increase the efficiency of the functioning of machine tools using different material removal techniques?

1.5 What is the function or purpose of machine tools?

1.6 What are the kinematic principles involved in the operation of machine tools?

1.7 Which kinematic principle is normally employed on numerically controlled machine tools?

Chapter Two

Control of machine tools

2.1 Levels of control

Control of machine tools involves:

(a) The applying of power
(b) The guiding or limiting of the movement of the tool or work
(c) The selection of the rate of movement of the tool and/or work
(d) The selection of manufacturing facilities, the clamping of work or tool, and so on.

There are different levels of control of machine tools, ranging from fully manual to fully automatic control. It is extremely difficult to clearly define the difference between successive levels of control; one level blends into the next, and there are overlapping areas.

In order to establish a hierarchy of the varying levels of control, it is intended in this text to designate an increase in the level of control where there is a major change in the operational characteristics of the machine tool or the introduction of a new feature which considerably improves the capability of the machine.

2.2 Zero level of control

The first machines used for cutting were completely manually controlled. As mentioned in Chapter 1, the bow drill is reported to have been used over six thousand years ago, and tools of similar design are still used today in some parts of the world. In terms of control, the brace and bit used by carpenters today is comparable with the bow drill; the power required to remove material, the positioning of the cutting tool and the advancement of the tool are all completely manually controlled. The first treadle lathe was also completely manually powered. Again, the relative positions of the cutting tool and the work are completely dependent on manual dexterity (control).

Therefore it can be considered that for both the bow drill and the treadle lathe, since there is no external power source to be controlled and there are no mechanisms for controlling movement, there is *zero level of control*.

The few treadle lathes in existence today are used only as a hobby by model-makers.

2.3 First level of control

With the introduction of the water wheel, then the steam engine and later the electric motor to drive machine tools, the on/off control of power was obtained through the use of clutches or switches.

The on/off application of power is designated the *first level of control*. The simple pedestal drill or wood turner's powered lathe are machines using the first level of control. The ability to have different speeds with the use of stepped pulleys or gear trains does not change the level of control.

2.4 Second level of control

The development of Maudsley's screw cutting lathe with slide rest (around 1792) and similar machine tools created a new level of control. Maudsley's lathe was revolutionary in concept because for the first time the movements of the tool and work were synchronized, using gear trains to drive a screw and it is these that define the second level of control. The screw was referred to as the leading screw or leadscrew. The gear trains were driven from the main spindle on which the work was mounted, and a nut on the leadscrew caused the saddle to move. The axial and rotary movements were thus synchronized. The tool was clamped to a holder on a slide rest positioned on the saddle. The machine enabled screw threads to be produced which were considerably better than anything made before that time. Figure 2.1 shows Maudsley's original lathe. It is interesting to note that the headstock is on the right of the operating position. The driving pulley and gear trains linking the spindle to the leadscrew have been removed.

There is no fundamental difference in the level of control existing on Maudsley's lathe and present-day manually operated sliding, surfacing and screw cutting centre lathes. The work/tool relationship and the selection of speeds and feeds are still dependent on the expertise and interpretation of the machinist. There has obviously been an improvement in the material removal capability of the lathe through the use of motors with higher power. It is possible to produce work consistently to a much higher degree of precision or accuracy because of the greater accuracy of the leadscrews and the better quality of the materials of which the machines are made.

Horizontal and vertical milling machines which are manually operated come within the second level of control because the work/tool relationship and the choice of speeds and feeds

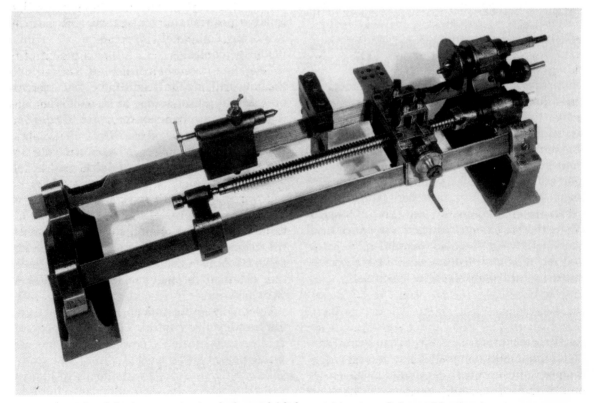

Figure 2.1 Maudsley's screw cutting lathe with slide rest (courtesy Science Museum)

are dependent on the discretion of the operator. Centre lathes or milling machines operated by a skilled machinist can produce very complex components. With this level of control the machines are economically suitable for producing components if:

(a) The shape is fairly simple.
(b) Only a small number of changes in the work/tool relationship are needed.
(c) Just a few components are required.

2.5 Third level of control

Machines which operate on a fixed cycle of movements introduce the *third level of control*. Cam-controlled automatic machine tools provide an example of this level of control. The shape of the work is 'contained' in the shape of the cam lobes. On lathe-type automatics the diameter of the work is dependent on the ability of the setter to position the tool in special holders mounted on turrets or special slides. Because of the time taken in setting machines with this level of control, they are likely to be the most economic only for producing large quantities of components (5000 plus).

Machines that have a 'plug board' for controlling the sequence of operations also come within this level of control. The size of the work produced on these machines is dependent on the setting of stops and tools. Microswitches, which have to be set with the stops, signal the end of operations so that the sequence can continue.

Tracer-controlled copy turning or milling machines are also within the third level of control. The work size is dependent on the depth of cut and the final distance between the tool point and the centre line of the work, both of which are under the control of the operator. It is not possible to determine the actual work size being produced until after the machining has been completed; that is, size control is 'post-process'.

For the continuous production of very large numbers of components requiring a wide range of milling and drilling operations, the most economic technique is with transfer machines.

Transfer machines consist mainly of individual milling, drilling and other types of tooling heads capable of carrying out all the machining operations required on one particular component, such as a cylinder block for a motor car. The work is mounted on pallets which are automatically moved (transferred) between the machining heads. All the machining heads operate on a fixed cycle of movements. The tool movements are activated either by cams or by screw and nut, driven by independent electric motors. The machining operations on all the heads are designed so that they take approximately the same time. There are transfer machines that are over 100 metres long with 100 machining stations arranged either side of the line.

2.6 Fourth level of control

In the *fourth level of control* the actual size of the work being produced during machining is monitored with the use of measuring transducers. The work/tool relationship is controlled so that the work is produced to the desired size; that is, size control is 'in-process'.

For economic reasons, in-process control is mainly used when producing components to small dimensional tolerances on certain specialist machines with abrasive tools such as cylindrical or surface grinders and honing machine. The self-sharpening effect of the abrasives ensures that their cutting ability does not deteriorate. To allow for wear of the tool which may have occurred, it is necessary to adjust the position of the tool to produce components of the correct size. The size of the work is monitored with specialized measuring instruments. Information from these instruments is transmitted to the power units. The feed movement of the cutting tool is stopped automatically when work of the correct size is produced. Generally this technique is used for controlling only one dimension of the work at one setting.

2.7 Fifth level of control: numerical control

Numerical control uses the *fifth level of control*.

As indicated in Chapter 1, numerical control is the name given to a fundamental concept of control in which information for controlling the machine is input to the control unit of the machine in the form of symbolic numerical values. The numbers represent such features as dimensions of workpiece, tools required, coolant, spindle speed and feed rate.

Numerical control of machine tools is reported to have been first developed between 1947 and 1952 at the Massachusetts Institute of Technology, in conjunction with the Parsons Aircraft Corporation. The development of numerical control techniques is accredited to the need for producing very accurately sized and intricately shaped parts for aircraft, principally space vehicles. Because of the complex shapes, considerable time was being spent in ensuring that the work/tool relationship was correct before machining took place; this led to long manufacturing times and hence high costs. In order to reduce costs, attempts were made to control automatically the work/tool relationship. This resulted in the development of positional control using numerical principles on a Cincinnati Hydrotel vertical mill.

The evolvement of numerical control was only possible because of the development of a

Year	Machine tools	Computing facilities	Controls	Input media	Tool materials
1800	Constant speed facing lathe		Jacquard loom	Jacquard punched cards	
1810					
1820	Milling machine				
		Babbage difference engine		Morse code	
1840					
1850	Capstan lathes				
1860	Cam-actuated machines		Cams	Perforated paper rolls (pianos)	
1870					Alloy tool steel (mushet)
1880	Copy turning		Copy circuits	Photocells	
1890				Hollerith punched cards	High-speed steel 18%W, 4%Cr, 1%Va
1900					
1910					Stellite
1920	Keller die sinker				Tungsten carbide
1930					
1940		ENIAC valve computer	Servo mech (military)	Stencil (for holes)	
1950	First NC machine	EDSAC stored program computer Electronic hand calculator		Punched tape	Ceramics
1960	2nd generation NC	LSI transistors		Magnetic tape	
1970	3rd generation CNC	Transistors		Magnetic discs (floppies)	
		VLSI transistors			
1980					
1990		Artificial intelligence?			

Figure 2.2 *Developments in machine and related technologies since* AD *1800*

number of separate features in a variety of technical subject areas, namely:

(a) Machine tools
(b) Computing facilities
(c) Control systems
(d) Input media
(e) Cutting tool materials.

Figure 2.2 shows developments since AD 1800 in some of the subject areas involved. Before this date, developments in some of the important related areas were as follows:

BC	5000	bow drill
	1000	abacus-type counting beads
	1000	simple lathe
AD	1250	treadle lathe
	1568	mandrel lathe
	1617	Napier's calculating machine (introduction of decimal point)
	1650	rotating drums with pins to ring bells
	1700	fin-controlled windmill steering
	1740	carbon steel (Huntsman's crucible process)
	1772	Wilkinson's boring machine
	1788	Watt's centrifugal governor
	1792	Maudsley's screw cutting lathe

The time saving and hence the cost reduction to be gained from the introduction of numerical control has led to its rapid introduction into a wide range of manufacturing processes. National surveys have found that on manually controlled machine tools the average time on actual material removal is of the order of 40 per cent of the time spent in producing a component. The rest of the time is spent on work and tool manipulation, gauging and so on. The same surveys found that on some manually controlled machines producing very complex and accurate components, where the tool or work positioning has to be very critical, material removal may take as little as 10 per cent or less of the overall time. On numerically controlled machines the time spent on material removal is frequently in excess of 70 per cent of the floor-to-floor time (see Section 2.12).

First-generation numerical control

The control systems and machines used for numerical control have varying complexities. The first machines, which are frequently referred to as being of the first generation, had been previously designed for manual or fixed cycle operation. These machines had numerical control systems added, but only for numerical control on positioning the work relative to the tool. Considerable time was saved, but the operator had to select the tools, speeds and feeds. The machines were development models and very few are still in general use.

Second-generation numerical control: NC machine tools

Second-generation machines are those on which material removal occurs at the same time as control of the work/tool relationship. A number of special mechanical design features were developed, such as recirculating ball screws and hydrodynamic slideway bearings (see Chapter 3 for further details). The use of these items enabled the machine tools to be more reliable and to have a longer useful life. There are a number of these tools still in use, and they are frequently referred to as NC machines. They were also termed *tape-controlled machines*, because the information was stored on either punched tape or magnetic tape. (see Chapter 7). With the second-generation machines the information to control a single operation is input to the machine, and when that operation has been performed the information for the next operation is input. It is not possible to edit the information at the machine; any change in the program can only be carried out at the place where the program was originated, and a fresh program is then produced.

The original second-generation machines had only a very limited memory capacity, and one development that took place was the introduction of a buffer storage. With this provision, while one block of information is being carried out, the next block is being 'read' into the buffer from where it is transferred to the active section when required. One advantage to be gained from the use of the buffer is that, because the tool does not stop moving while the next block is read in from the tape, there are no tool marks generated while a tool dwells in contact with the work.

Third-generation numerical control: CNC and DNC machine tools

The development of computers has created the the third-generation machines; these machines are capable of an extended range of machining operations.

Machining centres

It has become customary to refer to the machines that are capable of milling, drilling, boring and tapping operations, and can work on more than one face of a component as *machining centres*. Figure 2.3 shows a CNC machining centre which has the facilities normally provided on these machines. The machining envelope is 810 by 660 by 660 mm on the X, Y and Z axes respectively. The machine has a moving column configuration with automatic tool changing facilities, the tools being stored in a chain-type magazine. There are two work tables; work can be loaded on one table, while a component is being machined on the other table. Using a rotary table, machining operations can be carried out on four faces of the work. Swarf removal units are also part of the machine. The control unit has a keyboard which can be used by the setter/operator for inputting a program directly into the control unit. It is possible to interface the control unit with a punched tape reader or a desk-top computer and to load or save part

Figure 2.3 A CNC machining centre (courtesy Cincinnati Milacron)

programs on magnetic discs. Other examples of machining centres will be found in Chapters 3 and 5.

Turning centres

Machines which are capable of a wider range of turning operations than are normally available on lathes are known as *turning centres.* There are now CNC turning centres available for work of all sizes. The following machines are examples of design for different ranges of work.

For machining small diameter work a sliding head type machine with guide bush support is to be preferred, because the bush supports the work and ensures the work does not deflect during machining. The principle of operation of a sliding head turning centre is shown in Figure 2.4. All movements, turret indexing, collet opening etc. are programmed, and at the start of each component the headstock, with its collet open retracts to the left a distance equal to the length of the component. The collet closes and grips the bar which protrudes beyond the collet face through a bush which provides support against cutting forces. The bore of the bush is the shape or diameter of the bar. The bush is mounted in a supporting bracket on which a tool turret slides radially towards the work. The tools mounted in the tool turret are fed forward so that work of required diameter can be machined. The headstock is moved at feed rates along the bed amounts equal to the lengths to be turned. If a taper or curved section is required the movements of the headstock along the bed and the radial movement of the tool turret are synchronized. When the headstock has moved a distance equal to the length of the component, the component is parted off. The turret positioned on the right of the machine is used for end-cutting tool operations such as drilling.

Sliding head turning centres are used for bar work of up to 32 mm diameter. One of the reasons why the sliding head principle is not used for bars of larger diameter is the weight of the bar, and also because bars above this diameter are very rigid and if there is the slightest bend or distortion in the bar problems arise when the bar is fed through the guide bush.

There are turning centres for small diameter work where the headstock is stationary and the bracket with the bush and transverse sliding tool turret moves longitudinally.

Mill-turning centres

CNC machines known as *mill-turning centres* are capable of what is termed *complete machining*. These machines have small power-driven rotary spindle heads mounted on the tool turret so that cross drilling and milling operations can be completed while the component is still mounted on the turning centre. The tool turret on the guide bush bracket of sliding head CNC machines can have revolving spindle tool heads

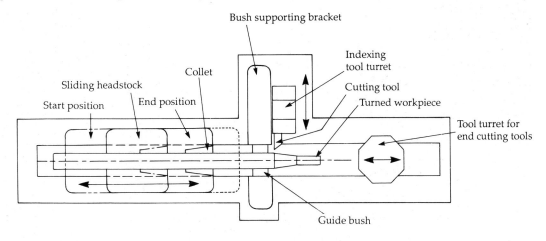

Figure 2.4   Sliding head turning centre

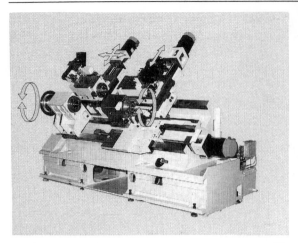

Figure 2.5   Mill-turning centre (courtesy Mills Marketing Services Ltd)

if required. A sliding head twin turret mill-turning centre is shown in Figure 4.6. Examples of work possible on a sliding head mill-turning centre are shown in Figure 2.23.

A slant bed, twin spindle, twin turret mill-turning centre with all the guards, control unit and turrets removed to show the essential construction is shown in Figure 2.5. The standard machine has the capability of machining bar work of up to 64 mm diameter by 300 mm long with automatic bar feeding. Additionally chuck held work of 250 mm maximum diameter can be machined. The various movements of the spindle and turrets are indicated in Figure 2.5. The left-hand headstock is fixed. The right-hand headstock can slide along the bed and can be moved in order to pick up parts directly from the left-hand headstock. The rotation of the spindles can be synchronized at the time of transfer to maintain the correct orientation of the work piece. The indexing of both spindles can be controlled from within the part program for drilling and milling operations on the end faces or radially or axially. The left-hand turret can move axially and radially, while the right-hand turret can only move radially relative to the work. An example of the sequencing of operations possible is shown in Figure 2.6. A vertical bed mill-turning centre is shown in Figure 2.7; the standard version of this centre is capable of

machining components of up to 356 mm diameter by 1066 mm long. Additionally bar work of up to 80 mm diameter with automatic bar feeding can be machined. The single turret has twelve stations, alternate stations being capable of providing power for driving rotary spindle heads. The main spindle can be indexed for milling and drilling operations. Work supported by a steady can be seen in Figure 2.7; the steady is fully programmable and the jaws automatically adjust to suit the work being machined. The tailstock can be positioned along the bed and can then be program controlled; the movement of the quill can also be program controlled. Work loading can be program controlled and Figure 2.8 shows a gantry loader in operation. The workpiece in the right-hand gripper has been machined and is being replaced with a work blank in the left-hand gripper.

The mill-turning centres shown in Figures 2.5 and 2.7 are typical of this type of machine in that they have movements that are parallel to the main spindle axis (Z axis) or at right angles to that axis (X axis). The turret on the CNC mill-turning centre shown in Figure 2.9 is capable of an additional vertical movement (Y axis) of ±55 mm at right angles to the Z and X axes. This machine is capable of turning components of up to 200 mm diameter by 400 mm length. The turret has twelve stations and rotary spindle heads can be mounted on any station. At the right-hand end of the machine there is a tool magazine which can hold twenty tools with an automatic tool changer capability.

More components are produced by milling, drilling and turning operations than any other cutting operations, consequently there are more CNC machines for these operations than other machining operations. However, as stated in Chapter 1, numerical control is a technique of controlling the operation of a machine, and there are many other types of machine tool applications where computer numerical control has proved to be very effective.

Sheet metal working
One application is on machines for cutting and pressing sheet metal. Figure 2.10 shows a CNC

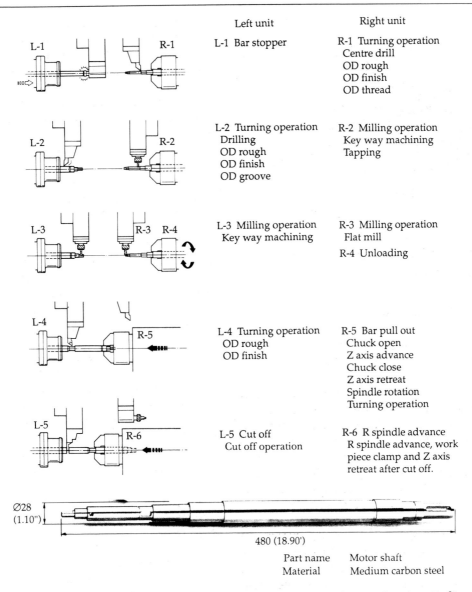

	Left unit	Right unit
L-1 R-1	L-1 Bar stopper	R-1 Turning operation Centre drill OD rough OD finish OD thread
L-2 R-2	L-2 Turning operation Drilling OD rough OD finish OD groove	R-2 Milling operation Key way machining Tapping
L-3 R-3 R-4	L-3 Milling operation Key way machining	R-3 Milling operation Flat mill R-4 Unloading
L-4 R-5	L-4 Turning operation OD rough OD finish	R-5 Bar pull out Chuck open Z axis advance Chuck close Z axis retreat Spindle rotation Turning operation
L-5 R-6	L-5 Cut off Cut off operation	R-6 R spindle advance R spindle advance, work piece clamp and Z axis retreat after cut off.

Ø28 (1.10")

480 (18.90')

Part name	Motor shaft
Material	Medium carbon steel

Figure 2.6   Twin spindle machining example (courtesy Mills Marketing Services Ltd)

punch press. With this machine, one edge of the sheet being punched is held by grippers which move the plate over the coordinate table, to the required punching position. There is a model of this machine that can punch patterns of holes and complex internal and external shapes in sheet steel of 12 mm thickness. The complicated shapes are produced using a combination of different shaped punches and dies; together with a nibbling technique coupled with linear and circular interpolation movements of the sheet. The displacement of the punch during its cutting and return strokes is monitored. The positioning of the plate being punched is synchronized with the movement of the punch and movement of the work commences as soon as the punch has cleared the work. Positioning speeds of 40 m/min are possible. The table is driven by a rack and pinion mechanism with a rotary encoder for position monitoring. The

Figure 2.7   Vertical bed mill-turning centre (courtesy Monarch DS&G Ltd)

Figure 2.8   Gantry work loader (courtesy Monarch DS&G Ltd)

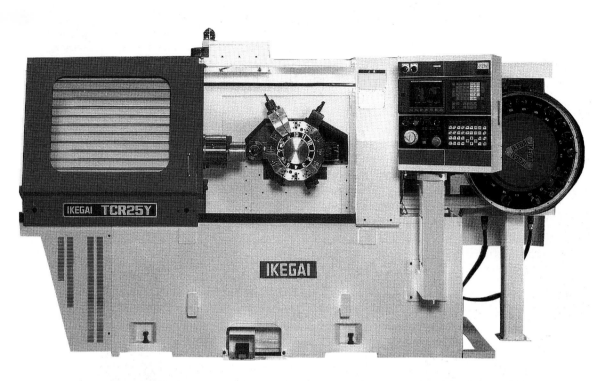

Figure 2.9   Mill-turning centre with Y axis control (courtesy N.C. Engineering Ltd)

Figure 2.10   CNC punch press (courtesy Pullmax Ltd)

speed of punching and nibbling is dependent on the material and thickness of the plate. Some examples of typical shapes that can be produced are shown in Figure 2.11.

Other cutting techniques using lasers, plasma arc and water jet are being used for cutting of sheet materials and tubes on CNC machines. The machines using these cutting techniques are essentially flat bed machines; normally the work is stationary and the cutting head moves along two rectangular axes. Flat work is located against end stops and rests on a series of beams which are replaced as required. Since there is no side force created during cutting the flat plate work does not have to be clamped. A CNC laser cutting and welding machine is shown in Figure 2.12. Various aspects of the laser beam are controlled. The laser is transmitted through the machine frame from the laser generator to the cutting head by mirrors.

Control is also maintained on the height of the cutting head and on the acceleration and deceleration of the speed of movement. Positioning

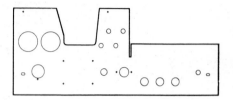

(a) Plate thickness 2 mm. Size of workpiece 555 × 230 mm. Number of tools 14. Number of strokes 70. Machine time 32 sec.

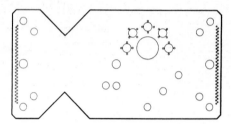

(b) Plate thickness 6 mm. Size of workpiece 1000 × 2000 mm. Number of tools 6. Number of strokes 354. Machine time 2 min 28 sec.

Figure 2.11   Examples of punch press work (courtesy Pullmax Ltd)

Figure 2.12   CNC laser cutting and welding machine (courtesy Pullmax Ltd)

(a)

Figure 2.13   Examples of the application of lasers (courtesy Pullmax Ltd)

(b)

(c)

and cutting speeds are of the order of 80 and 20 m/min respectively. Holes and openings in the walls of tubes can be cut; the tubes being held in a chuck mounted on the end of the bed the tube is supported with a type of tailstock. The indexing of the tube for positioning or during cutting can also be controlled. Figure 2.13 shows various applications of lasers for cutting. Figure 2.13a shows work being held in the end mounted chuck rotated through 90° – circular arcs can be cut by circular interpolation of the cutting head or by rotation of the chuck. Figure 2.13b shows a square tube held in an end mounted chuck – the rotation of the tube can be programmed to present different sides for cutting. Figure 2.13c shows helical slots being cut in a circular tube by a combination of linear interpolation of the head and synchronized rotational movement of the tube held in the end mounted chuck.

CNC grinding

Grinding is another machining process where CNC positional control (circular and linear path) is used very efficiently, and, in addition, control of process parameters such as rate of infeed, small changes in wheel and work speed and work traverse feed rate facilitates more efficient material removal. Figure 2.14 shows a CNC cylindrical grinding machine. The rate of infeed of a wheel during cylindrical grinding is normally quite slow and to ensure that a minimum of time is spent cutting air (advancing the grinding wheel into contact with the work) the wheel is advanced at rapid infeed rate and wheel/work contact is monitored by an acoustic transducer mounted on the tailstock. When the vibration set up as the wheel contacts the work is detected the programmed feed increment is activated and the grinding cycles are commenced. The reciprocation of the table is driven through a ball screw instead of the hydraulic cylinder and ram traditionally used on manually-operated grinding machines. To overcome the problems of wheel wear and thermal gradients on the CNC cylindrical grinding machine shown in Figure 2.14 the diameter of the work can be directly monitored with a calliper gauge as shown in Figure 2.15. The gauge is programmed to measure a particular diameter and when the desired size is reached, infeed stops and the datum of the wheelhead is reset so that other diameters of multi-diameter work pieces can be ground to increased accuracy. A particular shoulder position on multi-diameter work can also be directly monitored to ensure correct positioning of the longitudinal axis. A shoulder monitoring probe can be seen to the left of the calliper gauge in Figure 2.15. A fuller explanation of the use of calliper gauge for diameters and shoulder probes on CNC cylindrical grinding is given in Chapter 5.

An example of the sequence of movements of the wheel and work when grinding a multi-diameter shaft is shown in Figure 2.16.

CNC spark erosion

Machines removing material by spark erosion (EDM) using wire electrodes can utilize CNC positioning control (circular and linear path), feed rate control and also control of pressure and

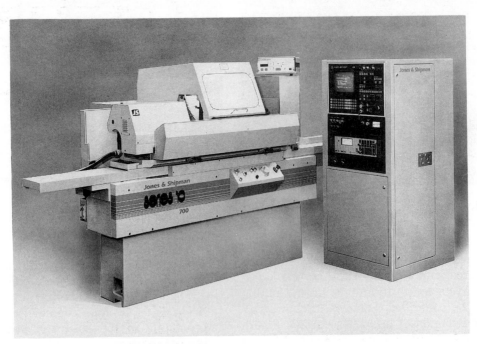

Figure 2.14   CNC cylindrical grinder (courtesy Jones and Shipman PLC)

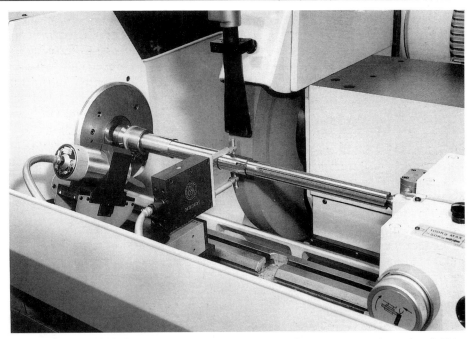

Figure 2.15   In-process gauging on CNC cylindrical grinding (courtesy Jones and Shipman PLC)

flow of the dielectric, with the conductivity of deionized water being measured and adjusted to suit conditions. CNC EDM machines using wire electrodes are very efficient – the wire electrode used by these machines is 0.1 to 0.3 mm diameter. Figure 2.17 shows a wire electrode machine with five axes control. The cutting zone is inside the cabinet in the middle of the machine on the right. The cabinet is required to prevent the dielectric from spraying and causing a hazard.

There is continuous development to improve the versatility of machine tools using computers, and two systems of computer control of machine tools have been designed: computer numerical control (CNC) and direct numerical control (DNC). These systems are described in the following sections.

Figure 2.16   Multi-diameter shaft (courtesy Jones and Shipman PLC)

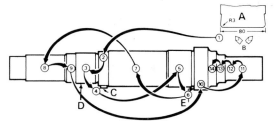

A Wheel   B Diamonds   C Steady   D Diameter gauge
E Shoulder location gauge

Sequence to achieve EP7 class bearing shaft tolerances on diameters 3 and 5

1	Profile dressing of wheel
2	Plunge grind steady $\varnothing$ with gap elimination (Program stop, apply steady)
3	Plunge grind with $\varnothing$ gauge
4	Shoulder grind with gap elimination
5	Plunge grind
6	Shoulder grind with gap elimination
7	Multiplunge grind
7	Transverse grind
8	Plunge grind
9	Plunge grind
10	Plunge grind
11	Plunge grind
12	Shoulder grind
13	Generate radius
14	Generate taper and blend radius

21

Figure 2.17   CNC wire cutting EDM machine (courtesy Agie (UK) Ltd)

2.8   Computer numerical control

There is no significant difference in the accuracy of the work produced on NC and CNC machines, because both machines use the same type of drive motors, carriage drive units (screws etc.) and monitoring systems etc. However, with CNC machines the control unit contains a dedicated computer which uses the data provided in the part program to control the machine tool. The complete program to produce a component is input and stored within the memory of the computer, and the computer can access information for calculation of work/tool movements. It is possible for cycles of movements (canned or fixed cycles) or repetitive movements (looping) to be activated with a single instruction. The reduction in the amount of input information is one of the big advantages of CNC. For example, only one line of input information is required for the drilling of 30 holes equally spaced on a pitch circle of 400 mm. On an NC machine, to drill this number of holes there would have to be 30 lines.

The information for each operation is fed from the control unit to the motors etc., when a signal indicating the completion of the previous operation is received by the control unit.

On some machines the memory of the computer can store a number of programs, and the particular program is selected as required by the setter/operator. With CNC machines it is possible to change (edit) any part of the program at the control unit. For future use a new punched tape can be produced at the machine or the amended program can be saved on a magnetic disc, tape cassette or other storage medium.

At one time it was considered that because the control was automatic, operating numerically controlled machines required a minimum of operator intervention and unskilled personnel could be used to operate the machine. However, to obtain efficient production with high utilization rates skilled setter/operators are regularly used which enable the facilities to be exploited to the full. CNC machines usually have the option for the manual override of programmed feed rates. The override is useful to move the work or tool at maximum feed for a 'dry run'. The override of programmed speeds and feeds is also an advantage if a particular work material is proving more difficult than expected and requires skilled judgement for its use.

With some CNC machines the setter can be inputting a new part program while machining is being carried out on a different component. Again on some machines, when a part program is being entered, the movements that would result after each block is entered can be checked by a graphical simulation shown on the visual display unit (VDU) contained on the control unit. When a program created at the machine has proved successful i.e. is verified, it should be stored on tape or disc for future use.

In the event of a tool failing it is possible, after replacing the tool, to move forward to the program block at the start of the operations where the previous tool failed, rather than having to repeat all the previous operations.

It is also possible with CNC systems to obtain information that is useful for management purposes such as:

(a) Number of components completed
(b) Number of different set-ups
(c) Time for setting up a particular job
(d) Time per component
(e) Time that material is actually being removed
(f) Time that a particular tool has been in use
(g) Time that the spindle has been running
(h) Time that a machine has not been working, and reasons
(i) Fault diagnosis for maintenance purposes
(j) Maximum load during machining.

As computer facilities become more available a firm will be able to develop a database of machining technology information from the experience gained during the use of the systems.

2.9 Direct numerical control (DNC)

DNC is a system containing a number of numerically controlled machines rather than a single machine. All the machines are linked to a mainframe computer which sends the information to the individual machines as required. There can be a number of different types of machine linked to the one mainframe computer. The computer is programmed to be able to select the order of manufacture of the components. DNC is capable of being integrated into the running of a complete factory, provided that the orders for equipment or parts are entered into the computer's memory, and that all the part programs have been written.

There are two main types of DNC: minimum cost or maximum flexibility.

Minimum cost With this system the facilities provided at each machine are the minimum required so that each machining operation can be completed. One problem with this system is that there could be delays in the main computer providing the details to the machine, while it is completing other work. One advantage of this system is that there is only one point of control.

Maximum flexibility This may also be referred to as satellite control. With this system each machine has its own dedicated computer, and the main computer provides the complete program for the component. The main computer is not actively involved while the machine is working. One big advantage of this system is that it is comparatively easy to link in machine tools and other plant. The machines can be used independently of the main computer. If the control unit of each machine has the facilities for conversational programming, the program can be transmitted to the main computer for future storage after it has been run successfully. It is also easy to edit programs at the machine in the event of some difficulty. Conversational programming will be explained in Chapter 13.

One of the first DNC systems was Molins System 24, developed in 1967, but the type of computers then available made the system extremely expensive. System 24 was a forerunner of what are now referred to as *flexible manufacturing systems* (FMS), and consisted of a number of specially developed machine tools. The machines provided all the machining functions required to produce quantities of components in small and medium batches. The components were clamped on to precision pallets which could be automatically loaded on to the machines by robots as required. A computer controlled the sequence in which the components were machined. The system was called System 24 because it was intended to be worked 24 hours a day, the loading of the pallets being carried out by manual labour during an 8 hour shift.

Flexible manufacturing cells

A complete factory-wide flexible manufacturing system is extremely expensive. A more economic application is a *flexible manufacturing cell.* The term flexible refers to the adaptability provided by the computer of economically machining a variety of single components or small batches of components to fulfil the production requirements of the firm. For minimum operator attendance a cell usually consists of:

(a) A computer with software for controlling the cell
(b) A CNC machining centre and a CNC turning centre (mill-turning centre) if required for the type of work being produced. These machines must be equipped with tool magazines, automatic tool changers (ATC) and preferably automatic tool setting facilities
(c) Automatic work loading and unloading facilities for machining centres such as an automatic pallet changer (APC) or automatic work carrying (AWC) vehicle; work changing facilities for turning centres such as automatic bar changing, robotic arm or gantry loaders. Preferably all the machines should have automatic work location setting and checking facilities
(d) A station for the removal of completed work and renewal of fresh work
(e) Automatic swarf removal.

The firm's production requirements will be entered and stored in the cell computer's memory, and the computer will decide on the order or sequence that the parts are to be machined. If there is more than one machine of the same type in the cell the most suitable machine to be used would be selected by the cell computer. The cell computer would send the required part program to the control unit of the machine tool selected upon receiving a signal from the machine when it has completed the previous component. If a cell contains a number of machining and turning centres, it is possible to machine a wide variety of work. If the items form part of an assembly it is good practice and economic to machine one-off of all the individual items required for one complete assembly, rather than produce batches of different components. This will ensure that all components are available when required and there will be no unnecessary waiting for batches of individual items to be machined, or needless use of expensive storage space and the associated record keeping. Obviously it is impractical to expect that all items for one assembly can be produced at exactly the same time but because of the dependability of CNC machines, using these machines can be extremely beneficial in meeting the requirements of the precepts of the 'just in time' (JIT) philosophy with respect to the machining aspects.

It can be seen that the control system of the CNC machine tools in an FMS cell must have the capability of accepting the changing of the part program for the component machined with a part program for a different component.

A work holder for machining centres in flexible manufacturing cells should have some method of being identified by automatic reading heads before the holder is loaded on to the machining station. This should ensure that the part program loaded into the control unit by the cell computer matches the work on the holder. One method of identification uses inductive element capsules similar to those used for tool identification explained in Chapter 3. The work is loaded on to the machining station automatically as also explained in Chapter 3.

Tool facilities

The tool magazines of the machine tools must store all the tools that will be required and the

Figure 2.18   Touch trigger probe checking tools (courtesy Monarch DS&G Ltd)

tools used must be either preset which can be time consuming; or preferably the machine should have facilities such as tool sensor probes for checking the presence, confirming the position and determining the offsets of the tools. A tool checking probe is mounted at a fixed location in the machining area on the machine. As will be explained in Chapter 5 the probe does not measure, but when contact is made between the probe and the tool or work a signal is sent to the control and the reading of the position transducers at which contact is made is stored within the control. A touch trigger probe which is being used to check tools on a turning centre is shown in Figure 2.18. For tool checking, a fixed (canned) cycle is used to program the advancing of the probe from a recess to a fixed datum position and each tool is moved in turn to contact the probe and establish its position in relation to the machine datum. During machining operations the probe is protected within the recess. The cycle can be activated before different components are machined or after a number

of the same components have been machined. See Chapter 10 for details of tool setting using probes on machining and turning centres. A probe for checking the presence and size of work is mounted in the lower right-hand quadrant of the turret shown in Figure 2.18. A probe for checking work is mounted in the tool holding unit (turret) and moves to contact the work; see Figure 3.44.

Work mounting

For machining centres in an FMS cell it has been practice for considerable time and care to be taken in clamping the work in a precise position on pallets or sub-tables which can be automatically loaded into precise positions in the machining area. However, using a work contact probe the work does not have to be positioned so precisely on pallets etc. as it is possible to determine the relationship of particular features of the work (datum edges, bores etc.) on the machine table to the machine datums. The probe for checking the work is mounted in the spindle nose and is programmed to move to contact the work. A probe in a horizontal spindle checking the position of a casting is shown in Figure 2.19. The probe contacts the casting datum faces at a number of positions and the control software uses the reading of the position transducers at each location to calculate the average values to determine the actual relationship of the workpiece datum faces to the machine datum. The difference between the desired and actual relationship is stored as an offset value. This offset value provides a compensation for the part program values. Thus the need for accurate location of the work on the pallets is reduced.

Probes can also be programmed to check the size of work produced on machining centres as shown in Figure 2.20. The probe makes contact at four positions on the work and the readings of the position transducers at each contact are used by the control to calculate the bore diameter. If the diameter is incorrect and can be rectified an additional machining operation is activated. A similar method could be used for checking the positioning of the workpiece for initial machining or second operation machining. A probe for tool checking can be seen on the left-hand corner of the work table in Figure 2.20.

The only human participation required for continuous operation of the cell, is the loading and unloading of the pallets on a work loading station, and general tool maintenance i.e. loading tools into the tool magazines. A flexible manufacturing cell in operation is shown in Figure 2.21. This cell contains the following:

(a) A computer which controls the cell
(b) Two T-10 machining centres with tool magazines and automatic tool changers
(c) Two rotary work loading facilities
(d) An automatic mobile work carrier
(e) Washing plant
(f) Automatic swarf removal.

In the right foreground is one of the eight station rotary work loading facilities. The station where the work has to be removed and

Figure 2.19   Probe checking position of casting (courtesy Renishaw Metrology and Avery Hardoll)

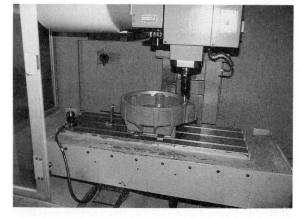

Figure 2.20   Probe checking size of bore (courtesy Renishaw Metrology and North West Engineering)

Figure 2.21 A flexible manufacturing cell with wire guided vehicle (courtesy Cincinnati Milacron)

loaded on to pallets or fixtures for this rotary work loader is in the centre foreground. The other eight station rotary work loading facility can be seen in the centre left. The correct pallet is automatically transferred from the loader on to the carrier on instructions from the control computer. An automatic mobile work carrier loaded with work blanks on a pallet is in the centre right. The path followed by the mobile work carrier is controlled by a wire embedded in the floor. The work carriers are also known as automatic guided vehicles (AGV). The machining centres are on the back right of the photograph. The boxed cabinet to the left of the machining centres is the washing plant. The cell control computer is on the centre left behind one of the work loaders.

A flexible manufacturing cell with a rail guided vehicle for transporting the loaded pallets to either of the two machining centres is shown in Figure 2.22. This type of installation can require a smaller floor area than the cell with AGV transportation. It must be noted that some of the guards have been removed for the purpose of taking the photograph. There are a minimum of twelve stations where the pallets are parked loaded with work ready for loading on to the machining centres as required. Six of the parking stations are on the front left of the installation. The other six are in the centre behind the

Figure 2.22   A flexible manufacturing cell with rail guided vehicle (RGV) (courtesy Cincinnati Milacron)

standing guards. The rail guided vehicle is between the pallet parking stations on the left. The station where the work is manually removed and loaded on to the pallets is next to the cell computer which is on the right edge of the figure. The double chain tool magazine and the automatic swarf removal unit for one of the machining centres are in the centre left of the cell.

2.10 Adaptive control

Adaptive control could herald the introduction of a sixth level of control. In adaptive control, sensors are mounted on the machine drive shafts and the tools. The sensors provide information on the cutting forces or temperatures generated during the actual material removal process. The information is analysed by the computer, which automatically adjusts the speeds and feeds during the actual cutting to maintain material removal at the highest rate of which the drive units (motors) fitted to the machine are capable. The machine is kept fully loaded as far as power consumption is concerned, but with protection against overloading. It is necessary to provide details such as:

(a) Maximum allowable forces exerted against the cutting tool
(b) Maximum permitted temperature at the tool tip
(c) Spindle torque.

The temperature generated during cutting is usually dependent on the cutting speed, and by measuring the temperature of the tool tip it is possible to maximize tool life by automatically varying the cutting speed to suit the conditions.

Another method, developed for use on turning centres, uses a force sensor placed under the tip of the tool. The feed rate is automatically adjusted until the optimum conditions are achieved. If the tool should fail the force will increase very rapidly, and this will cause the sensor to send a signal to switch off the main motor and apply a brake. The spindle will stop rotating within one revolution; this saves a possible serious incident.

Force transducers can also be fitted in the drive system of milling machines to monitor the force being created to remove material. The force transducers are used to provide feedback information. The rate of feed or possibly the depth of cut are adjusted automatically to maintain the maximum metal removal possible on the machine tool.

The control of diameters of ground components that have to be to precise sizes is achieved by measurement of the components in a temperature controlled area and feeding back information to the gauges controlling the infeed of the grinding wheel. The gauges are used for measuring the work-in-process. Chapter 5 gives more details of this method.

There could be a number of developments in adaptive control in the future, such as:

(a) Adaptive numerical control of surface finish. This would require the linking of a surface measuring instrument to the machine controls. As a component is produced it would be measured, and the feed rate would be automatically varied to produce the desired surface texture.
(b) Control of size by limiting the force on the tool during finishing operations to reduce the deflection of the tool.
(c) Reducing the vibrations (chatter) created during material removal by changing the dynamic characteristics with the use of controllable vibration dampers.

(d) Reducing the effects of alteration in the position of the tool tip due to thermal gradients existing in the machine during finishing operations. This would require in-process measurement of the workpiece.

The extra cost of the full implementation of adaptive control will only be justified by the advantages to be gained from the improved utilization of the bigger and more expensive machines.

2.11 Applications of numerically controlled machines

The decision as to whether a numerically controlled machine should be used for a particular application is mainly dependent on the cost of manufacture. A CNC machine can be very expensive and consequently the machine rate is usually quite high, but the preproduction time and set-up time can be quite low. (See Section 2.12 for definition of machine rate.) Other factors can also influence the selection of a CNC machine; such as when there is a shortage of the skilled personnel with specialized skills needed for setting and operating manually controlled or fixed cycle machine tools. As a consequence of the shortage of these personnel it may be more economic, for a firm which has only a few fixed cycle machines, to use CNC machines with higher machine costs rather than machines with higher labour costs. It must be noted that the mass production of large quantities of components which have fairly generous machining tolerances is more economic on cam-actuated fixed cycle or transfer machines than CNC machines. However, because of the accuracy obtainable with numerical control, it has been found that CNC machines can be economic to produce large quantities of components which have high quality standards such as pistons or work of up to 15 mm diameter. Another aspect to be considered is that since there is not the problem of designing and producing the specialized cams and form tools required for cam actuated lathes the developments in the design of CNC machines have made them more competitive for producing large quantities of the same component.

Numerical control is particularly suited for the manufacture of very small numbers of components needing a wide range of work, such as those with complex profiles or a large number of holes. It is possible that numerical control could be economic for only one component if a large number of tool or work positional changes were involved, particularly for components with close dimensional tolerances. CNC machines can be used very effectively for machining the cams required for the cam controlled automatics.

A single component requiring only one or two positional changes, and with large dimensional tolerances, could be produced more economically on a manually controlled machine than on a CNC machine. If special form tools were required for the manually operated machines then the numerically controlled machines would almost certainly be more economic. The important criterion is the amount of time spent in preproduction time, and in non-machining operations such as positioning the tools, compared with the machining time and overall time.

Once the part program has been written the setting up of CNC machines takes a minimum of time, which results in CNC machines being suitable for producing batches of small numbers. Because ideally, components should only be produced at the rate that they can be used; obviously this is not always practical. It must be remembered that large numbers of components held in store represent a sum of money which is not available until the components are sold or used. There is also the possibility that changes in design may result in finished components becoming obsolete, with the result that all manufacturing costs are lost.

The introduction of mill-turning centres which are capable of operations where there is a small amount of material removal such as the milling of flats or splines and cross drilling of components on the same machine and set up as

turning has further extended the applications of CNC machines. These mill-turning centres have reduced handling costs because most of the machining work is finished on one machine rather than having to set up a milling or drilling machine for second operation work. If the second operations are to be performed on manually operated machines a balance has to be achieved on the cost of using a machine with high machine rate against the handling time on high labour rate machines. If the second operations are to be performed on another CNC machine the use of the mill-turning centres will release the machines for operations where large volumes of material are to be removed. Figure 2.23 shows some examples of work machined per set up.

There are a number of reasons for the suitability of numerical control, such as:

(a) Faster production rates; less time spent on the positioning of tools or work; less work in progress.

(b) Quicker setting-up times, resulting in the viability of smaller batches of components. This fact, together with the shorter time spent on non-machining operations, means that the machine utilization is high.

(c) No jigs, specialized fixtures or special tools. The storage space previously required for the jigs and fixtures is available for production purposes. The skilled tool-makers can be used on setting numerically controlled machines.

(d) Reduced lead time, particularly with computer-aided graphical numerical control.

(e) Improved cutting efficiency; speeds and feeds are controlled.

(f) Improved quality of work. Components are more consistently accurate, resulting in less scrap and less reworking.

(g) Less inspection, because of improved quality. It is necessary that the first component produced from a new program is checked very carefully. Once the accuracy of the component has been confirmed, it is unlikely that other components will be faulty.

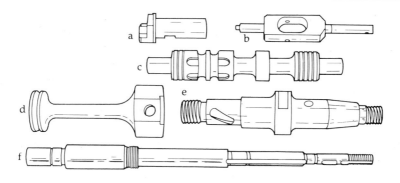

No.	Machining time (sec.)	Material
a	310	303 SST
b	420	AISI 12L13
c	410	303 SST
d	350	AISI 4135
e	320	303 SST
f	480	AISI 1035

Figure 2.23   Examples of work produced on mill-turning centres (courtesy N. C. Engineering Ltd)

The main possible faults are those created by tool wear.

(h) Less handling of components; more operations completed on one loading of the component.

(i) Ease of ability to change design of components, without expensive alterations to tooling equipment. The important alterations required would be achieved by the editing of the program.

2.12 Cost of manufacture

A major factor in selecting a particular machining process is that the cost of manufacture by that process is lower than any other.

There are two main types of costs: variable costs and fixed costs.

Variable costs   These may also be referred to as production costs or running costs. They are dependent on the number of components to be produced: the more components produced, the more the variable costs.

Fixed costs   These are the costs that have to be paid regardless of the number of components to be produced. They may be referred to as preproduction costs, lead time costs or capital costs. *Lead time* is the term used to describe the time leading up to actual manufacture. It is the time required to prepare tools and part programs; in some firms, set-up time may be included in lead time. If the work is to be produced in batches there will be set-up time costs each time a batch is produced. If the number of components to be produced is a fixed number then it is possible to allocate part of the fixed cost to each component. With these conditions, the greater the number of components to be produced, the smaller the fixed cost per component.

An example of the calculation of the various costs is given in Section 2.13.

Variable costs

There are four items in the variable cost of manufacturing a component:

(a) Material

(b) Plant or machine
(c) Labour
(d) Services or overheads.

In order to determine the manufacturing cost of a component it is necessary to calculate the cost of each of the above items for the component:

cost per component = material cost + machine cost + labour cost + overheads cost

Material cost

The material cost of a particular component is the same regardless of the machining process, and can be regarded as a constant for the component.

Machine and plant cost

Machine cost is the portion of the cost of the machine and equipment to be recovered on each component. The amount is normally dependent on the time the component takes to be produced, and only the machines etc. that are directly involved in the manufacture of the product should be considered. The term used to designate the time a component takes to be produced is *floor-to-floor time*. This is the time between the loading of one component ready for machining and the loading of the next component. The term 'floor-to-floor' comes from the action of picking up a component from a storage rack and loading it on to the machine, and then removing it after machining and placing it back in the storage space near the machine.

The following relationships are used in machine and plant costing:

machine cost per part = machine rate × floor-to-floor time (hours)

$$\text{machine rate} = \frac{\text{cost of new machine} + \text{insurance and maintenance}}{\text{estimated life of machine (hours)}}$$

$$\text{time per component} = \text{floor-to floor time} + \frac{\text{set-up time and tool reconditioning time}}{\text{number of components per set-up}}$$

Machine rule

In order to share the cost of the machine equally over all the components produced during its useful life, when calculating the machine rate it is customary to assume that there is linear depreciation in the value of the machine, as shown in Figure 2.24. The actual value of the machine follows the second curve in Figure 2.24.

The insurance and maintenance can be conveniently quoted as a percentage of the price of the machine. Insurance and maintenance are the minimum number of items considered; there could be other costs involved. Different firms have different values for different machines, but approximate values are insurance at 1.5 per cent and maintenance at 12 per cent of machine price.

Estimating the likely machine life is the critical factor in determining the machine rate. Past experience is the best guide, but frequently the machine tool manufacturers have evidence on which to base values.

Labour costs

Labour costs are the wages paid to the people directly involved in the manufacture of the component. The wages of the people concerned with the supervision or organization of the

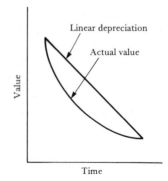

Figure 2.24   Depreciation of machines

manufacture are part of the overheads, and are often called salaries. The labour cost relationship is:

labour cost per component = labour rate per hour × floor-to-floor time per component

Overheads or services

It is difficult to calculate accurately how much of the wages of all the other various personnel in the firm, and the cost of the facilities provided, should be allocated to a component. In order that the costs are recovered it is standard practice to specify the overheads as a percentage of the major direct factor involved in the production of a component. Traditionally it is the productive labour cost to which the overheads are related; this is mainly because the labour cost is usually the largest and the most controllable cost. Where the machine costs are the largest single item, consideration is given to relating the overheads to the machine costs. The percentage for overheads varies considerably with different firms and range from 200 to 800 per cent.

Fixed costs

If computers and their software are used in producing part programs, the cost involved can be calculated in a similar way to the machine cost above. The difficulty is in knowing exactly how long it will take to produce a part program. An alternative method is to specify a sum for the creation of a part program, based on previous experience.

The cost of the tools is only for the special tools required for the component, and is commonly estimated.

The cost of setting up the machine can be calcualted from the amount of time taken and the cost per hour of setting up. The actual time allowed to set up can also be based on past experience.

2.13 Break-even charts

Break-even charts can help to decide which machine should be used to produce a particular component. A typical break-even chart is shown in Figure 2.25. The chart shows the comparison of only two machines, but the same technique would be used when comparing a number of machines (as in the example to follow).

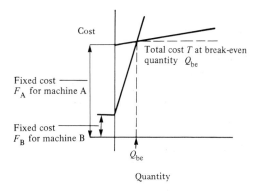

Figure 2.25 Break-even chart for two machines

The quantities of components are plotted along the horizontal axis. The variable costs and the fixed costs are plotted along the vertical axis. The break-even point is where the total cost of production is the same for both machines.

Let the quantity and the total cost for both machines at the break-even point be Q_{be} and T respectively. Let the fixed costs for machine A be F_A and for machine B be F_B, and let the cost per component on machine A be C_A and on machine B be C_B.

Then at the break-even point the total cost for machine A is given by:

T = fixed costs F_A + (break-even quantity Q_{be} × cost per component C_A)

Similarly, the total cost at break-even point for machine B can be specified as:

$$T = F_B + (Q_{be} \times C_B)$$

Since the total cost is the same,

$$F_A + (Q_{be} \times C_A) = F_B + (Q_{be} \times C_B)$$

Hence

$$F_A - F_B = Q_{be}(C_B - C_A)$$

and so

$$Q_{be} = \frac{F_A - F_B}{C_B - C_A} = \frac{\text{difference in fixed costs}}{\text{difference in variable costs}}$$

If there are a number of machines to be considered, it is necessary to compare each pair of machines.

Example of the use of break-even charts

This example is intended to illustrate a method that could be employed to determine the costs involved. For simplicity, only basic cost factors are considered. Other factors might include:

(a) Rent and rates for the floor area occupied by the machine
(b) Heat, light and power
(c) Inspection
(d) Safety
(e) Cost of preparation of components.

The actual figures used in the example are typical at the time of writing. These costs will increase with time, but the ratios of the costs should remain much the same.

It is possible to produce a component using (a) a hand-operated centre lathe (b) a cam-operated automatic lathe (c) a CNC turning centre. Given the following information, the range of work for which each machine is the most economic is to be determined. Overheads are to be taken as 300 per cent on productive labour.

	Centre lathe	Cam auto	CNC centre
Cost of new machine (£)	8 000	20 000	30 000
Estimated life (hours)	20 000	16 000	16 000
Planning, tool design or program preparation, i.e. lead time (hours)	2	100	24
Lead time cost (£ per hour)	3	4	4
Set-up time not included in lead time (hours)	1	8	0.25
Set-up rate (£ per hour)	3.50	3.50	3.50
Production time per component (minutes)	20	6	7
Labour rate (£ per hour)	3.50	3	3.50

The set-up time for the centre lathe is the time required to determine the readings of the index dials for the diameters required. The labour rate for the cam auto is half the setter's rate plus half the feeder's rate. The feeder is a semi-skilled person who is responsible for ensuring that the automatic lathe is kept loaded (fed) with the work material. An automatic lathe setter generally looks after a minimum of two machines.

Since it is the same component that is being produced, the material costs will be the same for all machines. Therefore it is intended to neglect the material costs as they will have no influence on the break-even quantity.

Fixed costs

fixed costs = (lead time × lead time costs) +
(set-up time × set-up rate)

Centre lathe: $(2 \times 3) + (1 \times 3.5) = 9.5$
Cam auto: $(100 \times 4) + (8 \times 3.5) = 428$
CNC centre: $(24 \times 4) + (0.25 \times 3.5) = 96.875$

Variable costs

variable cost per component =
machine cost + labour cost + overhead cost

Machine cost =
$$\frac{\text{cost of new machine} \times \text{time/component}}{\text{estimated life in hours}}$$

Labour cost = (labour rate × time/component
Overhead cost = (labour costs × 3)

Centre lathe =
$$\frac{8\,000 \times 20}{20\,000 \times 60} + \left(3.5 \times \frac{20}{60}\right) + \left(3.5 \times \frac{20}{60} \times 3\right) = 4.8$$

Cam auto =
$$\frac{20\,000 \times 6}{16\,000 \times 60} + (3 \times 0.1) + (3 \times 0.1 \times 3) = 1.325$$

CNC centre =
$$\frac{30\,000 \times 7}{16\,000 \times 60} + \left(3.5 \times \frac{7}{60}\right) + \left(3.5 \times \frac{7}{60} \times 3\right) = 1.852$$

Break-even quantities

The machines are compared in pairs to find the break-even quantities:

Centre lathe and cam auto: $\dfrac{428 - 9.5}{4.8 - 1.32} = 120.432$

Centre lathe and CNC: $\dfrac{96.875 - 9.5}{4.8 - 1.85} = 29.638$

Cam auto and CNC: $\dfrac{428 - 96.875}{1.85 - 1.32} = 628.32$

These break-even quantities can be interpreted as follows:

(a) For quantities up to and including 29, the centre lathe is the most economic.

(b) For quantities between 30 and 628 inclusive, the CNC turning centre is the most economic.

(c) For quantities of 629 and more the cam automatic is the most economic.

One of the clearest ways of displaying the range of work for which each machine is the most economic is by means of a graph, as shown in Figure 2.26. It is seen that the break-even point between the centre lathe and the cam auto is of no concern, because at that point the CNC is more economic than either. At the two break-even quantities of interest, the total costs are:

Centre lathe and
   CNC: $(4.8 \times 29.638) + 9.5 = £151.76$
CNC and cam auto: $(1.325 \times 628.32) + 428 = £1\,260.52$

The break-even quantities are plotted on the horizontal axis and the total costs on the vertical axis, and the cost lines are drawn as shown.

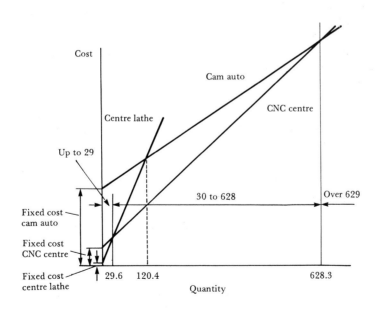

Figure 2.26 Break-even chart for three machines

Questions

2.1 What does the control of machine tools involve?

2.2 What is numerical control?

2.3 Explain the difference between NC and CNC as applied to the operation of machine tools.

2.4 What are the advantages of CNC compared with NC?

2.5 Explain how information obtained from CNC machines can be useful for management purposes. Specify the information that can be obtained.

2.6 Why are certain types of machines called machining centres or turning centres?

2.7 What advantages are to be gained by the introduction of DNC?

2.8 What is adaptive control?

2.9 Explain what instruments are required for the introduction of adaptive control on machines.

2.10 Specifying eight reasons, detail the type of work for which CNC machines are particularly suited.

2.11 Giving reasons, detail the type of work for which CNC machines are *not* economically suited.

2.12 Explain the difference between fixed costs and variable costs as applied to manufacturing costs.

2.13 Explain how fixed and variable costs can be calculated.

2.14 Explain how a break-even chart can provide information for selecting a machine tool to be used for machining a quantity of components.

Chapter Three

Construction of machine tools

3.1 Constructional features

The present design of machine tools has evolved over many years, as outlined in Chapter 2. It is not intended to present here a treatise on machine tool design; the subject is complex and forms a complete study in itself. Consideration will be given only to those features of machine tool construction that can influence the function of the machine and that have been found necessary for efficient operation under numerical control.

A machine tool can be considered to be a complex assembly consisting basically of the following:

(a) Power units
(b) Speed control units for both work and tool movements
(c) A means of controlling the line of movement of the tool and/or work
(d) Tool or work movement actuating mechanisms with control on the amount of movement
(e) Tool and work holding devices
(f) A structure linking and aligning the individual units together, so that the tool and work are maintained in a controlled relationship.

3.2 Power units

There are three types of power units required:

(a) For driving the main spindle (cutting speed)
(b) For driving the saddles or carriages (feed)
(c) Providing power for ancillary services.

The motors are of two kinds:

Electrical Alternating current (AC), Direct current (DC) or stepper motors.
Fluid Hydraulic or pneumatic motors.

3.3 Providing the cutting speed

The motors for this purpose must be capable of providing power and high speeds at the main spindle. For this requirement the majority of numerically controlled machine tools use electric rather than fluid motors. Electric motors can provide sufficient power and speed more economically for a wider range of applications.

The cutting speeds required for single-point or multipoint tools vary from 10 metres per minute (m/min) when turning cast iron with high-speed steel tools to 1000 m/min when turning aluminium alloys with cemented carbide or ceramic tools. The cutting speeds are provided by rotational speeds of the main spindle: the majority of machines provide rotational speeds ranging from 30 to 4000 revolutions per minute (rev/min). On some machines developed for more specialized applications, such as the use of diamonds as tools for turning, the top speed can be as high as 5000 rev/min. In the late 1980s, spindle speeds of 20 000 rev/min are being investigated; these speeds require special ceramic bearings. The search for higher speeds is all in the cause of increased production rates at lower costs. The maximum spindle speed provided is dependent on a number of factors, such as the power of the drive motor, the type of bearings used and the lubrication system.

The power available on a wide selection of machines is usually 5 kilowatts (kW) or less. To be able to machine components at high performance rates, the power required is of the order of 20 to 30 kW.

In the majority of applications it is possible for the actual cutting speed to vary within 10 per cent of a theoretical value without significantly reducing the efficiency of the cutting tool. Except at fairly low speeds of below 100 rev/min, it is not essential for the number of revolutions per minute to be an exact figure. It is more desirable that the actual rotational speed remains constant during machining. At all speeds when cutting screw threads it is essential that the rotational speed is synchronized with the rate of feed so that accurate threads are machined. For tapping operations it is frequently an advantage for the spindle rotation to be reversible.

It is difficult to separate consideration of speed control methods from consideration of the type of power units used. In many cases speed variation and control are obtained directly through the power units themselves.

AC motors

AC induction motors are more reliable, more easily maintained and less costly than any other electrical motor. It is comparatively easy to obtain reversal of direction of rotation of three-phase AC motors. It is also possible, through the use of pole change motors, to obtain four speeds of the order of 2800, 1400, 700 and 350 rev/min.

Apart from the pole change motors, AC motors have not been generally used for driving the main spindle directly. This is because specialized and expensive electrical equipment is required to provide high power with accurate steplessly variable speeds.

Where AC motors are used for driving the main spindle, it is necessary for a mechanical variable speed unit to be used in order to obtain the speed variation of the spindle required. However, developments in all aspects of engineering are continually taking place. In 1984 it became possible to obtain speed control and variation of AC motors by varying the frequency of the electricity. As a result of these developments, there may be a more widespread use of the AC induction motor with speed control by frequency variation.

DC motors

When variable speeds are required, the DC electric motor is normally used. DC motors are capable of transmitting sufficient power with steplessly variable speeds. At one time special equipment such as a Ward–Leonard system was used to obtain the speed variation and control. With a Ward–Leonard system it is possible to obtain a speed range of 10 to 1. Controlled rectification using thyristors has also been used to convert the alternating current to direct current and obtain the required speeds.

Fluid motors

Fluid motors are driven by oil or air under pressure. The fluid is pressurized by a pump, which is normally driven at a constant speed by an alternating current electric motor. As stated previously, fluid motors are not generally used for providing the cutting speed. But there are a few applications, such as for milling and drilling operations on aluminium alloys, which use fluid power for driving the main spindle. One of the effective applications was on the Molins System 24 machines. Machine units in that system had the cutter spindles driven directly by turbines mounted on the spindle shaft. Oil under high pressure was directed on to the blades, enabling rotational speeds of the order of 30 000 rev/min.

3.4 Power units for moving the carriages holding tool or work

On manually operated machines, the power for moving the carriages holding the tool or work is usually provided through a gear train driven by the main spindle of the machine. This method is not suitable for numerically controlled machines, because of the need to move tools or work exact distances at desired rates of movement (feed rates).

There are two separate requirements in the moving of the tool or work:

(a) The actual position of the tool or work at any time when material is being removed has to be within at least 0.01 mm of the desired position and for some applications to within 0.005 mm.

(b) Except when cutting screw threads, the rate of movement has to be controlled but does not have to be precise.

The power units that provide the feed movement do not have to be so powerful as those used for driving the main spindle, and motors of 1 kW are adequate. In addition the feed rates are much slower than cutting speeds, being normally from 5 to 200 mm/min during machining and at 5 m/min during rapid positioning. These rates of feed can easily be obtained with a screw and nut driving system, where the screw of 5 mm pitch rotates at 1 to 40 rev/min when machining and at 1000 rev/min when positioning. These speeds are considerably slower than the main spindle speeds, but it is essential that the movement provided by feed motors can be controlled very precisely and quickly, and it must be reversible. To obtain position control there are two systems:

(a) A closed loop system using feedback signals generated by position transducers

(b) An open loop system.

Chapter 5 provides an explanation of transducers, and Chapter 6 an explanation of closed and open loop systems.

DC motors

Generally DC motors are used in closed loop systems for positional control and for moving tools or work under precise control. The developments that have taken place in the

control of speed using frequency variation may result in AC motors being more widely used.

Stepper motors

Open loop systems use stepper motors. A very simplified drawing of a stepper motor is given in Figure 3.1, and is used to provide an explanation of the principle of operation of stepper motors. This principle is the magnetic attraction and repulsion of unlike and like poles respectively. The ends of the rotor section are of a fixed polarity but the polarity of the stator poles can be switched.

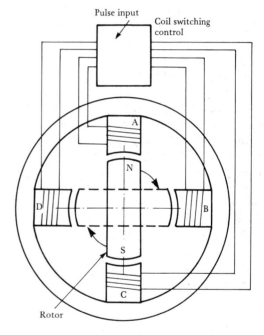

Figure 3.1 *Simple stepper motor*

The stepping action commences with a pulse of electricity being sent from the control unit to the coil switching control. The coil switching control causes poles A and D to be north polarity, and poles B and C to be south polarity. The rotor has permanent north and south poles; its north pole is repelled from the north polarity of poles A and D, and attracted by the south polarity of pole B. Simultaneously the south pole of the rotor is repelled from the south polarity of poles B and C, and attracted by the north pole D. When the rotor has moved a quarter of a revolution, the table which is driven by the rotor will have moved a quarter of the lead of the driving screw.

The next pulse will cause the coil switching control to change the polarity of the poles, and the next step will take place: A and B will be north polarity, and C and D will be south polarity.

The control unit generates the number of pulses which corresponds to the programmed amount of movement required of the tool or work. The rate of movement of the carriages is controlled by the speed with which the pulses are sent to the stepper motor. As soon as all the pulses programmed for that operation have been fed to the drive unit and the steps have occurred, the next action is initiated.

Stepper motors used on numerically controlled machines have rotors which can be controlled to move very small rotary steps, in either single or multiple steps. This results in the driving screws, which are on the same spindle as the rotors, moving the carriages to an accuracy of 0.01 mm or 0.001 in.

At one time stepper motors were not powerful enough to drive the carriages of machine tools directly, and had to use a type of hydraulic system to provide the required power. However, they are now able to move work tables weighing 50 kg over a distance of 500 mm at 5 m/min.

Stepper motors have a wide range of applications, such as driving robots or moving the pens on plotters used for draughting purposes.

3.5 Power units for ancillary services

Because of the dependability of AC induction motors they are generally used for such services as coolant pumps, swarf removal equipment and driving fluid motors. The only control required for these applications is on/off switching.

3.6 Speed control units

Before speed can be controlled, it has to be measured. Two techniques used for this purpose are:

Digitizer or pulse generator   See Chapter 5 for an explanation of working principles.
Tacho-generator   This is widely used. It is an electricity generator of small physical size, the strength of the voltage generated being dependent on the rotational speed of the rotor.

The pulse generator or tacho-generator is mounted directly on the end of the spindle or is driven from the spindle whose speed has to be controlled.

The variation in speed of DC motors is obtained by varying either the armature voltage or the field voltage. The output signal generated by the pulse generator or tacho-generators is compared with the input value of the voltage to be applied to the DC motor, which represents the desired speed, and any difference between these voltages is used to vary the voltage to obtain the required speed.

Motors with armature voltage control are used for drives requiring high power; motors with field voltage control are used for low-power applications. The control of the rotational speed of the main spindle to provide cutting speed poses different problems from the control of feed rates.

Control of cutting speed
In addition to the techniques of varying the cutting speed by electrical means which have been outlined above, i.e. using pole change motors for stepped speeds and DC or AC motors for steplessly variable speeds, there are mechanical systems used which can provide speeds which are either stepped or steplessly variable:

Stepped   Cone pulleys with belts; stepped cluster gears.
Steplessly variable   Expanding pulley; friction discs.

Stepped drives
Stepped drives have limited use on numerically controlled machine tools, and can cause problems. It is difficult for the speed change to be

programmed to occur automatically. The problem with programming these drives is that usually the spindle has to be stopped before the speed can be changed. It is possible to change speed using a series of clutches, but this may cause slippage and put movements out of phase. If the tools have to be changed manually then a stepped drive may be acceptable because, for safe working, the spindle has to be stopped for the tool to be changed, and any change in spindle speed can be carried out at the same time.

The cluster gear type of stepped drive is preferred to the cone pulley, as it is more convenient to change speeds by moving a few levers rather than belts.

Steplessly variable drives

One of the problems of both the steplessly variable systems listed is the transmitting of sufficient power. The expanding pulley is more efficient than the friction discs, and is used on a range of machine tools. A mechanical variable speed controller powered by an AC motor is used on a number of numerically controlled machine tools.

Milling and drilling machines have different speed requirements from turning centres. On milling and drilling operations, for a particular work material, there is a spindle speed at which each diameter of tool is at optimum efficiency for material removal and tool life. The same spindle speed set at the start of the cutting operation can be maintained throughout the operation without detriment to the efficiency of the cutting tool. Therefore it is possible for milling and drilling machines to have a drive unit that provides a stepped series of speeds; the speeds are usually arranged in a geometric progression.

On turning centres it is very desirable that the spindle speeds should be steplessly variable. It is important that as the diameter of the work being machined changes, the spindle speed in revolutions per minute should change. This is to maintain the actual cutting speed (peripheral speed) at a constant value so that the cutting action when removing material is at maximum

efficiency. This may not be of significance if the diameters being machined do not change more than 20 per cent, but is important when facing large-diameter workpieces.

To overcome this problem there are control systems which provide constant cutting speed at the point of cutting. With these systems the actual peripheral cutting speed (metres per minute) required is programmed; as the radius being machined changes, the spindle speed (revolutions per minute) changes so that the peripheral speed at the tool point is kept constant. These systems ensure that the tools are cutting at optimum efficiency and the desired tool life is obtained.

There is another problem associated with the rotation of the main spindle on turning centres, which is not of such great importance on milling and drilling machines. When cutting screw threads on a lathe it is essential that the rotation of the spindle is synchronized with the axial movement of the cutting tool. This is achieved by having one transducer (typically a pulse generator) mounted on the main spindle and another transducer (position transducer) which monitors the movement of the tool. When reference points on both transducers are coincident, axial movement of the tool is initiated; this is similar to the use of the screw cutting dial on manually operated lathes. Special tapping attachments are used on milling and drilling machines to cut screw threads; there is less of a problem with synchronizing the axial and rotary movements, as the threads are produced in one pass of the tool.

3.7 Control of feed rates

Only DC motors or stepper motors have been used for providing the power to move the carriages at the required feed rates, as the other methods (detailed previously for control of cutting speed) are not capable of sufficient precision.

Tacho-generators or pulse generators provide the feedback signal to control the rate of movement. The transducers are mounted on the screw which is used to move the carriages. The amount of movement is monitored by position transducers, whose principles of operation are explained in Chapter 5.

Fluid motors

The main application of fluid power was for driving cylindrical motors for linear movements. The hydraulic ram is capable of very smooth and slow movements. The rate of movement of the rams is controlled by flow control valves which regulate the rate at which oil is permitted to leave the cylinder. At one time there were a number of numerically controlled machine tools that used fluid motors to drive the saddles and carriages of the machine tools. The tendency now is to use DC motors or stepping motors for this purpose; it is easier and cheaper to provide positional control with electric motors. There is a problem with the use of cylindrical rams for long movements, related to the amount of oil that is required to fill the cylinders to move the pistons.

One advantage with using hydraulic rams is that when two are mounted in opposition they can be used to stop carriages very quickly by feeding oil to the ram opposing movement.

The pneumatic ram is only effective for quick movements, such as moving handles right or left. When pneumatic rams are moved at slow speeds considerable judder and vibration can occur, owing to the compression of the air.

3.8 Control of translational (linear) movements

There are two types of movement to be controlled: translational (linear) and rotational. This section describes translational movements, and Section 3.9 deals with rotational movements.

When machining is taking place, the rate of the translational movement (feed rate) can be as low as 20 mm/min. During non-machining operations such as positioning, the feed rate can be 5 m/min or more. For fine machined surfaces and for accurate positioning during machining, the movement must be smooth and continuous, and free of any stick-slip reaction.

Stick-slip reaction occurs when two bodies which are in contact tend to stick together until the applied force is large enough to cause one of them to slip over the other. Stick-slip happens at very slow rates of movement because friction at these speeds tends to be high. To cause movement, the force to overcome friction and any force resisting movement has to be correspondingly high. This force results in the drive mechanism, such as a screw, being elastically deformed. The energy thus stored in the screw, together with the applied force, causes the carriage to slip and move at a faster rate. At this higher speed the friction decreases and a greater amount of movement than that intended of the carriage or table results. There is a possibility of this cycle of events repeating, resulting in errors in positioning or changes in the direction of movement. To reduce the possibility of stick-slip there should be minimum but constant friction between the surfaces in contact.

Slideways

A slideway is used to control the direction, or line of action, of the translational movement of the carriage or table on which the tools or work are held.

In a closed loop control system the amount of movement is monitored by positional transducers which provide feedback signals. In an open loop system the movement of the carriage is dependent on the input commands to the stepper motors. Different shapes or profiles can be produced by a combination of translational movements of varying amounts occurring simultaneously on different slideways.

The alignment of the slideways to each other and to the axis of the spindle is critical. The shape and size of the work produced depends not only on the accuracy of the amount of movement, but also on the direction of the relative movements of the tool and the work.

As a safety measure, microswitches are positioned near the ends of the slideways. If the microswitches are activated (closed) by contact with the carriages, the power to the feed motors will be switched off. This will prevent the carriages travelling too far and striking the leadscrew bearing end housings, or the nut screwing off the leadscrew. On some control systems the information stored in memory may be corrupted or lost, and the datums and program may have to be re-entered.

Slideway forms

Slideways commonly used on machine tools have a number of different forms, such as cylindrical, vee, flat and dovetail. The cylindrical guideways are usually made of hardened steel and have to be bolted or fastened on to the main casting of the machine. It has been the practice for the other slideways to be machined out of the sides or top of the main casting of the machine. Whatever the form, the slideway must be rigid and not subject to any deformation under load. On most machines the guiding action for the movement of the table or carriage in one direction, X or Y or Z, is provided by one slideway only. The use of two slideways to provide the guiding action for movement in the same direction has been considered to cause unnecessary restraint, and the alignment is said to be overdetermined. Recent developments using a pair of bolt-on guideway rails for a single axis have eliminated this overdetermination.

At one time, after the slideways had been machined in the casting of the machine bed or column, it was common practice for them to be hand scraped to ensure efficient bedding of the carriages on the guideways. A limited amount of hand scraping may still be carried out on some machines, but the development of very efficient guideway grinding machines has reduced the amount of scraping required. A problem when the slideways are machined on the casting is that, after the machine has been in service for some time and the slideways have become excessively worn, it is necessary to strip the entire machine down so that the slideways can be remachined. While the slideways are wearing the machine has been losing its accuracy. To overcome this difficulty there has been an introduction of premachined hardened steel guideways which are bolted on to the main casting. The slideways have special forms and, to reduce wear and maintain accuracy, balls or rollers are used to provide rolling contact rather than the sliding contact which previously occurred.

The position of the drive mechanism relative to the guiding faces of the slideway is very important. Ideally the drive mechanism should be placed midway between the guiding surfaces. If it is not placed centrally the friction on the guiding surface furthest from the drive tends to cause the carriage to twist or skew. For efficient operation and to prevent the skewing or 'cross-winding' occurring the slideways are made as narrow as practicable and the slideway in the contacting member as long as possible.

Cylindrical guideways

The bore in the carriage housing provides support all round the guideway, and therefore one cylindrical guideway can effectively provide guiding action in two planes without creating unnecessary restraint. For relatively short movements or light loads, cylindrical guideways are very efficient. A limitation with the use of these guideways for long translational movements is that, if the guide bar is only supported at each end, it may sag or bend in the centre of the span under load. One way of overcoming this problem has been to provide a support underneath the guideway over its complete length; the sleeve unit in the carriage then does not have a complete outer casing, as shown in Figure 3.2. Cylindrical guideways are well suited for the use of linear ball bearings. The bore in which the balls travel must be highly finished and accurately cylindrical.

One widely used application of a cylindrical guideway is in the head of a drilling or vertical

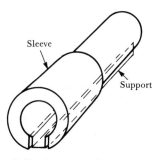

Figure 3.2 Cylindrical guideway

milling machine. In this application the cylinder moves axially, and the head with the
mating bore remains stationary. The main spindle is located in bearings that are mounted
within the cylinder. The spindle is driven at the
top and has to be fairly short and torsionally
rigid to prevent it twisting under load. The tools
are mounted at the opposite end to the drive.

The axial movement of the tools, mounted on
the end of the spindle, is usually of the order of
150 mm. This is because the lack of support at
the extremity of the travel will result in a tool
tending to be deflected by any sideways acting
forces on the end of the tool.

Vee guideways

The vee or inverted vee has been widely used on
machine tools, especially on lathe beds. Maudsley's original screw cutting lathe had inverted
vee guideways (see Figure 2.1). One of the
advantages of vee or inverted vee is that the
parallel alignment of the slideways with the
spindle axis is not affected by wear. There is a
closing action as the upper member settles on
the lower member, and this automatically main-
tains the alignment. Jib strips are therefore not
required with the vee slideways to take up the
clearance caused by wear. On some machines
the angles of the vee are different to reduce the
possibility of the sides of the vee wearing
unevenly. The majority of lathes that have a vee
slideway also have a flat slideway to prevent the
carriage twisting, as shown in Figure 3.3. Provision also has to be made to prevent the carriage
from lifting off the slideway.

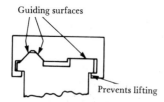

Figure 3.3 Vee and flat guideway

Flat and dovetail guideways

Although the vee type has certain advantages, it
is the flat or dovetail forms (shown in Figures

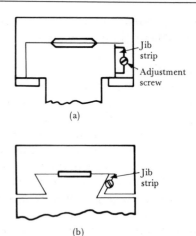

Figure 3.4(a) Flat and (b) dovetail guideways

3.4a and b) which have been used on numerically controlled machine tools. The flat
slideway has better load-bearing qualities than
the other slideways. After a period of use, wear
may have occurred owing to the sliding of the
surfaces over each other. Jib strips are used to
ensure the accurate fitting of the carriages to
both the flat and dovetail slideways. The jib
strips are tapered and can be adjusted to reduce
excessive clearance caused by wear.

The metal-to-metal contact on the vee, flat and
dovetail types of guideway is normally cast iron
to cast iron. The cast iron may have been heat
treated (flame hardened) to increase its hard-
ness, and the surfaces ground to obtain the
accuracy required. A problem with the metal-to-
metal contact is the relatively high coefficient of
friction, and this inevitably results in wear and
an increase in the power required to move the
carriage.

Various design techniques are used on the
slideways of numerically controlled machine
tools to:

(a) Reduce the amount of wear
(b) Improve the smoothness of the movement
(c) Reduce friction
(d) Satisfy the requirements for the movements
      stated above.

The techniques used include hydrostatic
slideways, linear bearings with balls, rollers or

needles, and surface coatings. These are described in the following sections.

Hydrostatic slideways

Hydrostatic slideways use essentially the same principle of operation as a hovercraft. Oil or air is pumped into small cavities or pockets machined in the faces of the carriage which are in contact with the slideway of the machine (Figure 3.5). The pressure of the fluid gradually reduces to atmospheric as it seeps out from the pockets, through the gap between the contacting faces of the carriage and slideway. An almost frictionless condition exists for the movement of the carriage. One of the problems with hydrostatic slideways is that there has to be a relatively large surface area to provide adequate support.

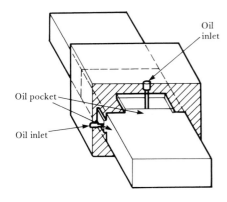

Figure 3.5 *Hydrostatic slideway*

Machines may use pressure-balanced slideways where the pockets for the fluid are machined in all the faces of the carriage contacting the slideway. If the carriages are heavy and the cutting forces created are high, the pockets may only be required in the faces supporting the load.

The hydrostatic principle can be used on slideways of vee or inverted vee, flat and dovetail forms. For efficient operation it is very important that the fluid and the slideway surfaces are kept clean.

Linear bearings with balls, rollers or needles

As indicated previously, one of the problems when there is metal-to-metal contact is the relatively high coefficient of friction which exists between the faces in contact, typically 0.15 for lubricated steel sliding on steel. A number of machines have flat roller bearings fitted to the carriages to provide a rolling motion rather than a sliding motion. The rollers are in contact with the guideways machined on the bed of the machine. These have been very effective in providing smooth and easy movement, but still require an accurate form to be machined on the bed. The surfaces in contact with the rollers have to be hardened and must have a smooth surface texture.

To reduce the problem of machining an accurate form on the bed of the machine, hardened steel rails with special guide forms (Figure 3.6) may be bolted on to the casting of the machine tool (as mentioned earlier). Special blocks with recirculating spherical balls bearings can move along the rails. The balls provide rolling motion and, because the contact form on the rail is a mating form of the balls, there is a line contact

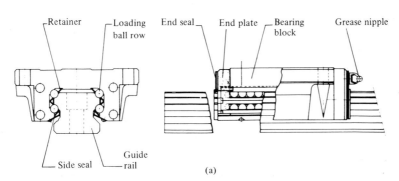

Figure 3.6(a) *Construction of linear bearing guide rails (courtesy Unimatic Engineers)*

Figure 3.6(b) Set of rails and blocks (courtesy Unimatic Engineers)

between the balls and the rails. The coefficient of friction is reduced, and is reported to be of the order of 0.005; there is no stick-slip, and the slowest and smallest movements can take place efficiently. At least four blocks are fitted to each carriage; there is a pair of blocks along each guide rail. Each pair is placed to provide the greatest possible separation between blocks.

Surface coatings

On some machines the guiding surfaces on the bed and columns are coated with low-friction material such as polytetrafluoroethylene (PTFE). Alternatively replaceable strips of low-friction material are used. The material used is a type of plastic with particles of graphite embedded into the surfaces that take the heaviest loads. When the strips wear to such an extent that the alignment is in error, they can be replaced relatively easily and the accuracy restored.

There are machines where the guideways are plated with chromium, which is very hard wearing. The thickness of the plating is controlled; when the plating is eventually worn away, the original surface shows through. The accuracy of the machine can be re-established by replating the surfaces.

3.9 Control of rotational movements

In a similar way as the translational movement

of the carriage is controlled by the slideways, rotational movements of spindles are controlled by circular bearings. The accuracy of the roundness of the rotation is dependent on the quality of the bearings.

There are two types of rotational movements required: those providing feed movements, and those providing the cutting speed. The rotational speed of feed movements is generally very slow, between 20 and 200 mm/min, and for the majority of feed movements the rotation is usually less than a full 360 degree rotation. Consequently there is not such a problem with the guiding of the tools or work with respect to the amount of movement. On large machines, the circular tables on which the work is mounted can be at least 2 metres in diameter.

Rotating spindles

As stated previously, material removal using single-point or multipoint tools requires rotational speeds of the order of 30 to 4000 rev/min and even higher. All work or tool carrying spindles rotating at these speeds are subject to deflection (twisting) and thrust forces depending on the nature of the work being performed. To increase stability and minimize torsional strain on the spindles they are designed to be as short and stiff as possible, and the final drive to the spindles is located as near to the front bearing as possible.

When a work holder (such as a chuck) is mounted on the spindle, the accuracy of rotation is extremely important as it affects the roundness of the components produced. The rotational accuracy of the spindle is dependent on the quality and design of the bearings used. The bearing should support the spindle radially and axially. There are various types of bearings used, namely plain, ball or roller, and fluid.

Plain bearings

Plain bearings require accurate fitting, but efficiently fitted plain bearings have certain advantages over the other bearings in vibration damping. The main objection to a plain bearing is that a definite clearance must be provided for the oil film to be maintained between the spindle and the bearing. This clearance may result in the relative position of the centre of a spindle in the bearing changing its position owing to variation in the position of the applied force. Clearances normally provided between the spindle and bore of the bearing for the oil film are given in Table 3.1.

Table 3.1 *Plain bearing clearances*

Diameter of spindle (mm)	Clearance (mm)
10 to 25	0.05
26 to 65	0.075
66 to 85	0.1
86 to 100	0.125
Above 100	0.025 per 10 mm increase in diameter

Ball or roller bearings

These are suitable for high speeds and high loads. They are often used in preference to plain bearings because of their low friction, moderate dimensions, lesser liability to suffer from wear or incorrect adjustment, and ease of replacement when necessary.

For efficient service it is essential that all the components of the ball and roller bearings, particularly the rollers and the inner and outer bearings (tracks), are of the highest accuracy. An error on one component can affect the quality of the work produced.

Fluid bearings

Both oil and air are used for machine tool spindle bearings, but generally oil is used for the spindles of numerically controlled machine tools. There are two main types of oil bearings:

Hydrostatic With the hydrostatic bearing, the spindle is supported in the bearing by a relatively thick film of oil supplied under pressure, similar to that used in the bearings for linear movements. The oil is pressurized by a pump external to the bearing. The principle of the hydrostatic bearing is illustrated in Figure 3.7a. The load carrying capacity of this type of bearing is independent of the rotational speed.

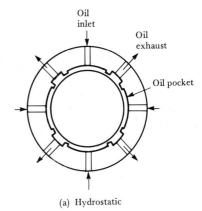

(a) Hydrostatic

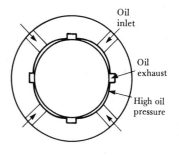

(b) Hydrodynamic

Figure 3.7 Fluid bearings: (a) hydrostatic; (b) hydrodynamic

Hydrodynamic With the hydrodynamic bearing, the rotation of the spindle helps to maintain the position of the spindle in the bearing. The principle of the hydrodynamic bearing is illustrated in Figure 3.7b. The pressure of the oil is created within the bearing by the rotation of the spindle. As the spindle rotates, the oil in contact with the spindle is carried into wedge-shaped cavities between the spindle and the bearing. The oil pressure is increased as the oil is forced through the small clearances between the bearing and the spindle.

3.10 Actuating mechanisms

Of all the constructional features that make up a numerically controlled machine tool, the efficiency and responsiveness of the actuating mechanisms (the drive unit) have the greatest influence on the accuracy of the work produced.

For an efficient drive unit there are a number of essential requirements:

(a) The drive must be stiff and responsive.
(b) There must be virtually no backlash in the drive.
(c) The drive must be free running with low temperature rise.
(d) There should be freedom from high-frequency vibrations.

The actuating mechanisms for the carriages for tool or work found on numerically controlled machine tools are: screw and nut; rack and pinion; and ram and piston.

Screw and nut

These are effective for short to medium length (100 mm to 8 m) movements. With longer movements there is a problem with the screw sagging under its own weight. Conventional vee, acme or square thread forms are not suitable for the drive mechanisms. This is because although it is possible to produce the screws to the required accuracy, the sliding action of the contacting surfaces of the thread on screw and nut results in rapid wear, and the friction is high. The efficiency of these screws is of the order of 40 per cent.

There are three types of screw and nut used on numerically controlled machine tools which provide low wear with continued accuracy over a long life, reduced friction and smooth action, higher efficiency and better reliability. These are recirculating ball screws, roller screws and hydrostatic screws.

Recirculating ball screws

For both open and closed loop systems, recirculating ball screws are widely used. The thread form used with these screws is shown in Figure 3.9 and is known as the 'Gothic arch'. The balls rotate between the screw and the nut and at some point they are returned to the start of the thread in the nut. Two types of recirculating ball screws are shown in Figures 3.8a and b. In Figure 3.8a the balls are returned through an external tube after three threads. In Figure 3.8b the balls return to the start through a channel inside the nut after only one thread. Obviously

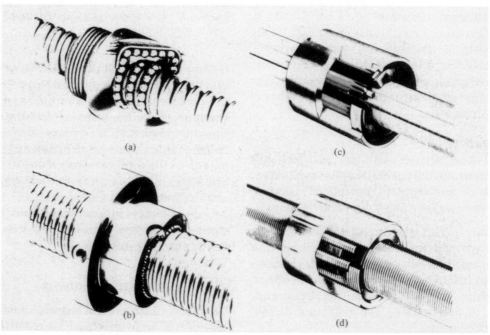

Figure 3.8(a) Recirculating ball screw with external return; (b) recirculating ball screw with internal return; (c) planetary roller screw; (d) recirculating roller screw (courtesy Unimatic Engineers)

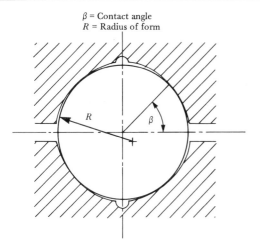

β = Contact angle
R = Radius of form

Figure 3.9 Gothic arch thread form

there must be balls in each thread within the nut. To enable the movement of the carriages (work tables) to be bidirectional without any significant errors in position occurring, there must be a minimum of backlash in the screw and

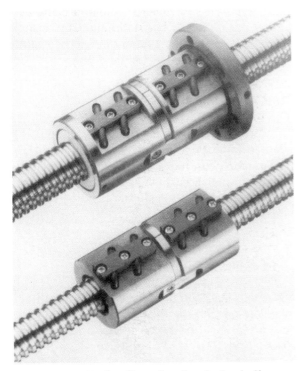

Figure 3.10 Preloading of recirculating ball screws (courtesy Unimatic Engineers)

nut. One method of achieving virtual zero backlash with these screws is by fitting two nuts, as shown in Figure 3.10. The nuts are preloaded by an amount which exceeds the maximum operating load. The nuts are forced apart, or alternatively squeezed together, so that the balls in one nut contact one side of the threads in the nut and screw, and the balls in the other nut contact the opposite side of the threads.

The efficiency of recirculating ball screws is of the order of 90 per cent and is obtained by the balls providing a rolling motion between the screw and the nut. A problem with recirculating ball screws is that in order to transmit reasonable power a minimum diameter of ball must be used. The diameter of ball used is 60 to 70 per cent of the lead of the screw; this results in the minimum pitch of the screw being of the order of 3 mm. So that the balls can return to the start there is a limitation on the rate of movement of the screw.

Roller screws

There are two types of roller screws used: planetary and recirculating. Both types provide backlash-free movement and their efficiency is of the same order (90 per cent) as the ball screws. An advantage of roller screws is that because the pitch of the drive screws is smaller than the minimum of the ball screws, the less complex electronic control circuitry will provide more accurate positional control. Roller screws are, however, three times the cost of ball screws. The rollers of both types of screw are positioned between the nut and the screw, and engage with the thread form inside the nut and on the outside of the screw. The thread form is triangular with an included angle of 90 degrees.

The planetary roller screws are shown in Figure 3.8c. The rollers in the planetary type are threaded with a single start thread as shown in Figure 3.11a. At each end of the rollers, gear teeth are cut. The gear teeth mesh with an internally toothed ring on the nut, which drives the rollers to provide a rolling motion between the nut and the screw. The rollers are equally spaced around the shaft and are retained in their circumferential positions by spigots which

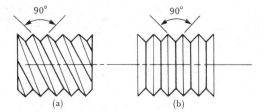

Figure 3.11(a) Planetary roller; (b) recirculating roller

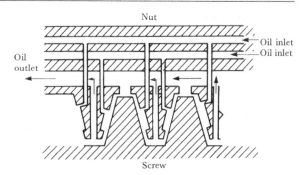

Figure 3.12 Hydrostatic nut and screw

engage in locating rings at each end of the nut. There is no axial movement of the rollers relative to the nut. The planetary roller screws are capable of transmitting high loads at fast speeds.

The recirculating roller screws are shown in Figure 3.8d. The rollers in this type are not threaded but have circular grooves of the thread form along their length, as shown in Figure 3.11b. The rollers are equally spaced around the shaft and are kept in their circumferential positions by a cage. In operation the rollers move axially relative to the nut a distance equal to the pitch of the screw for each rotation of the screw or nut. There is an axial recess cut along the inside of the nut, and after one rotation of the drive screw the rollers pass into the recess and disengage from the thread on the screw and nut. While they are in the recess, an edge cam on a ring inside the nut causes them to move back to their start positions. While one roller is disengaged, other rollers are providing the driving power. The recirculating roller screws are slower in operation than the planetary type, but are capable of high loads with greater accuracy.

Hydrostatic nut and screw
These were developed to provide an alternative to the recirculating ball screws and are reported to be more accurate for high loads. The oil pressure required is of the order of 7 MN/m². The principle of operation is shown in Figure 3.12. Oil at the high pressure is pumped through the nut into the gap between the threads in the nut and the screw, and provides virtual frictionless movement with no backlash and no wear. The gap between the threads is about

0.025 mm, and the positional accuracies obtainable are reported to be better than 0.0025 mm per 300 mm.

Rack and pinion
These are only used for the longest drives. The pinion is made as large as practical in order to have as many teeth in contact with the rack as possible. There are special pinions which provide minimum backlash. These pinions are in two sections across the width; teeth on one side of the pinion mesh with one side of the rack teeth, and teeth on the other side of the pinion mesh with the other side of the rack teeth.

Ram and piston
Obviously this drive is only possible with hydraulic cylinders, and these are generally limited to fairly short movements (under 1 m). Pneumatic cylinders are not used as drive units for positional control because, as stated previously, air compressed under high loads at low speeds creates 'judder'.

3.11 Tool holders

The materials used for cutting tools for milling, drilling and turning must have special physical and mechanical properties and are very expensive. Therefore the cutting tools used are comparatively small, and the tools have to be held in holders. The tool holding method used will be dependent on the machining technique.

Grinding wheels

For example, the most common method of holding wheels for cylindrical or surface grinding is between flanges and once mounted it is unusual for a grinding wheel to be removed from the flanges except when it has been so reduced in diameter as to be unusable. The changing of a grinding wheel is an event of significant importance, and normally occurs fairly infrequently. It must be noted that a grinding wheel is the only tool that does not have to be removed from a machine for resharpening. A wheel complete with flanges is only removed from the machine for such reasons as when either a different work material is to be ground, or when a wheel with a different grit size is required for roughing or finishing operations, or a wheel with a basically different shape is to be used. Because of the structure of grinding wheels and the high speeds involved in grinding, the major consideration in the design of the wheel holding method, and when changing a grinding wheel is safe working practice.

Spark erosion

The method of holding tool electrodes for spark erosion (EDM) must provide electrical continuity. The wire electrode used on EDM wire cutting machines is normally made of copper and for the majority of machining operations is 0.1 to 0.3 mm diameter. The wire is wound on a reel and a 6 kg reel is sufficient for at least 100 hours of continuous cutting. The wire can be automatically fed through the support guides being carried by a jet of water from the upper to the lower support guides. In Figure 3.13 the wire can be seen entering the pivot head at the top centre of the photograph and shows the wire and jet between the upper and lower guides. After passing through the cutting zone the wire is wound on to a second reel and discarded after one pass through the work. The changing of a reel takes very little time. For both grinding and EDM the tools do not have to be changed as frequently as tools on machining or turning centres.

Machining centres

The tools used on machining centres are invari-

Figure 3.13   EDM wire guide head (courtesy AGIE (UK) Ltd)

ably round and rotate at high speeds (rev/min) so the tool holder has to be capable of holding the tools securely and provide positive driving of the tools.

Single point tools used for turning operations are mainly rectangular and the holder has to accommodate this section.

The individual tool holders for both machining and turning centres are designed so that the tools are held securely within the holder and supported sufficiently to prevent the tools twisting or bending under the action of the cutting force. The individual milling cutters and single tools themselves must be strong enough not to deflect under the action of the cutting force.

Generally, for a particular component, more than one tool has to be used and it is more convenient to change the holder with the cutting tool already mounted rather than to change the individual tool in the machine. An advantage to be gained from the use of tool holders is that tools can be preset, which assists in ensuring repeatability of position when the tool is changed. With presetting the time the machine is out of operation is reduced. (See Chapter 10 for details of presetting of tools.)

3.12 Tool changing arrangements

For efficient operation of numerically controlled machine tools it is essential that the correct tools are available at the time required. The tool change should be capable of being carried out as quickly as possible because, although it is essential to change tools, time spent on carrying out these operations does nothing positive to the work; only when work material is being removed, or there is any change in the shape or size of the work is any economic benefit being obtained.

The design of the mechanism for securing the holder on the machine should ensure that it is possible to locate the tool holder in only one position on the machine; due attention being paid to the safety of the operators.

For situations where there may be a single machine or small numbers of CNC machines, manual tool changing is widely used. However, for continuous operation of machines such as in flexible manufacturing cells, various programmable automatic tool changing techniques have been developed which have a number of advantages such as:

(a) Unattended operation of the machine
(b) Provision of information for management on the number of times a tool has been used
(c) More accurate estimation of machining costs because the time of changing the tool is known
(d) Generally tool changing times are reduced.

Manual tool changing

This is the least sophisticated method and an important consideration with manual tool changing is that efficient operation is dependent on the ability of the operator to select the correct tool to be loaded. There are, however, some advantages with this method of tool loading:

(a) When changing the tool with quick clamping action tool holders, the overall time from stop to restart of cutting can approach that of the automatic methods.
(b) There is no theoretical limit to the number of tools from which the selection can be made.
(c) The tools can easily be checked after each change to decide if they require regrinding. The checking will not increase the machining time as it can take place while the tool that has been loaded is cutting.

Tool identification

For numerically controlled machines it is usually necessary for a number to be allocated to every tool. The number may have to be entered in the part program and be input into the machine control unit; the number must be either on the tool holder itself, or on the location in which the tool is mounted on the machine. Tool identification is very important as a wrong tool can cause considerable damage.

During manual tool changing the tool has to be selected by the operator. A skilled operator should be capable of recognizing the tools to be used for a particular machining operation especially when an operation schedule and tool detail sheet has been provided. For operators' convenience there are various tool identification aids that can be used. One technique is to have a special stand in which the tools are held in numbered locations – the stand is linked to the control unit. The number of the tool required for a particular operation being called up from information stored in the part program, and a light at the required tool location in the stand is illuminated, or the location number displayed on the control unit's screen. With this arrangement it is essential that the tools are placed in the correct location at the setting up stage, and also replaced correctly after use.

Automatic tool changers use a number of systems to identify the tool. One system used on machining centres of identifying the tools themselves has a series of rings or discs fitted on the body of the tool holder. The rings are all of equal thickness but are two different diameters. The larger-diameter rings activate microswitches at the reading head; there is a microswitch for each ring on the tool holder. The smaller-diameter rings are referred to as 'spacers' and do not activate a microswitch. A signal in binary number format is created by placing activating and spacer rings at selected places, as shown in Figure 3.14. The number created by certain of the microswitches being activated is sent to the control unit. There is usually a screwed collar which can be removed so that the rings and spacers can be arranged in the desired sequence to create the tool number. It is essential to have a tool library which lists all the tools available. There are various ways in which the different types of tools can be identified. In Figure 3.14 the binary number 000 to 111 (denary 0 to 7) created by the top three rings can be used to indicate the type of tool, and the remainder of the rings (which create denary numbers 0 to 31) can be used to indicate the size of the tool. An example is shown in Table 3.2.

The number generated by the setting of the rings shown in Figure 3.14 is 101 01101 = 5 13. If more rings are provided, a more elaborate numbering method can be employed for the tool identification system.

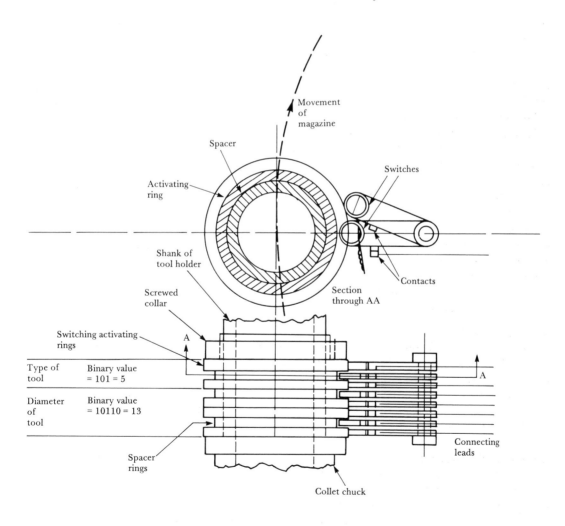

Figure 3.14   Tool identification

Table 3.2 *Tool identification*

First number	Type of tool	Second number	Diameter of tool (mm)
1	Drills	01	2
2	End mills	02	4
3	Slot drills	15	10
4	Shell mills	25	20
5	Taps	30	50
	etc.	etc.	

Another tool identification system uses small sealed capsules embedded in the shank of the holder. The capsule contains semi-conductor circuits. The housings of the capsules are waterproof and are made of stainless steel; the ceramic cover of the sensing face is resistant to hot metal chips.

Figure 3.15   Tool identification using inductive elements (courtesy Euchner (UK) Ltd)

There are two forms of this system; in one form there a number of capsules embedded in the shank of the holder. Figure 3.15 shows a milling cutter holder with five capsules fitted into the shanks. Each capsule generates a single decimal digit (0 to 9) at the reading station. The capsules which are 7 mm diameter and 5 mm thick can either be glued into holes in the shank or can be held in element holders which can easily be changed if necessary. The tool number is 'read' when the tool passes in the proximity of a sensing head. The reading head contains an electrical circuit and the number is generated by inductive coupling. It is reported that the reading of the numbers can be at speeds of up to of 30 m/min when the distance between the capsules and the head is 1 mm. It is possible to use capsules for identifying the location rather than the individual tools.

In the second form there is only one capsule into which data can be programmed and read. The capsule can be inserted at any covenient position that is accessible to the reading head. Figure 3.16 shows a capsule mounted in the shank of a milling cutter holder. The maximum sensing distance is 7 mm at reading speeds of up to 30 m/min. An individual capsule has a memory capacity to store up to 1 K bit of data. The information 'written and read' can be the tool number of up to 16 digits, and also details on the tool setting length and diameter offsets, tool life etc.

For CNC centres there are a number of tool holding systems available, but the contrast between the type of tools used on machining and turning centres results in different turret and magazine tool changing arrangements.

Tool holding on machining centres

On milling and drilling machines there are a number of quick clamping action tool holders. One type has a clamping ring on the spindle nose that needs less than a complete turn to secure the tool holder into the spindle nose. Another type has a draw bar which passes through the centre of the spindle and screws into the end of the shank to pull the holder securely into the spindle nose. The draw bar can

Figure 3.16   Single capsule identification (courtesy Euchner (UK) Ltd)

be rotated rapidly by a special mechanism. Figure 3.17 shows an air-actuated hydraulic-powered draw bar, which has a safety interlock to prevent the spindle rotating if the tool is not correctly located. For manual loading the tool holder can be released when the operator presses a button on the head of the machine. For automatic tool loading the releasing of the tools can be programmed.

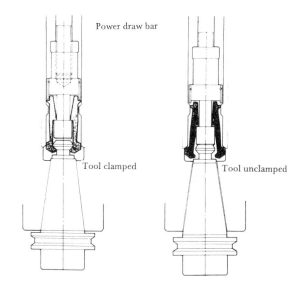

Figure 3.17   Power draw bar (courtesy Bridgeport Machines)

A common type of shank has a non-stick taper which has an included angle of the order of 16 degrees; rather than a friction type taper such as a Morse taper which has an angle of approximately 3 degrees. There are standard sizes of tool holder shanks, ranging in size from no. 10 ISO to no. 60 ISO; these have tapers with maximum sizes of the order of 16 mm and 108 mm respectively. This type of holder derives the driving power from keys on the spindle nose engaging in slots in the tool holder. The taper section is to ensure that when the tool holder is drawn into the spindle nose, the centre of the tool is coincident with the centre of the spindle. The tool holder can be removed quickly and easily because the taper does not lock, and the tool holder is free when the clamping mechanism is released. The tool holders with the Morse taper shanks derive the driving power from the friction between the shank and the spindle, and there has to be some form of ejection device to remove the tool holder.

Tool turret

One problem with the use of tool turrets is the limited number of stations (tools) available; a small machine can have a four-station turret, while large machines can have up to twelve stations. An advantage of the use of a turret is that the time taken to change the tool during the program is only the indexing time of the turret. Indexing times range from one to five seconds. The turret indexes automatically so that the station required by the program is in the cutting position. Because of the limited number of tools available in the turret the programmer has to give very careful consideration to the tools selected for use in a particular program. It is the turret station which is used as the tool identification and not the tool itself. During the setting-up stage, the setter has to ensure that all the tools required for the component are correctly loaded into the respective stations in the turret. On turret type milling and drilling machines the tools are usually held in the spindle nose in the turret by means of some form of clamping ring, as it is not possible to use a draw bar. For safe working conditions the tools

Figure 3.18   Cross-sectional view of an eight-spindle turret (courtesy of Cincinnati Milacron)

do not rotate during indexing; in addition, only the tool actually cutting rotates. The rotational speed of each tool can be individually programmed. Figure 3.18 is a cross section of an eight-station turret, showing the method of driving the spindles.

Tool magazines

Tool magazines are more commonly found on milling and drilling machines than on turning centres. This is because milling and drilling operations require a larger variety of cutting tools than does turning. The number of tools held in magazines has to be very carefully considered. The greater the number of tools in use, the more problems there are in tool maintenance and the higher the tooling costs. In addition, as the capacity of the magazine is increased to accommodate more tools, the larger and consequently heavier the magazines have to be, and the more maintenance is required. More power is required to move the larger magazine. There has to be a balance between cost of tools readily available and increased time required for tool replacement. An advantage of tool magazines is that tools subject to excessive wear can be duplicated so that, when a tool has reached the end of its life, another is ready for

use. This facility is particularly useful when the machines are under computer control in flexible manufacturing systems.

Identification of the tool required in the magazine may be either by the number of its magazine location, or by individual marking of the tool holder. When the magazine location is used as the identification the tool specified by the programmer has to be placed in the programmed location, or the tool's location has to be entered at the setting-up stage through the keyboard of the control unit.

There are a number of different types of tool magazines: chain, circular and box.

Chain magazines

The chain type holds the greatest number of tools; some can store up to 200 tools. A chain type magazine is shown in Figure 3.19. This magazine holds up to 45 tools but there are double chain models that can hold up to 170 tools. An automatic swing around tool changer can be seen at the right-hand end with one tool held ready for a tool change. Tools are normally manually loaded into the magazine as shown in Figure 3.20.

Circular magazines

The circular magazines should not be confused with tool turrets; with circular magazines the tools have to be transferred from the magazine to the spindle nose. The most common type of circular magazine is referred to as a carousel, and is of a flat disc construction with one row of tools around its periphery. Carousels can be configured at different angles, as shown in Figures 3.21a to d. Circular magazines have different tool holding capacities; there are machines with circular magazines such as that shown in Figure 3.22 that can hold up to 30 tools. The transfer of tools with this machine uses a swinging-arm automatic tool changer which can be seen in the 'park' position. (The guards have been removed for the purpose of taking the photograph.) A circular magazine capable of holding 12 tools can be seen in Figure 3.23.

This machine has an automatic tool changer

Figure 3.19   Chain type magazine (courtesy Cincinnati Milacron)

Figure 3.20   Loading of tools into a chain magazine (courtesy Cincinnati Milacron)

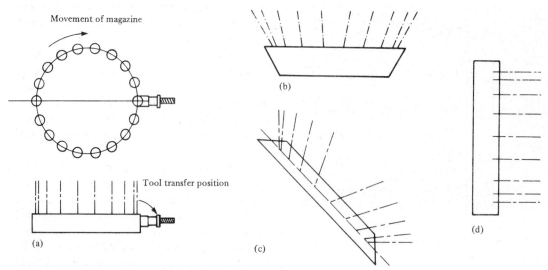

Figure 3.21 Configuration of circular magazine: (a) vertical axis; (b) vertical axis, angled tools; (c) angled axis; (d) horizontal axis

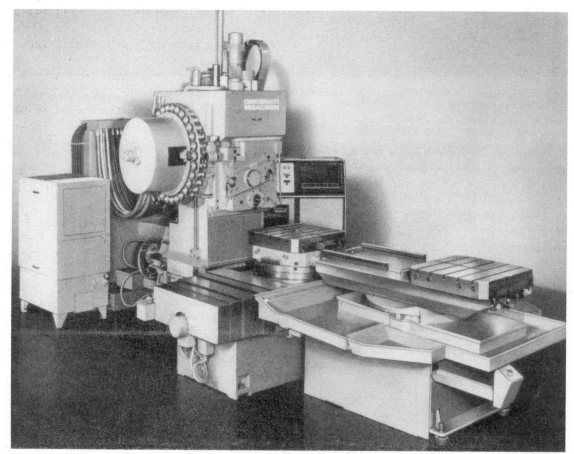

Figure 3.22 CNC machining centre with circular tool carousel (courtesy Cincinnati Milacron)

Figure 3.23   Twelve-tool circular magazine with Geneva mechanism tool change (courtesy Bridgeport Machines)

which uses a Geneva mechanism for changing the tools.

There is another circular magazine with very limited application – the drum type. It is more like a cylinder, and has rows of tools around the periphery.

Box magazines

The box magazine also has limited application. In this magazine the tools are stored in open ended compartments similar to pigeonholes. It is not as convenient as chain or circular magazines. The head with the spindle nose has to move to the tool storage position rather than the tool to the spindle nose. It is also necessary for the tool holder to be removed from the spindle nose before the new tool can be loaded. This results in the tool changing times being longer, because the new tool cannot be positioned ready for the change until the previous tool has been removed from the spindle nose.

Automatic tool changers

With CNC machines the control unit sends a signal to the magazine driving motor to move the magazine. The control unit selects the shortest distance to move the magazine so that the next tool programmed is at the transfer position ready for the tool change when called up by the program. The movement occurs during the

machining operation prior to the tool change operation. Either the tool's identification or the magazine location is read as the tools pass a sensing head. The tool to be loaded is removed from the magazine by the automatic tool changer ready to be transferred into the tool spindle while machining is still proceeding. When the tool to be replaced has completed its machining operation, the tool spindle stops and the spindle head moves so that the tool change can take place. On some control systems when the tool change command is input the spindle nose is placed in the correct position automatically, but there are machines which require the spindle nose to be programmed to move to a suitable position for the tool change to take place. The automatic tool changer removes the tool from the work spindle and loads the new tool. While machining continues the tool removed from the spindle is held in the automatic tool changer while the tool magazine moves until the original location of the replaced tool is in position ready to receive the tool.

The tool holders are automatically transferred from the magazine to the spindle nose using various mechanisms. One changing mechanism is a rotating arm which has facilities at each end for gripping tool holders. In Figure 3.22 shown previously the tool changer is in a 'park' position. A tool changer with tools in both grippers is shown in Figure 3.24. The stages in the tool change are as follows:

(a) In preparation for a tool change to take place, the magazine moves so that the tool holder to be loaded is suitably positioned for the transfer to take place. The movement of the magazine occurs during the machining operations prior to the tool change.

(b) The rotating arm moves so that the gripping mechanism at one end of the arm grips the tool and removes it from the magazine. This operation also takes place during machining. The arm then remains at rest until the tool change is called for and the tool spindle has stopped rotating.

(c) The rotating arm housing swings through 90 degrees, and the arm rotates and grips the tool in the spindle nose.

Figure 3.24   Swinging arm automatic tool changer (courtesy Cincinnati Milacron)

(d) The tool holder clamping mechanism releases the tool in the spindle, and the arm moves axially forward sufficiently to remove the tool from the spindle nose. A jet of air cleans out the spindle nose.

(e) The rotating arm, with tools in both grippers, rotates through 180 degrees. It then moves axially backward to position the tool to be loaded in the spindle nose, which is secured by the tool holder clamping mechanism. The gripper then releases the tool.

(f) The gripper arm rotates through 90 degrees, the rotating arm housing swings through 90 degrees, the spindle starts rotating and machining recommences. The gripper arm waits while the magazine moves until the original location of the tool is in position, and then moves backwards to place the tool removed from the spindle nose into its original location in the magazine.

With the swinging arm mechanism the removal of a tool from the spindle nose and loading a different tool takes 4 seconds, and non-machining time due to a tool change is only 10 seconds. A rotating arm mechanism used on a number of machines does not require a swinging arm housing. On these machines the tool holder in the magazine moves into a tool transfer position which is coplanar with the spindle nose as shown in Figure 3.25a and b. The tools in the magazine and in the spindle are gripped and removed simultaneously, from the spindle and the magazine, and interchanged. Figure 3.25a shows a holder with a touch trigger probe being removed from the magazine and a holder with an end mill being removed from the spindle nose; Figure 3.25b shows the holders transferred ready for replacing in the tool magazine and the spindle nose on the right.

Problems can arise if a tool is not replaced in its original numbered location, especially if the tools are of large diameter and overlap into adjacent locations. Additionally, if the tools are replaced at random or replaced in the location vacated by the next tool to be used, it will be necessary to check the identification of each tool at each location at every machine start-up. When the tools themselves are individually

Figure 3.25   Rotating arm automatic tool changer (courtesy Renishaw Metrology and Matchmaker Machines Ltd)

identified they may be placed in any location, provided there are no problems created with large tools.

Tool holding on turning centres

Less tools are required for use on turning centres

than machining centres; there is no limit to the number of diameters that can be machined by one tool such as an external turning tool with zero approach angle, whereas a different drill is required for each hole diameter.

Although manual tool changing is used on turning centres, the most common method of holding tools is on turrets, although there are magazines used on some turning centres.

Tool identification

Because turrets are mainly used on turning centres, identification of tools is not as much a problem as on machining centres. The turret station is used as the tool number. If required, the sealed capsule technique of tool identification described for milling tool holders can be used with tools used on turning centres.

Manual tool changing

On CNC lathes with manual tool changing facilities the tools are clamped in holders that can be interchanged quickly. The tool posts used have special locating features to ensure that the holders can be replaced in position accurately. The tools can be stored in numbered locations in stands linked to the control unit as explained previously. The tool post shown in Figure 3.26 has two vee-type locations and a quick acting cam clamp to secure the tool holder to the tool post. It is also possible to position tool holders on the tool post in a right-angle relationship, so that tools for turning outside diameters, and tools for facing or boring can be located correctly for use as required. The centre height of the tool is set and maintained with the aid of adjustable screws that are locked when the tool is first set up. There are other tool posts which use splines to locate the holders instead of the vee-type locating ribs.

Tool turrets

As stated previously, the majority of turning centres use tool turrets and on some of the smaller turning centres it is possible to remove the entire turret to preset the tools. (See Chapter 10 for details of presetting.) Where the turret cannot be removed from the machine, the cut-

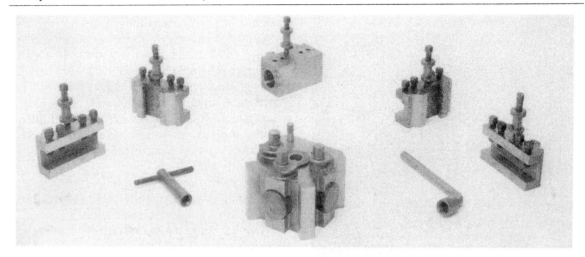

Figure 3.26   Quick change toolpost (courtesy Dicksons Engineering)

ting tools are mounted in special holders so that the position of the actual cutting point of the tool can be preset. Figure 3.27a and b shows special block holders that use indexable inserts rather than rectangular shank type tools. The block holders fit on to clamping units to suit the particular turret or tool post. The design of the shanks of the clamping units varies with the different machine tool manufacturer. Figure 3.28 shows a range of shanks found on turrets of turning centres. At the top left of the figure can be seen a pair of cassettes for mounting on tool magazine as shown in Figure 3.28. As with turrets used on machining centres, during the setting-up stage the setter has to ensure that all the tools required for one component are correctly loaded into the respective stations in the turret specified by the programmer.

The turrets on mill-turning centres can hold turning tools and, additionally, small heads with revolving spindles that can be used for milling or cross drilling operations. Figure 3.29 shows a tool turret on a mill-turning centre. As can be seen it is possible to mount two tools at each station if required. Figure 3.30 shows a twenty-station turret that is used on a sliding head mill-turning centre.

There are three main types of revolving spindle heads used on CNC mill-turning cen-

tres. These are shown in Figure 3.31a to c. The heads shown in Figure 3.31a and b can be used for milling or drilling using circular shank tools held in collets. Tools of up to 20 mm diameter can be held in the axial head (Figure 3.31a) which has a maximum speed of 4000 rev/min, and tools of up to 16 mm diameter can be held in the radial head (Figure 3.31b) which has a maximum rotational speed of 3000 rev/min. The head shown in Figure 3.31c can only be used for milling using a shell mill with a bore diameter of 16 mm at a maximum rotational speed of 3000 rev/min.

The revolving spindle head is driven only when the head is indexed to the position for the machining operation. Obviously the main head-stock spindle with the work is stationary when any cross drilling operations are being carried out, but it is possible to synchronize the rotation of the headstock spindle with the axial or radial movement of the rotary head to cut spiral grooves. During the machining operations using the revolving heads any indexing or rotational movement of the main headstock spindle is driven by a different motor than the motor used for rotating the main spindle at the required cutting speed. See Figure 4.7 in Chapter 4 for a drive for the angular control of the headstock spindle.

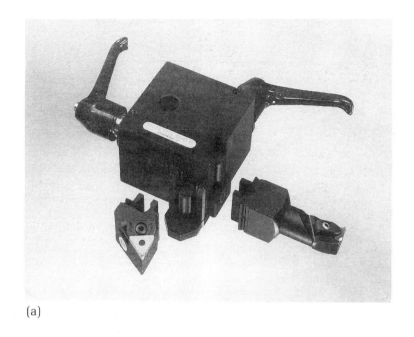

(a)

(b)

Figure 3.27(a)   Mounting of block tools; (b) block tool holders (courtesy Sandvik Coromant)

Figure 3.29   Eight-station turret with revolving spindle heads for a mill-turning centre (courtesy N.C. Engineering Ltd)

Figure 3.28   Tool holder shanks (courtesy Sandvik Coromant)

Figure 3.30   Twenty-station turret with rotary heads (courtesy N.C. Engineering Ltd)

Figure 3.32   Tool magazine (courtesy Sandvik Coromant)

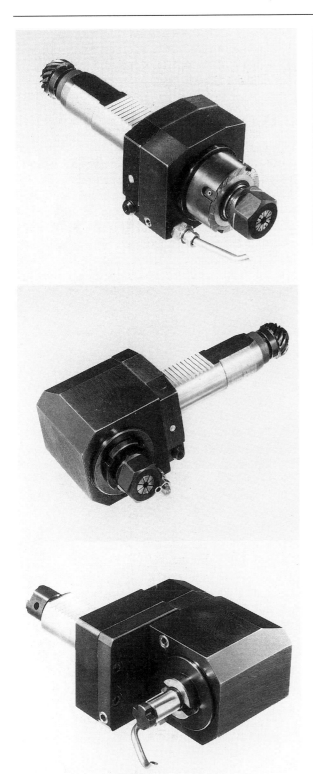

Figure 3.31   Revolving spindle milling/drilling heads (courtesy DBC Machine Tools)

Tool magazines

Tool magazines that are used on turning centres are mainly used as tool storage, tools being transferred from the magazine to the turret as required. The tools transferred can be either duplicate tools to replenish worn tools, or tools with a different geometry that may be needed especially for a particular operation on small batch production schedules. Figure 3.32 shows a rotating magazine capable of holding up to 10 cutting units (block tools) in each of 10 cassettes. Each cassette being able to store one type of tool. The rotating magazine shown in Figure 3.32 would need to be located near to the machine tool and interfaced with the control unit. This would enable the tool gripper in the automatic tool changer to remove a block tool from the magazine and replace a worn tool or exchange a tool in the tool turret. The tools are exchanged automatically as called up in the part program. The used cutting units are stored in the magazine as they are returned from the turret. There are models of these tool magazines which can hold up to 240 cutting units.

Tool holding on sheet metal working CNC machines

Tools for pierce and blank operations, as well as nibbling and step punching, invariably are required in pairs – a punch and a die. On many CNC presses the tools are mounted in a turret but as shown in Figure 3.33 there are CNC punch presses that use tool magazines. The individual tools (punch and die) are held in standardized

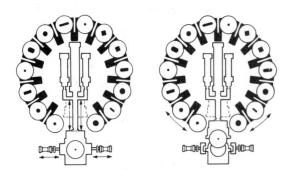

Figure 3.33   Tool magazine for punch tool work (courtesy Pullmax Ltd)

holders which are automatically transferred to the punching station under program control. The shape of the individual tools can be seen in the centre of the holders.

3.13   Work holders

Similar methods are used for holding work on numerically controlled machines as on other machines. Careful consideration should be given to the method chosen because the holder is the link between the work and the machine – if the holder is deficient it is impossible to produce consistently accurate work. The design of the work holders should embody the principles of the restraint of the six degrees of freedom within their construction; in addition, for numerically controlled machines the following details must be considered:

(a) It should be possible for the work to be located precisely and secured quickly such that the datum edges are in the correct relationship.
(b) The work has to be well supported.
(c) The work has to be held securely to resist the cutting forces generated during machining, but must not be deformed by the clamping forces.
(d) The work should be released quickly without any danger to the operator.
(e) The work holding method should be as simple as practicable without creating a hazard.
(f) The work holder should be capable of being adapted to hold different components without major changes.
(g) It is an advantage if it is possible to automatically load and unload the work under program control.
(h) If the work has to be manually loaded on to the machine, the work holding method must not be too elaborate or specialized and it should be possible to load the work quickly, in only one position.

In order that the work blanks, castings and forgings can be loaded correctly, they must be premachined to stipulated sizes and have the datum edges, locating faces and any holes used for holding purposes prepared before the work is loaded on to the numerically controlled machine. This premachining is normally carried out on manually controlled machines. The design and type of work holder will be influenced by the type of machining system being used.

Obviously, machining cannot be taking place on the numerically controlled machine while work is being loaded or unloaded from the actual machining area, therefore it is essential that work loading times should be kept to the minimum; for this purpose automatic loading techniques have been developed.

One technique, which takes up the minimum of machine time and hence increases machine utilization, is to load the work on special holders away from the machine, and transfer the loaded work holder on to the machine table. Time can be spent on ensuring that the work is accurately located and clamped securely without taking any machine time. The loaded work holder can be precisely located and clamped in position quickly. The use of touch trigger sensor probes has made the need for accurate work positioning less important.

Milling and drilling work

It is possible to secure work directly to the machine table, but the aligning and clamping of the work frequently takes appreciable time. However, this method may be practical if only a very few components are required and are not likely to be repeated, especially if the work is large and bulky.

The technique of clamping the component has to be considered. For example, hand operated toggle clamps may be suitable for holding work for drilling operations but are not strong enough for milling operations. Work loading directly on to the machine table can be carried out automatically, using pneumatically or hydraulically operated clamps or vices. The program can contain the information for the changing of the work and sending a signal to the pneumatic or hydraulic control valves to activate the clamps or the vice.

There is a design of machine which has a very wide base and work can be set at one end of the table while machining is being carried out on components mounted at the other end of the table. This technique requires special attention to be paid to safety to ensure that no quick movement of the machine table endangers the machine operator.

The size of the components and the number to be machined at any one set-up will also influence the work holding method employed. For the majority of work the holders generally used are vices, grid plates, fixtures, pallets and subtables. For specialized applications such as holding thin plates or heavy components, vacuum clamping or magnetic chucks are used. Cylindrical work requiring milling or drilling operations can be efficiently held in table mounted, non-rotating three jaw and collet chucks.

Vices

A stepped jaw vice (shown in Figure 3.34) is one of the easiest methods of holding work that is to

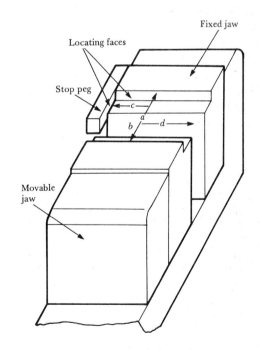

Figure 3.34   Stepped jaw vice

be manually loaded on drilling and milling machines. Care has to be taken by the operator to ensure that the work is held securely but is not overstressed. Work can be effectively located against a stop peg fitted in the jaw. It is desirable that the cutter movement should be programmed so that any large cutting forces are directed towards a solid part of the vice (in directions *a* and *b* in Figure 3.34) and not parallel to the vice jaws where the work is only held by friction (direction *d*).

One of the limitations in the use of vices is that the maximum work size is dependent on the amount of jaw opening and the width of the jaws. Another limitation is that it is impossible to cut around the full depth of the work. Although it is comparatively easy to arrange for vices to be opened and closed automatically (using pneumatic or hydraulic power) it may not be economic to use automatic mechanisms (robots etc.) for removing and loading small quantities of components from the vice jaws.

Grid plates

Grid plates are suitable for holding work of a wide range of size – up to 500 mm long by 250 mm wide. The height of components held will be dependent on the availability of the clamping bolts. It is more convenient to use a grid plate to locate a number of small components all requiring the same machining operation rather than to clamp the components on the machine work table.

Grid plates are rectangular blocks of cast iron with the base and top face machined flat. The top face of a grid plate has a series of holes that are spaced in rows and lines. A typical grid has a selection of holes as shown in Figure 3.35. The holes are alternately plain and tapped. The plain holes are reamed or ground to a precise size so that dowel pins can be inserted against which the work can be located; the tapped holes are used for clamping. The dowel pins can be plain cylindrical, headed or stepped. Headed pins are placed in holes underneath the work, but clear of the holes being drilled, to support the work and stop it deflecting when subjected to thrust forces during drilling. Stepped pins can be used

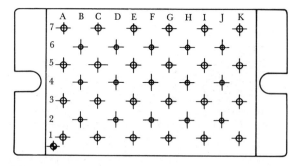

Figure 3.35   Grid plate

to support the edge of a plate and provide a location for the side of the plate. Right angled plates can also be used for locating the corner of the work.

The rows and lines of holes are identified by letters and numbers; the programmer can identify the holes to be used for a particular component by giving their grid reference. On the underside of the plates, holes are precision machined (ground) and positioned accurately so that the plates can be precisely located on the machine work table by using plugs in the ground holes. The plugs locate in the tee slots of the machine work table. The location holes in the top surface are precisely positioned relative to the ground holes in the base of the grid plate. The components are clamped to the grid plate with straps and bolts.

Fixtures

A fixture is a special device which holds work during machining operations. However, for numerically controlled machining it is not economic to design and make a fixture which can only be used for a particular workpiece.

Fixtures used on numerically controlled machines are constructed from base plates (also known as mounting plates) and a wide range of support blocks, clamping bolts and clamps as shown in Figure 3.36. There are different sizes of base plates which are available as flat rectangular sections, upright or knee sections and cubes. The surfaces of the plates have a series of tee slots which are used for clamping bolts, location plates and pins. Various attachments, such as round plugs of differing diameters, can be used

Figure 3.36   Base plate with blocks, clamps and bolts (courtesy George Kuikka)

with the plates to locate components with premachined bores etc. The blocks and clamps can be used to provide the location and clamping of awkwardly shaped components ready for machining; Figure 3.37 shows such a component being held securely. It is advised that a photograph be taken of the method of holding the component for record purposes. A camera that gives 'instant' photographs is very useful for this purpose.

Figure 3.37   Example of work holding (courtesy George Kuikka)

A fixture can be mounted on a pallet or directly on the machine table. There are rotary tables on which fixtures can be mounted. The tables can be indexed to present a different face of the work for machining operations. Some of these rotary fixtures can be programmed.

Pallets

A pallet is a block of steel or more usually cast iron of the order of 400 mm square by 100 mm thick. The sides of the pallet are machined square to known size. Pallets are similar to grid plates except that the surface and base of the pallet are machined flat, and where grid plates have a series of holes, a pallet has a number of tee slots cut at right angles across the top surface. The sides of a pallet usually have slots for the holding bolts on the machine table. Pallets can be located on the machine table with dowel pins.

Pallets are useful for holding components that are heavy or of complex shape, and where positioning of them on the machine table can be a problem. To save time in loading these components and adjusting their position on the machine table, use can be made of two pallets; one component is set up and clamped on one pallet off the machine, while another component is being machined on the second pallet. The pallets with the work are exchanged as required. The component must be accurately located on the pallet, and the pallet positioned on the machine to the part programmer's specification. A problem that can occur with the use of pallets can be their weight, and mechanical handling techniques may be required for efficient and safe usage.

If the components are small (up to 150 mm square) it is possible to load more than one on to a pallet. The pallet shown in Figure 3.38 has a capacity for holding four components on each face; the pallet is clamped to a work table, which rotates to present each face in turn to the tool spindle.

A method of automatically transferring pallets loaded with small or medium-size workpieces (up to 300 mm cube) to the machining area is with the use of an automatic pallet

69

Figure 3.38   Pallet holding four components per face (courtesy Maskin Fabrick Gerni)

piece to be machined is at the transfer position. If the machine is manned the operator uses a handle to move the pallet holding the machined component to the work loading/unloading station. A pallet loaded with work can be moved on to the magazine.

When automatic clamping is used, if a pallet is not held securely on the machine table a signal is sent to the control unit to stop the program.

Subtables

A subtable is of similar shape to a pallet, but where a pallet can be used on any machine that has a table large enough, a subtable can only be used on a particular machine. Subtables are of one size for each machine, and the underside of the subtable is machined to a form that permits it being transferred on to a mating form on only one particular machine.

Figure 3.39   Automatic pallet changer (courtesy Matchmaker Machines Ltd)

changer (APC) shown in Figure 3.39. With this facility ten pallets 300 mm square are held on a chain type magazine. Each pallet has an identification number which can be called up in the machine control program so that the particular pallet to be selected can be transferred to the machining area. The pallet change takes 17 seconds. Figure 3.40 shows a schematic layout of the method of transferring the pallets. The magazine moves to present the pallet required to the transfer position, and the pallet changer arm automatically transfers a loaded pallet to the machining area for machining to take place on the work, and simultaneously transfers a pallet with machined work to the magazine. The pallet and the pallet holding table in the machining station are automatically cleaned with compressed air before the pallet is loaded.

If the machine is unattended the magazine moves so that the pallet holding the next work-

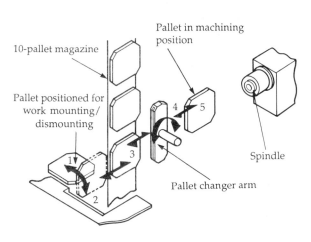

10-pallet magazine

Pallet in machining position

Pallet positioned for work mounting/ dismounting

1

2

3

4

5

Spindle

Pallet changer arm

Figure 3.40   Pallet change system (courtesy Matchmaker Machines Ltd)

Figure 3.41   Shuttle-type work tables (courtesy Cincinnati Milacron)

Figure 3.42   Six-station automatic work changer (courtesy Cincinnati Milacron)

Figure 3.41 shows two work tables, referred to as shuttle-type work tables, in front of a machine. The spindle nose without a tool can be seen in the background. (Guards have been removed from the machine so that the photograph can be taken.) Work clamped in fixtures could be loaded on to the work table, or work can be loaded directly on to the machine table. The work table on the left would be the first to be loaded. The work loading stage rotates to transfer the loaded work table to the actual machining station, which can be seen on the right-hand side. While machining is proceeding on one set of components, the empty work table is loaded with the next workpiece. When the machining has been completed, the work table holding the machined component is transferred back to the work loading unit, which rotates and transfers the second work table to the machining station. The interchange time of work tables is only 26 seconds. In Figure 3.22 shown previously one of the subtables has been transferred to the machining station.

Rotary automatic work changers can have 6, 8, 10 or 12 work stations. A six-station work changer is shown in Figure 3.42. Once fully loaded the machine can be left to operate with minimum supervision. The particular work is selected by the computer in the control unit as required to programmed instructions. The subtables must contain a form of identification which can be read by sensors such as the inductive elements used for tool identification described previously; these transmit the information to the computer so that the correct part program is input to the machine's control unit.

The use of subtables is particularly beneficial in flexible manufacturing systems (FMS), and if the machine is part of a DNC system the components are loaded on to the subtables at a work loading station away from the machine. An FMS cell with two rotary automatic work changers is shown in Figure 2.21 (see Chapter 2 for fuller details of this installation).

Work handling
When the work to be machined consists of castings, forgings and billets for one assembly it can be more productive to load automatic work changers with the different items rather than produce small batches of the same component. This procedure will ensure that all the related parts are completed at one set up.

Work pieces have to be manually loaded on to grid plates, pallets and subtables, at the machine itself or at a work handling station away from the machine. Work which has had all the machining operations completed has to be removed manually from the machine or at the work loading station. Figure 3.43 shows an operator loading work on to a pallet mounted on a rotary automatic work changer. The pallets can contain a number of components; the two pallets in the foreground of Figure 3.43 each hold two components. When these pallets are loaded on the machine they are rotated under program control to present the faces that have to be machined to the cutting tool. The loaded pallets are transported from the rotary automatic work changer to the machining centre by a mobile carrier. Figure 3.44 shows a loaded pallet being transferred on to a mobile carrier which transports it to the machining centre. Automatic guided vehicles (AGV) are used for transporting the loaded base plates or pallets to machining centres in fully automatic flexible manufacturing cells – an AGV can be seen in Figure 2.21. For smaller flexible manufacturing systems rail guided vehicles are being used as shown in Figure 2.22.

The pallets or subtables contain a form of identification such as the inductive capsules used for tool identification which can be read by sensors to ensure that the work loaded and the part program in the control are matched. The majority of work machined is of a size that can be handled without a great deal of trouble. If the work is heavy or cumbersome either overhead gantry cranes or robots are used for handling the work while it is being loaded on to the machine.

Turning centres
The holding of work to be machined on turning centres is to some extent simpler than on machining centres because the majority of the work is cylindrical at the start. However, be-

Figure 3.43   Loading work on to a rotary automatic work changer (courtesy Cincinnati Milacron)

cause the work may have to revolve at high speeds (rev/min) the work and the work holder has to be balanced so that there are no out of balance forces creating vibrations or eccentric

Figure 3.44   Work being transferred on to a guided vehicle carrier (courtesy Cincinnati Milacron)

running of the spindle that can effect the roundness of the work and cause excessive wear of spindle bearings.

The work blanks should be prepared or conform to the programmer's specification, as problems can arise if the diameter or the length of the work varies. Irregularity in blank size can also cause difficulties for the work holding method. Additionally, if the program has been created to cater for the largest work blank (maximum metal condition) when a work blank of minimum size is loaded it is possible for a significant amount of time to be wasted with the tool 'cutting air' i.e. unnecessary roughing cuts. It is possible to accommodate variation in work size by programming a sensor probe to contact the work before commencing machining. (See Chapter 10 for details of sensor probes.) Figure 3.45 shows a contact probe which is mounted in the turret and can be programmed to advance to contact the work when machining is completed and check the size of the work. Alternatively, the probe can measure the amount to be removed on a work blank and provide a feedback signal to

Figure 3.45   Component gauging (courtesy Monarch DS&G Ltd)

the control unit. The probe could also be used to detect the presence of work that has been automatically loaded. Sensor probes are particularly necessary on CNC machines in flexible manufacturing cells. The holding methods used are: mainly chucks for bar work or short length work; between centres or supported at one end by a centre; or turning fixtures for work of complex shape.

Chucks

There are two main types of chucks used – jaw chucks and collet chucks – either of which can be manually or power operated.

Chucks with two or three self-centring jaws are commonly used on turning centres and are capable of holding a work with a wide range of size. However, as the maximum diameter of the chuck increases, the safe operating speed of the jaw chucks decreases. Particular attention must be given to the maximum speeds used with chucks which have a scroll to move the jaws. Figure 3.46 shows a cross-section through a

chuck and spindle designed for high speed work on a CNC turning centre. These chucks have counter-balanced jaws which overcome the possibility of the jaws opening at high speeds. The tapered cones in the cross-section of the chuck in Figure 3.46 are part of the mechanism for opening and closing the base jaws. The jaws which hold the work are bolted to the base jaws.

Figure 3.46   High speed quick change jaw power chuck and hydraulic cylinder mounted on a machine spindle (courtesy Pratt Burnerd International Ltd)

Typically chucks for high speed CNC turning centres to grip work of 160 mm and 320 mm maximum diameter have safe maximum speeds of 6000 and 3000 rev/min respectively. Jaw chucks can be manually loaded, but when chucks are loaded by robots special care has to be taken to ensure that the work is gripped correctly and securely so that it runs concentrically, especially for second operation work. Back-stops fitted in the spindle bore can be used for positioning the work to ensure that the protrusion of the work is constant and to the programmer's specification.

Collet chucks

Collet chucks are very efficient and convenient for use on turning centres, particularly with automatic operation for bar feeding which can be programmed. The work is generally bar stock of convenient length (up to 4 m long) and preferred sizes in diameters. The tolerance on the diameter of the bar is usually small enough so that it is only necessary to adjust the collets when a bar is first loaded. Figure 3.47a shows a range of different types of collets, and Figure 3.47b shows a range of collet chucks. The collet chuck with two collets provides double grip which has greater resistance to pushback, particularly required for increased speeds and feed rates and when rotary driven spindle tools are used in the tool turret.

The individual collets can only accommodate a limited variation in the size of work, but with a range of collets bars up to 80 mm diameter can be held. Dead-length collet chucks are frequently selected because they hold the work accurately and efficiently. Bar stops mounted in the tool post or turret can be used to ensure that the correct length of work protrudes from the collet and is maintained to the programmer's specification. When the chuck is under automatic control, and there is no work in the chuck or the work is not clamped securely (lack of pressure), a signal is sent to the control unit and the program is stopped.

If the length of the work that protrudes beyond the end of the collet or chuck jaws is more than five diameters and large depths of cut or fast feed

rates are used, it should be supported by centres or steadies, as shown in Figure 3.48. It is not practical to use steadies for supporting work smaller than 32 mm diameter; for this size of work a sliding head turning centre with a supporting bush is used, see Chapter 2 and

(a)

(b)

Figure 3.47(a)   Collets; (b) collet chucks (courtesy Crawford Collets Ltd)

Chapter 4 for figures of sliding head turning centres.

Between centres

Work greater than 25 mm diameter and more than 10 diameters long that has to be machined over its complete length or to have further machining, such as cylindrical grinding, or other machining carried out on the outside diameter is generally mounted between centres. It is important that the length of the work blanks is to the programmer's specification, and also that the centreholes are a consistent size.

One method of driving the work is with the use of a unit which has a number of teeth which 'bite' into the end face of the work. The work is pressed against the teeth by a running centre which is located in a hydraulically powered tailstock. For this type of work driver the diameter of the work has to be larger than the drive unit and stiff enough not to bend under the axial pressure of the running centre. The pressure exerted by the running centre is monitored and a warning signal is created if the pressure reduces below the desired level.

Steadies

For very long work greater than 25 mm diameter, fixed and travelling steadies are normally used to ensure safe working practice. Figure 3.48a shows a steady and Figure 3.48b shows the steady supporting a multi-diameter shaft. The steady can be programmed to be at any position along the work, and the jaws of the steady automatically adjust to any diameter.

Turning fixtures

Turning fixtures are essentially faceplates with clamping bolts and locating plates or pins. Turning fixtures are mainly used to hold castings or forgings. The work has to be prepared to suit the location provided on the fixture. It is essential that the fixture is balanced correctly to rotate without any eccentricity, as vibration problems can easily arise if there is any out of balance of the rotating masses.

Work handling on turning centres

As explained in Chapter 2, Section 2.7 there are

(a)

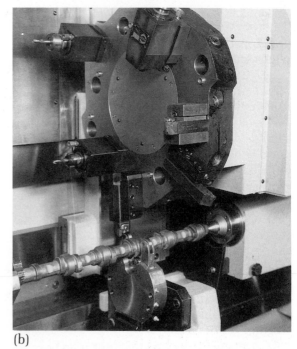

(b)

Figure 3.48(a)   Steady; (b) steady supported work (courtesy Monarch DS&G Ltd)

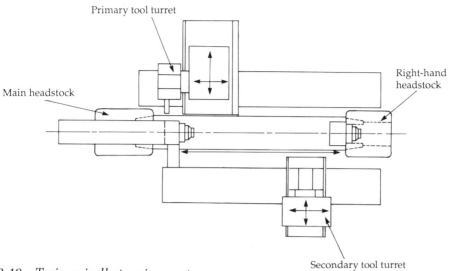

Figure 3.49   Twin spindle turning centre

Figure 3.50   Part catcher (courtesy Monarch DS&G Ltd)

turning centres where the transferring of the component is achieved by the right-hand headstock sliding along the bed until the end of the workpiece can be held in its collet; the rotation of the spindles in both headstocks is synchronized, and the component is gripped as it is parted off, the right-hand headstock sliding back to its machining position. The advantage of this method is that it is possible for the angular orientation of the workpiece to be maintained for any second operation cross drilling etc. Figure 3.49 shows the principles of operation of the twin spindle machine. It is possible to control machines with two turrets and headstocks (twin spindle, dual turret) so that different workpieces held in each headstock can be machined simultaneously.

When all machining has been completed the chuck opens and the components are ejected into a chute, or removed by a parts catcher which is programmed to move into position and carry the work to a location outside the machining area so that the operator can remove the component for checking if required. A part catcher is shown in Figure 3.50 ready to receive the component when it has been parted off.

The majority of the work machined on turning centres originates from bar stock, and safe working practice requires the bars protruding from the rear of the headstock spindle to be protected by a non-rotating tube. Automatic bar loaders

and automatic bar feeding can be used on turning centres. Figure 3.51 shows a bar feeder that can hold six bars at a time ranging in size from 5 to 80 mm diameter. The bars are supported hydrodynamically in oil to eliminate vibration and will ensure noiseless feeding of round, hexagonal or asymmetrical sections.

Workpieces originating from bar stock greater than 80 mm are generally cut off to the required length on a power saw and loaded on to the turning centre as work blanks. These work blanks, and also castings or forgings, are loaded manually or on fully automatic manufacturing cells loaded by robotic work loaders. The automatic loading of brake drums is shown in Figure 3.52. When the machining is completed on one drum an overhead loader moves into position.

Figure 3.51   Bar feeder (courtesy N.C. Engineering Ltd)

Figure 3.52   Automatic loading of brake drums (courtesy Monarch DS&G Ltd)

In Figure 3.52a a machined brake drum is being held on the outside by the grippers of an overhead loader, while the brake drum to be machined is held in the bore by the grippers of a second overhead loader. The jaws of the chuck holding the machined drum are released and the first overhead loader lifts the part clear. Figure 3.52b shows a loader holding the drum to be machined ready to be moved into position to be held by the chuck jaws. After changing the work the loaders move to the work loading/unloading station, and the work is replaced.

Overhead gantry cranes are commonly used for handling extremely large workpieces on large turning centres.

3.14 Linking structure

Two of the major constructional features of a machine tool are:

(a) The means of holding and moving the work
(b) The means of holding and moving the tool.

These units have to be assembled so that the movements required take place in the correct relationship. The structure which joins the two is referred to in this text as the linking structure. All the motors, mechanisms and other assemblies which constitute the machine tool are rigidly fastened to the linking structure or are aligned to each other by the linking structure.

The nature of the cutting action results in milling machines and to some extent turning centres being subject to fluctuating and variable forces during material removal operations. It is essential that the linking structure does not bend or move in any way under the action of these forces. All parts of the machine must remain in the correct relative relationship regardless of the amount and direction of the forces being applied. It is also essential that when required all slides move simultaneously.

Numerically controlled machine tools are involved with actual metal removing operations for a greater proportion of the time than manually controlled machines. Numerically controlled machines are therefore subject for longer periods to the dynamic forces that result

when actually removing materials. The linking structures therefore generally tend to be of more substantial proportions. It is advantageous that the structure should combine high stiffness with low weight and provide vibrational rigidity; in particular, it should give freedom from low frequency structural vibrations. High stiffness gives long-term dimensional stability and results in increased accuracy being maintained. The majority of the linking structures are of a closed box section, which has the greatest bending strength and torsional rigidity for equal weight when compared with tee, H and tubular sections.

The linking structure can have two main configurations: a column (vertical orientation), usually for milling and drilling machines; or a bed (horizontal orientation), the common form for turning machines.

Milling and drilling machines

The column structures are mainly milling and drilling machines. These machines have at least three axes of numerically controlled movement. A knee and column type of milling and drilling machine is shown in block format in Figure 3.53. The knee can be moved vertically on slideways machined on one side of the column;

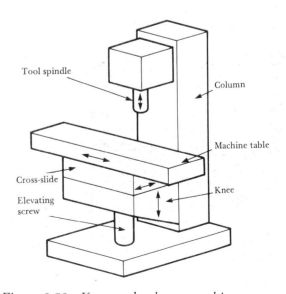

Figure 3.53   Knee and column machine

on most machines this movement is not programmable. Vertical raising or lowering of the knee enables work of different heights to be accommodated on the work table. Generally the knee is moved manually during the setting-up stage, and clamped in position. If the work is heavy, consideration has to be given to the design of an elevating screw and nut mechanism to provide mechanical assistance. The maximum weight that can be loaded on a machine which has a work table of 500 mm by 300 mm is of the order of 130 kg. Figure 3.54 shows a knee and column CNC machining centre (the guards have been removed from the machine for the purpose of taking the photograph). The small round handwheel on the right of the machine table is an electronic handwheel used for positioning the tables during the setting up of the machine. The button in the middle of the top row on the spindle head is for actuating the power draw bar.

The work table normally has numerical control on only two axes of horizontal movement. The third numerically controlled axis is usually provided by the vertical movement of the tool spindle. This movement is usually limited to about 150 mm; if the spindle extended further it would not be adequately supported, and there would be a danger of the tool being deflected sideways under the action of the cutting force. As stated previously, there is a limitation on the weight of the work that can be loaded on to the work table because of the problems of manually moving the knee, and also to reduce the possibility of the knee assembly deflecting under the weight of the work.

For large work the machine configuration commonly adopted is a solid bed at the base of a column, as shown in Figure 3.55. A work table which may have at least two numerically controlled axes for translational movement is positioned on the solid bed. A third axis of numerically controlled translational movement is obtained by the tool head moving on vertical slideways on the column. On many machines, numerically controlled rotary tables can be mounted on the work table. It is also possible for the tool spindle to be mounted on a numerically controlled rotary head for machining holes or faces at an angle. On these machines there can be up to five numerically controlled axes.

Figure 3.54   Knee and column CNC milling machine (courtesy Bridgeport Machines)

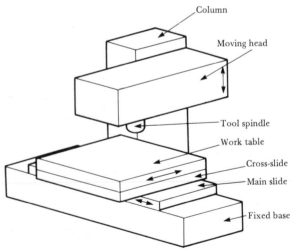

Figure 3.55   Solid base, fixed column vertical spindle machine

There are machines where the work table has only one numerically controlled axis of translational movement. The two remaining numerically controlled axes for translational movement are provided by moving the tool head in two directions on the column, as shown in Figure 3.56. With this configuration the base has to be twice the length of the movement required of the work table, to ensure that the work table is supported at all times.

On an alternative configuration the base is fixed and does not have any translational movement. All translational movement is provided by the column moving in two horizontal directions and by the head moving vertically on slideways on the column, as shown in Figure 3.57a. Figure 3.57b shows an actual machine body with a fixed bed and the tool head moving in the three translational directions. With this type of configuration the length of the base and the amount of horizontal movement of the column only has to be the maximum length of the work. Figure 3.58 shows a moving column machine with a work table with two rotary tables, both of which can have programmed rotary movements. (The guards have been removed for the purpose of taking the photograph.)

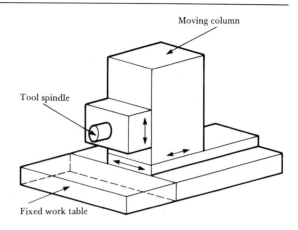

Figure 3.57(a)   Moving column machine

Turning centres

The configuration of centre lathes has traditionally been a horizontal bed on which a saddle moves longitudinally as shown in Figure 3.59. The cross-slide moves transversely on the saddle, and there can be tool posts mounted at both the front and rear of the cross-slide. The head-

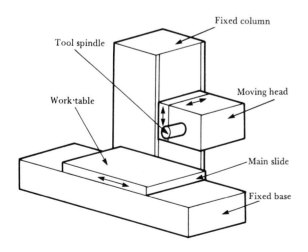

Figure 3.56   Solid base, fixed column horizontal spindle machine

Figure 3.57(b)   Constructional elements of moving column machine (courtesy Bridgeport Machines)

81

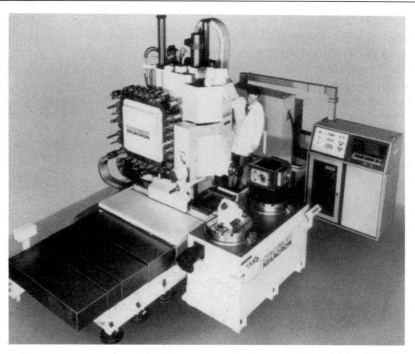

Figure 3.58   Travelling column machining centre (courtesy Cincinnati Milacron)

stock with the work spindle is fixed at one end of the bed. This arrangement is suitable for manual control because the tool post is fairly easily accessible to the turner. One problem is that although the tools are near enough to be seen clearly, so that they can be positioned accurately, the turner has to lean over the tool post. One

other problem with the horizontal bed is that, in order to provide the bed with the strength and rigidity to support the saddle and resist cutting forces, the bed has to have cross-webbing; this results in the swarf tending to collect on the bed slideways. Concentrations of swarf create a source of heat; this causes temperature gradients

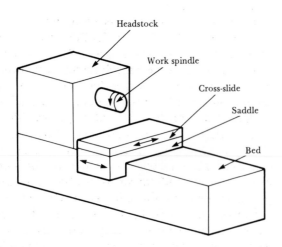

Figure 3.59   Traditional centre lathe

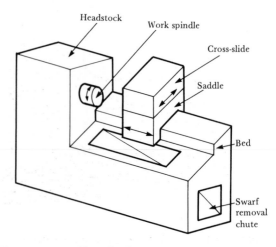

Figure 3.60   Slant bed turning centre

through the machine, which affect the alignment of the various slideways. In addition, the swarf can cause the bearing faces of the slideway to deteriorate and become worn.

One development in the design of a linking structure for turning centres is to position the saddle and cross-slide behind the work spindle on a 'slant' bed, as shown in Figure 3.60. Figure 3.61 shows a slant bed casting with the machined slideways. One effect of the slant bed is that the swarf can fall freely to the base of the machine, from where it can be removed automatically. A slant bed machine is more ergonomically acceptable; for example, the operator can see the tools cutting more easily than on the traditional horizontal arrangement. Changing of tools is not a problem because during the setting-up stage the tool slides can be brought under manual control to the front of the machine within reach of the turner. An additional tool turret can be provided on a secondary saddle and cross-slide mounted on the end of the bed. Figure 3.62 is a slant bed lathe with the guards open showing the tool turret.

There are turning centres which have vertical beds, as shown in Figure 3.63. One version of this configuration has the saddle moving on two cylindrical guideways of approximately 160 mm in diameter.

For work of 2 m diameter and larger, the bridge-type configuration shown in Figure 3.64 has been used; on these machines the work spindle is vertical. Because of the large diameter of the work the spindle speeds are usually quite slow, the maximum being of the order of 300 rev/min.

Cast iron structures

Traditionally the linking structure has been made of cast iron, which has excellent vibration damping properties. Another advantage of cast iron, as mentioned in Section 3.8 for flat guideways, is that any slideways can be surfaced hardened and ground to provide efficient and long-life bearing surfaces. One problem with cast iron is that the molten iron must flow quickly and fill the mould when the structure is being cast; it is thus difficult and sometimes impossible to place strengtheners where required in the structure, as these may prevent full casting. On some machines with complex cavities, the sand cores that were used to provide the cavities are sealed in the casting. It has been found that this can be an advantage, as the sand may increase the vibration damping qualities of the structure.

Concrete structures

There are machines where the structure is partly made of concrete. This results in a cost saving compared with cast iron, especially for one-off machines. Concrete for machine tool structures is not a new idea; it is reported that tests were carried out in 1917 when concrete was used to replace cast iron for the bed of lathes. More recently, in order to reduce the cost of machines, attempts have been made to use reinforced concrete for the bed with metal plates embedded in the concrete. Metal strips are bolted to the plates to provide the slideways. One of the advantages of this design is that if the slideways wear it is comparatively easy to replace them. Another important reason for the use of concrete is that it has excellent vibration damping

Figure 3.61 Casting for slant bed turning centre (courtesy Cincinnati Milacron)

Figure 3.62   Slant bed turning centre with conversational programming facilities (courtesy Cincinnati Milacron)

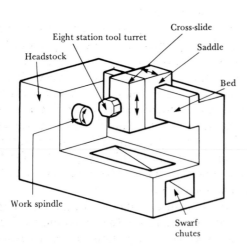

Figure 3.63   Vertical bed turning centre

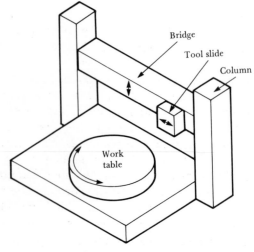

Figure 3.64   Bridge-type turning (boring) centre

properties. Concrete is a poor conductor of heat and so does not transfer heat from the swarf to the metal of the machine structure.

There is a machine available which is a combination of a cast iron box section mounted on a concrete block. All the drive units are secured to the box section to ensure alignment, and items such as the swarf removal unit are fastened to the concrete base.

Welded frameworks

An increasing number of machine tools have been constructed as welded frameworks. The light-weight welded steel construction (with cross-webbing) has great advantages, as it is possible to distribute the material most favourably in the structural elements and to make full use of the material properties. The distribution of material in the best positions cannot always be obtained with cast materials.

3.15 Overall considerations

There are a number of factors which have to be considered:

(a) Swarf removal
(b) Positioning of transducers
(c) Accuracy of machining
(d) Accuracy of the machine
(e) Safety of the machine
(f) Ergonomic design.

Swarf removal

Because of the high rates of material removal possible with cemented carbide and ceramic tools, a large quantity of swarf is created; its removal can be a major problem. As stated previously, concentrations of hot swarf cause temperature gradients through the machine which affect the alignment of the various slideways. Another problem is that if the chippings collect in the cutting area they will provide an obstruction to tool or work changing. To overcome these difficulties special openings in the machine structure are provided where the swarf is directed via chutes to swarf disposal units. The openings in the castings for the clearance of swarf are shown in Figures 3.60 and

3.63. These units remove the swarf as quickly as practicable. There are various designs of swarf removal mechanisms such as conveyer belts and screw mechanisms which transport the swarf to bins that are separate from the machine. If different materials are being machined it is relatively easy to have separate bins for the different materials. There is a particular problem when turning materials which result in a continuous ribbon type swarf. The use of formed chip controllers on the indexable inserts of the tool help to break up the swarf.

Cutting fluid is used as a coolant to remove some of the heat created during material removal. The fluid continually flows over the work and tool and absorbs heat, and so will gradually get hotter. As the heated cutting fluid runs over the machine structure, heat will be transferred into the material of the structure, and temperature gradients will be created through the machine. Special cooling systems can be used to reduce the temperature of the cutting fluid and therefore the thermal gradients created in the structure. It is also important that the temperature of the lubricating oil should be kept steady to reduce any thermal effects created by hot oil on bearings and spindles. Cutting fluids are also used to lubricate the cutting tool, and to wash away small particles of swarf. There are different types of cutting fluids for different materials.

Positioning of transducers

The location of the position monitoring transducers is very important, and they should be mounted according to Abbe's principle of alignment. This requires that the line along which the measurement is taken is coincident with the line of movement at the cutting tool. It may not be practical for the principle to be completely implemented, as the work is in contact with the tool and it is not physically possible for the measuring transducer to be in contact with the work at that point.

Linear transducers should be mounted so that they monitor (measure) the actual movement of the carriage. The rotary transducers are usually mounted on the end of the leadscrew that moves the carriage, and so they monitor the rotation of the leadscrew and not the actual linear movement of the carriage. See Chapter 5 for further details on the mounting of transducers.

Machining accuracy

The accuracy of the work produced – the machining accuracy – is dependent on the work/tool relationship when material is being removed. This is dependent on a number of factors, such as:

(a) The machine accuracy
(b) The measuring accuracy of the transducers (their resolution)
(c) Any deflection of the tool when cutting
(d) The size and amount of any wear of the tool
(e) Temperature variations in the work and tool.

The accuracy with which the transducers measure the movement of the carriages is therefore only one factor relating to the accuracy of the work produced. The biggest factor is the machine accuracy. To obtain maximum accuracy it is possible to calibrate each axis of the machine under different operating conditions, and any inaccuracies can be stored in the control unit's memory. When a command to move to a particular coordinate is input it is compared with the stored information, and the value of the coordinate is modified to the correct value. As a result of the calibration the numerically controlled machine can produce components that are more accurate than the individual parts of the machine itself. The calibration can be repeated as often as results show that it is necessary. The results of the calibration, stored in the control unit's memory, are the electronic equivalent of the mechanical error correcting device that has been fitted to the leadscrews of jig borers for a number of years.

Machine accuracy

Machine accuracy is the accuracy of the movement of the carriages and tables. This is influenced by:

(a) The geometric accuracy of the alignment of the slideways
(b) The deflection of the bed and other features due to load
(c) Any temperature gradients existing through the machine
(d) The accuracy of the screw thread of any drive screws, and the amount of backlash (lost motion)
(e) The amount of twist (wind-up) of the shaft, which will influence the measurement of rotary transducers (lost motion)
(f) The responsiveness and accuracy of the control system
(g) Wear of moving parts.

Accuracy of the final positioning of point-to-point machines (drilling and boring machines) is of more importance than that the actual path followed to arrive at the position is a straight line. With continuous path machines it is important that the work/tool relationship is correct at all times.

It has been practice for a number of years that, when a new machine is being installed, it is first levelled up and secured to suitable foundations, possibly using anti-vibration mountings. Alignment and other acceptance tests are then carried out to check that the machine has the capability of accurately performing the movements required. Numerically controlled machines are now very sturdy and many have a three-point support, so that securing and levelling up are not so critical; however, they should still be subject to acceptance tests. Practical tests using the maximum movements of the carriages, and taking light finishing cuts with sharp tools, will prove beneficial. Special test programs can be developed which can be used at periodic intervals to test quickly and efficiently that the machine is maintaining its capability. Laser measuring equipment is available for checking the response and accuracy of the amount of movement of the carriages to input signals. Figure 3.65 shows laser testing equipment being used in the acceptance tests concerned with the movement of the work table of a machining centre.

The accuracy of the work produced on manually controlled machines is influenced by a number of factors additional to those given above for numerically controlled machines:

Figure 3.65   Laser testing of machine alignments (courtesy Bridgeport Machines)

(a) The pressure exerted against stops
(b) Setting of tools
(c) Operator's interpretation of the reading of the index dial on the cross-slide screw.

To some extent these factors also apply to those machines with the third level of control that operate on a fixed cycle of movements (see Chapter 2). On both the manually controlled and fixed cycle machines the setting of the tools can also vary with each set-up and it can be very time consuming to adjust the position of the tools or work. Numerically controlled machines have the advantage that exactly the same information is input for every component, and the positional relationship is continuously monitored. As a result the repeatability of the positional relationship of the work and tool is more consistent, and consequently the work produced is more consistent. It is also comparatively easy to edit the program to adjust the position of the tools or work on numerically controlled machines.

Manufacturers of machine tools have paid considerable attention to the mechanical design aspects of numerically controlled machine tools. However, as a result of the high utilization of the machines, and in order to maintain their accuracy, it may be necessary for the bearings, the actuating mechanism or other constructional features such as slideways to be replaced at periodic intervals. Developments in control systems have resulted in otherwise serviceable machines becoming uneconomic in operation, and there are cases where new control systems have been fitted to machines under a retrofit scheme.

Safety of machines

When the machine is under continuous automatic control the operator has very little involvement in the actual running of the machine, and three different problems are created:

(a) Overloading of the machine
(b) Guarding of the machine guideways
(c) Safety of the operator.

The difficulty with the *overloading* of the machine is that, since the operator has very little if any physical contact with the machine, there is no tactile (feel) sense which enables the operator to judge how the machine is reacting to the applied loads. The operator has to depend on the sensors fitted to critical points of the machine to provide the information. Adaptive control will considerably help in preventing the machine being overloaded, and will also ensure that the machine is working at maximum capability (see Section 2.10).

The *guarding* of the guideways is absolutely essential for the efficient long-term life of the machine. Various types of collapsible guards and covers are used to protect the slideways, drive screws and measuring equipment (transducers etc.). Special jets of cutting fluid are used to wash away swarf and clear the tool work area. The wipers placed directly in contact with the slideway faces must be given continuous attention and replaced at the required intervals to prevent them becoming contaminated. To prevent contamination of the electronic circuit boards by dirt particles, cutting fluid fumes etc., air filters are mounted on the sides of the control unit; these must be replaced at regular intervals.

The *protection of the operator* is of prime importance and, where it is not possible to install efficient metal or plastic guards around the machine, proximity protection has to be provided. A problem to be overcome is that there must be good access for setting up, but safe working conditions when the machine is operating.

In addition to the table mounted guards shown in Figure 3.66, which provide a limited and localized protection against flying swarf and cutting fluid, there are a number of techniques of guarding numerically controlled machines, such as overall guards, pressure mats and light barriers.

Overall guards

These are also known as perimeter guards. The difficulty of access for setting up is generally overcome by the fitting of large sliding doors. The doors have various types of electrical interlock switches fitted; during the setting up, the signal from the interlock switches can be cancelled. If the doors are opened when the machine is operating under program control, this activates some form of warning signal, either visual such as a flashing light or auditory such as a buzzer. On some machines the power may be switched off if the doors are not closed within a short interval. Windows made of high-strength plastic are fitted to the guards so that

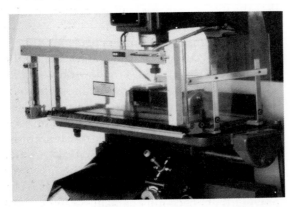

Figure 3.66   Table mounted guards (courtesy Bridgeport Machines)

the machining area is visible. One advantage with the overall guards is that they prevent excessive splashing of the cutting fluid and restrain flying swarf, which can be a hazard. When machining some materials, particularly plastics, it may be beneficial to have extractor fans to clear the fumes from within the enclosure. Overall or cubicle guards are particularly suitable for machines which have carriages moving over a fixed bed or base, such as turning centres. Figure 3.67 shows a machining centre with overall guards that have been designed so that the work on a pallet can be changed while machining is proceeding with other work. The swarf removal mechanism can also be seen.

Pressure mats

These are commonly found around milling and drilling machines where the tables can extend either side of the machine. The tables can move quickly and quietly during positioning operations and can be a hazard if the operators are standing too close. The mats are generally placed around the machine, and if anyone treads on them a visual or auditory warning signal is generated. There is no difficulty in providing access during the setting up of the machine, as the signal from the mats can be cancelled with the use of a switch on the control unit. Usually there is a limitation on time or on the number of components produced with the pressure mat override switch active, and it is not possible to run the machine continuously.

Light barriers

These are also mainly used on milling and drilling machines. The light barrier consists of a source of light, usually infra-red, sending a beam to a light-sensitive cell. If anything obstructs the light beam a warning signal is generated. The light barriers are placed around the machine, and they can be made inactive with a switch on the control unit. On the machine in Figure 3.68, an infra-red light beam is provided between the two guide rails which are used for the loading of pallets on to the work table. If an operator moves between the guide rails when a pallet change is due, the machine

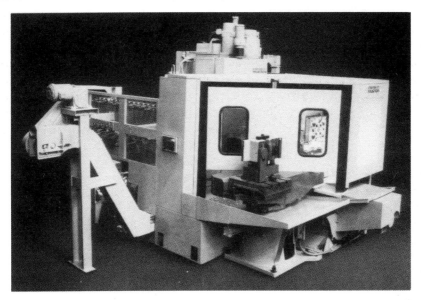

Figure 3.67   Overall (cubicle) guarding for a CNC machining centre (courtesy Cincinnati Milacron)

stops. It is possible to restart the machine without loss of position.

Ergonomic design

Since there are very few operating levers and hand wheels on numerically controlled machines, their positioning is not as critical as on manually operated machines. A fairly large cabinet is required for the control unit and electrical transformers, rectifiers etc.; a typical

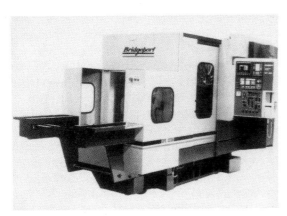

Figure 3.68   CNC machining centre with infra-red light barrier (courtesy Bridgeport Machines)

size is 600 mm wide, 600 mm deep and 1 m high. The cabinet is usually placed a short distance away from the machine to allow adequate access to the machine. The operating panel containing the input reader, keyboard, switches, buttons etc. is commonly inclined at a small angle and placed on top of the cabinet. Taller control cabinets have the control panel mounted vertically, with the visual display unit located on one side. The layout of control panels has had to be given very careful consideration, and there are a number of symbols that are recommended for indicating certain control features. British Standard 3641:1980 provides designs of recommended symbols.

As a help during setting up, and to avoid the setter continuously having to move between the machine and the control cabinet, certain of the various switches and buttons are duplicated and arranged on an overhead pendant box which can be swung into position close to the machining area. The pendant provides the setter with access to the essential control features required when setting up the machine. The reduction in the physical size of computers has meant that a complete control unit can now be placed on a pendant (Figure 3.69).

Figure 3.69 CNC machining centre with pendant control unit (courtesy Cincinnati Milacron)

Questions

3.1 Describe the principle of operation of a stepper motor.

3.2 For what type of NC machines are the following electric motors used: (a) servo motors and (b) stepper motors?

3.3 Detail the different types of slideways that are used on machine tools.

3.4 What advantages are to be gained by the use of linear bearings using rollers or balls?

3.5 Describe the difference between (a) hydrostatic and (b) hydrodynamic spindle bearings.

3.6 Specify the advantages to be gained by the use of recirculating ball screws or roller screws.

3.7 What is the main difference in the method of use of a tool turret and a circular tool magazine?

3.8 Explain the principle of operation of an automatic tool changer.

3.9 Detail the essential requirements for work holders on NC machines.

3.10 Name the various types of work holding arrangements that can be used on NC machines, specifying the type of work for which each method is particularly suited.

3.11 Describe with the aid of line diagrams four different machine configurations for machining centres, and four different configurations for turning centres.

3.12 What problems arise due to the high rates of material removal that are possible on NC machines?

3.13 What are the safety problems related to the actual running of NC machine tools?

3.14 Describe three different methods of guarding used on NC machines.

3.15 Explain the difference between (a) machine accuracy and (b) machining accuracy, detailing the different factors that can affect both.

3.16 Why is the work produced on NC machines generally found to be more consistently satisfactory than work produced on other types of machine tools?

Modes of operation

4.1 Designation of axes on NC machine tools

In Chapter 1 it was explained how the generating, forming or copying principles can be used in the production of various geometric shapes on machine tools. Numerically controlled machine tools invariably use the generating principle in the machining of components because it is easier and more economic to control the movement of the tool or work rather than to make special tools or templates for a particular workpiece.

Numerically controlled machines are frequently used for making specialist tools, templates and cams for use on other machines, especially for machines used for mass production. The range of engineering components produced by machining on numerically controlled machines can be broadly classified into:

(a) Those that can be contained within a cuboid. The profile of the component consists mainly of straight or curved lines, and may contain holes in some of the faces. This type of component normally requires three views to be drawn in orthographic projection to show the overall length, height and width or depth.

(b) Those whose shape is basically cylindrical. The profile consists of a circle and a line that is straight or curved. There may be holes running axially through the component.

The work blanks for the components in class (a) could be solid blocks, castings or forgings depending on the total quantity to be produced. A typical component is shown in Figure 4.1. This type of component is generally produced on machine tools where the tools rotate to create the cutting speed. The plane (flat) or profiled surfaces are generated as a result of the movement of the work. The machines used to produce these components normally have a minimum of three independent axes of movement.

The work blanks for components of class (b) could be bar stock, castings or forgings depending on the total quantity required. The components are produced on turning centres, and as explained in Chapter 1 there are four different techniques for generating cylinders. Numerically controlled machines generally produce the cylindrical type of components where the cutting speed is created by the rotation of the work, and the rotation of the work also generates circles of revolution. The movement of the tool along and across the axis of work generates the profile of the work. A large number of machines have a minimum of only two independent axes of movement.

There are turning centres with at least four independent axes of movement. When programming these machines, care has to be taken to ensure that tools on different axes are not programmed to be on a collision course. At one time a simulator mounted on a drawing board

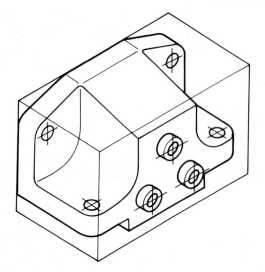

Figure 4.1 *Component contained within cuboid*

using the Cartesian system defines the X, Y and Z axes, as shown in Figure 4.2. The Z axis is always considered to be vertical. On a numerically controlled machine the direction of the axes does not always conform to the Cartesian system.

For numerically controlled machine tools, a programming convention for the designation of the axes of movement has been recommended by ISO. For convenience and standardization the convention is generally adopted, but the designation of the letters to be applied to the various axes (spindles and slideways) of the machine tool is dependent on the manufacturer of the machine. The programmer has to use the axis designation specified by the manufacturer. It is generally found that the Z axis has to be defined first.

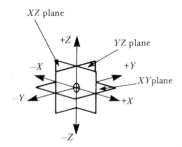

Figure 4.2 *Cartesian coordinate system*

was used to check for the possibility of a collision. The simulator contained tool templates fixed to special heads which could move along two pairs of two axes to provide coordinate movement. The heads are moved manually and the amount of movement is indicated on a two-axis digital readout. The heads are moved to program coordinates and the possibility of a collision can be seen and predicted. The simulator has now tended to be replaced by computer systems with graphic display units.

The machine tools used for the production of both cuboids and cylinders must be capable of performing movements that cause the required geometric shapes to be reproduced accurately and consistently.

On numerically controlled machines the geometry of the components is contained within the numerical data which is input to the control unit of the machine tool. This programmed data is used to control the amount and direction of movement of the tool or work so that components are produced to the desired shape and size. For this reason it is convenient and has become practice for the shape of the components to be dimensioned by coordinates using X, Y and Z planes. It is logical that the axes of the machines should be similarly designated. The customary designation of the axes and planes

Z axis

As stated above, this must be the first axis to be defined. The convention adopted for numerically controlled machines is that the Z axis is identified with the axis of the main (i.e. principal) machine spindle. This provides the cutting power or cutting speed.

If there are a number of spindles that provide cutting power on a machine tool, then the one which is used most is designated the principal spindle, and this is the Z axis. The other spindle or slideway, which may be parallel to the main spindle (Z axis) and on which the movement can be programmed, is designated as the W axis. If there is a third spindle or slideway parallel to the Z axis, this is designated the R axis.

For those machines that do not have a spindle that can be considered to be the main spindle, the Z axis is in a direction at right angles to the work table surface.

On some machines (vertical milling and drilling machines) the cutting tools rotate and are held in the main spindle; on other machines (turning centres) the workpieces rotate and are mounted on the main spindle.

On a vertical milling and drilling machine the Z axis is vertical. On the majority of vertical milling machines it is the tool movement (up and down) which is controlled. However, in some instances it is the table which moves vertically up and down. In either case the Z axis is vertical. If both tool and work table can be moved vertically under program control, then it is practice for the axis that is the most conveniently controlled to be the Z axis and the other axis to be the W axis. Generally the spindle on which the tool is held is the Z axis and the table is the W axis.

On a turning centre the work is mounted on the main spindle, the axis of which is usually horizontal and parallel to the bed, and therefore movement of the tool along the bed parallel to an extension of the spindle axis is the Z axis movement. This condition also applies to the 'slant' bed and vertical bed lathes, the axial movement of the tools being parallel to the axis of the spindle.

X axis

The X axis is generally horizontal and parallel to the work holding surface. If there are a number of carriages mounted on separate parallel slideways, one of the slideways is termed the primary slideway and designated the X axis; the secondary and tertiary slideways are designated U and P axes respectively.

When one is standing at the front of vertical milling and drilling machines, the movement of the table to left and right is an X axis movement.

On a turning centre the X axis movement is radially oriented to the work, i.e. it is the in and out movement of the cross-slide.

Y axis

On a three-axis machine the Y axis is at right angles to the other two axes, because the three axes X, Y and Z are always mutually perpendicular to each other. If there are a number of slideways at right angles to the X and Z axes, one is designated the Y axis and the others are designated V or Q axes.

Examples of nomenclature of axes

Figures 4.3a–e show the recommended nomenclature of the axes for a variety of machine tools.

A traditional centre lathe is shown in Figure 4.3a; this machine has two axes of movement. The main spindle axis is horizontal, and therefore by convention the horizontal axial movement of the saddle parallel to the main spindle axis is the Z axis. The X axis is at right angles to the Z axis and is horizontal as shown. The tools on these lathes are usually clamped in position at the correct centre height and do not move vertically; a Y axis is therefore not required.

Figure 4.3b shows a turning centre with four translational movements. The vertical slide is considered to be the primary or principal slide. Therefore the Z axis is accredited to the horizontal axial movement of the saddle, parallel to the main spindle axis on this slideway. The transverse movement on the principal slideway is by convention designated the X axis; this means that the actual movement is in a vertical direction. The horizontal axial movement of the secondary slideway parallel to the main spindle axis is designated the W axis, and the transverse movement on this axis is designated the U axis.

Figure 4.3c shows a traditional vertical mill of the knee and column type. The axial movement of the main spindle is designated the Z axis, and the horizontal movement of the machine table is designated the X axis. The cross-traverse of the machine table is designated the Y axis. The vertical movement of the knee has not been designated with an address because this movement is not programmable.

Figure 4.3d shows a bridge-type vertical turning (boring) centre which has four translational movements, Z and W being axial movements parallel to the main spindle axis and X and U being radial movements. The tool head nearest the control panel is generally used most frequently, and the vertical movement of that tool

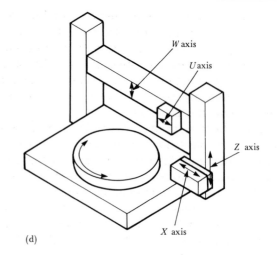

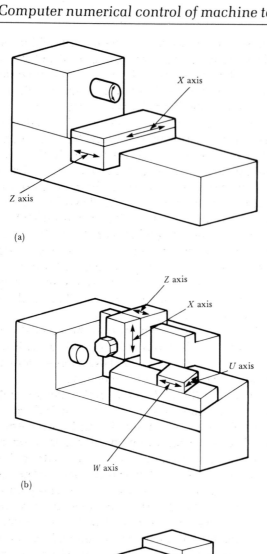

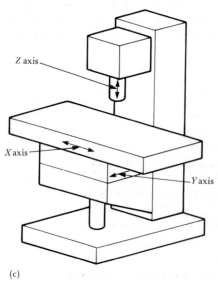

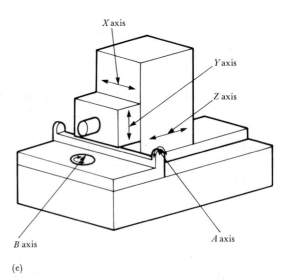

Figure 4.3(a) *Centre lathe; (b) turning centre with two tool slides; (c) knee and column machine; (d) bridge-type turning (boring) centre; (e) five-axis machining centre*

head is designated the Z axis, with the associated radial movement designated the X axis. The vertical movement of the bridge is designated the W axis. The radial movement of the tool head across the bridge is parallel to the X axis movement and is designated the U axis. The cutting speed is provided by the rotation of the work table, which only requires speed control and does not have to be designated any axis.

Figure 4.3e shows a moving-column horizontal spindle machining centre with three translational movements, Z, X and Y and two rotational movements. The Z axis is the horizontal movement of the column parallel to the main spindle axis, and the X axis is the other horizontal movement of the column. The vertical movement of the tool spindle head is designated the Y axis. The rotational movement of the rotary work table is designated the B axis because its axis is parallel to the Y axis. The tilting of the machine table is designated the A axis because its axis is parallel to the X axis. An explanation of the designation of programmable rotary movements is given in the next section. The rotation of the tool spindle provides the cutting speed; it only requires speed control and does not have to be designated any axis.

A travelling column machining centre of different configuration is shown in Figure 4.4. This machine has three translational movements and one rotary movement. The horizontal movement of the work table provides the Z axis, the horizontal transverse movement of the column provides the X axis and the vertical movement of the spindle head provides the Y axis. The rotary movement of the work table is the B axis because its axis is parallel to the Y axis. This configuration permits for easier loading of the pallets on to the work table from an automated pallet loading system.

A machining centre of compact design with a vertical fixed column of box construction is shown in Figure 4.5a. The standard machine has three translational movements X, Y and Z and one rotary movement the B axis. The essential constructional details of the machine are shown in Figure 4.5b. The X axis carriage travels on vertical slideways and the Z axis headstock travels on horizontal slideways. Both the X and Z axes have square slideways, while the horizontal movement of the pallet table on the Y axis has cylindrical guideways. The rotation of the pallet table is designated the B axis because its axis is collinear with the Y axis. It is possible to mount an indexing table (not shown in Figure 4.5) on the pallet table of the B axis which provides another rotary movement whose axis is parallel to the X axis and therefore this rotary movement provides the A axis. This table is known as the fifth plane table.

Having all the slideways on the vertical column enables the swarf to fall directly away from the machining area, and reduces any thermal distortion of the machine members.

Although the designation of the axes generally conform to that recommended by ISO, some manufacturers have found it propitious to deviate. Figure 4.6 shows a twin spindle, dual turret sliding head mill-turning centre with guide bush support which has five translational movements and two rotary movements. All the guards and control unit have been removed to show the essential features of a slant bed construction. The horizontal movement of the main spindle has been designated by the manufacturers as the

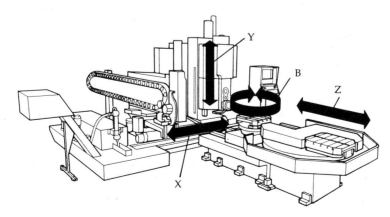

Figure 4.4   Travelling column machining centre (courtesy Cincinnati Milacron)

(a)

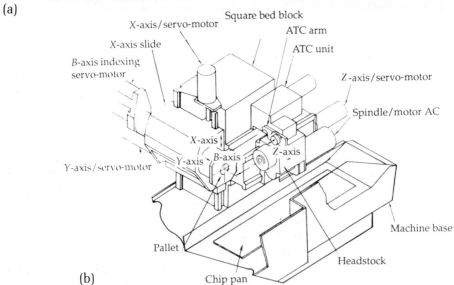

(b)

Figure 4.5(a) Fixed column machining centre; (b) schematic layout of fixed column machine (courtesy Matchmaker Machines Ltd)

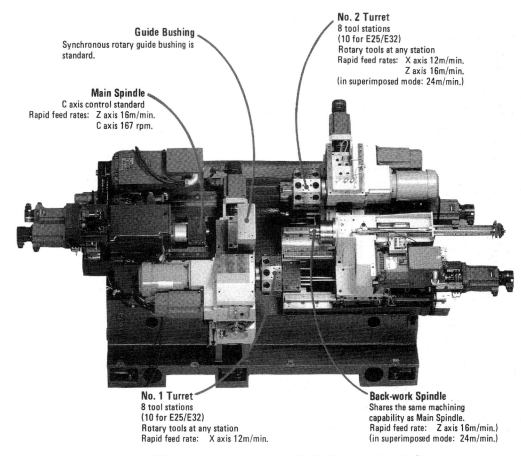

Guide Bushing
Synchronous rotary guide bushing is
standard.

No. 2 Turret
8 tool stations
(10 for E25/E32)
Rotary tools at any station
Rapid feed rates: X axis 12m/min.
Z axis 16m/min.
(in superimposed mode: 24m/min.)

Main Spindle
C axis control standard
Rapid feed rates: Z axis 16m/min.
C axis 167 rpm.

No. 1 Turret
8 tool stations
(10 for E25/E32)
Rotary tools at any station
Rapid feed rate: X axis 12m/min.

Back-work Spindle
Shares the same machining
capability as Main Spindle.
Rapid feed rate: Z axis 16m/min.)
(in superimposed mode: 24m/min.)

Figure 4.6   Sliding head mill-turn centre (courtesy N.C. Engineering Ltd)

$Z1$ axis; the parallel horizontal movement of turret No. 2 is designated the $Z2$ axis. The right-hand headstock (back work spindle) can slide along the bed and this third parallel horizontal movement is designated the $Z3$ axis. The transverse horizontal movement of turret No. 1 mounted on the guide bush bracket is designated the $X1$ axis and the parallel transverse movement of turret No. 2 is designated the $X2$ axis. The indexing of the main spindle for cross drilling and milling operations is designated the $C1$ axis; the indexing of the backworking spindle for similar drilling and milling operations is designated the $C2$ axis.

Angular positioning of headstock

During the machining operations using any indexing or rotational movement of the spindles, the spindle is driven by a different motor from the motor used for rotating the main spindle at the required cutting speed. Figure 4.7 shows a drive for the angular control of the headstock spindle, positioning accuracy is reported to be ± 0.01 degree using rotary position transducers. The motor and drive causing the angular movement of the headstock spindle are disengaged during any normal turning operations.

Axes designation is also applicable to all machine tools. Figure 4.8 shows the designation of axes on an EDM wire cutting machine. The X and Y axes provide the guidance of the work table along the actual cutting path. The U and V axes control the position of the upper wire guide in relation to the lower guide. This generates the slope or clearance angles of the cut surfaces. The

U and V axes are collinear with the X and Y axes respectively. The Z axis provides the vertical adjustment of the wire guide heads for different heights of work.

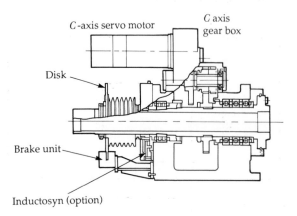

Figure 4.7   Drive arrangements of the spindle for angular control (courtesy N.C. Engineering Ltd)

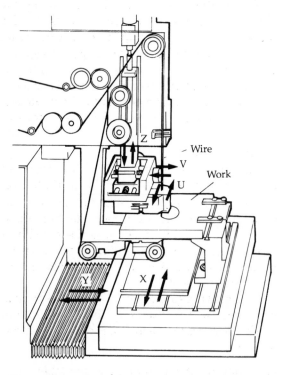

Figure 4.8   EDM wire cutting machine (courtesy AGIE (UK) Ltd)

4.2 Linear and rotary motions

On all machines there has to be both a linear and a rotary movement. The amount and direction that the tool or work moves have both to be accurately controlled.

Movement on Z axis

When working with absolute dimensions for milling and drilling operations, if the surface of the work is designated the Z datum (zero), a Z coordinate with a positive $(+)$ value will result in the tool being above the work surface, as shown in Figure 4.9a. Conversely a Z coordinate with a negative $(-)$ value will result in the tool penetrating the work. On turning operations, if the end of the work furthest from the spindle is designated the Z datum, a Z coordinate with a positive $(+)$ value will result in the tool being along the bed away from the work, as shown in Figure 4.9b. A negative $(-)$ Z coordinate will result in the tool moving axially into the work.

Movement on X axis

If the Z axis is vertical when looking at the machine from the operating position, absolute X coordinates with positive $(+)$ values will result in the tool being right of the datum, as shown in Figure 4.9a.

On turning centres where the Z axis is horizontal, the centre line of the component is usually selected as the X axis datum (X zero). If the tools are mounted to the left of the centre line when looking towards the spindle from the workpiece, absolute X coordinates with positive $(+)$ values will be on the left of the datum as shown in Figure 4.9b. If the tools are mounted on the right of the centre line, positive $(+)$ values will be on the right of the centre line.

There are systems that require the centre of the tool post to be the X axis datum.

Movement on Y axis

On those machines which have three axes, absolute Y coordinates with positive $(+)$ values will result in the tool being away from the datum in a direction towards the machine from the operator when looking at the machine from the operating position, with the positive X axis being towards the right, as shown in Figure 4.9a. On the majority of turning centres the height of the tool is fixed and it is extremely unlikely that there will be any programmable change in the height of the tool; consequently there is no Y axis movement on turning centres.

Reversed movement of work

In the above description, reference has been made to the movement of the tool. However, on some milling and drilling machines it is the work that moves on the X and Y axes; the tool only moves on the Z axis. The above conventions concerning the movement for positive coordinates still apply. The control system ensures that on those machines where the work moves, if a positive X coordinate of any value is input the resulting movement will position the work to the right of the datum. With reference to Figure 4.10, if the tool is at position A $(X10, Y10)$ and the next position required is B, the absolute X and Y coordinates to be input are $(X30, Y20)$.

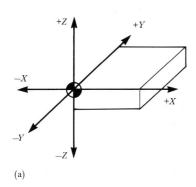

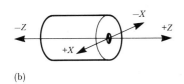

(a)

(b)

Figure 4.9   Linear movements: (a) on three-axis machine with vertical spindle; (b) on lathes

The table will move to the left on the X axis and outwards or away from the column on the Y axis. When the coordinates of point C $(X25, Y5)$ are input, the table will move from B to the right on the X axis and inwards on the Y axis.

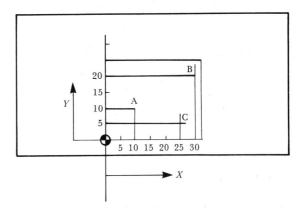

Figure 4.10   Reversed movement of work

If difficulty is experienced in following the above explanation of the movements, hold a marker such as a pencil over point A in Figure 4.10 but not touching the paper. Keep the marker stationary and move the book under the marker so that point B comes under the marker, and observe the movement of the book. Repeat the movement so that the marker is over point C.

Rotary movement

For milling or drilling machines it is essential for the tools to rotate to cut the work, but on turning centres it is essential for the work to rotate. Both these types of rotary motion are at the required cutting speed; it is the speed and direction of the spindle rotation which has to be input and controlled, and not the amount of circular movement.

Control of the amount of circular movement is important when generating the shape of the work. There are two techniques of cutting circular arcs:

(a) Mounting the work on a circular table.
(b) Simultaneously moving the work different amounts on two axes so that a circular arc is

generated; this action is termed *circular interpolation*.

When the rotary movement of the work is obtained through the use of a circular head or rotary table the angular rotation has the designatory letters A, B and C as shown in Figure 4.11. The axes of the rotary movements A, B and C are always parallel to the *X*, *Y* and *Z* axes respectively, regardless of the orientation of these axes. Positive A, B and C rotational movements would cause a right-handed screw to advance in the direction of positive *X*, *Y* and *Z* movements respectively.

The radius of the circular arc cut in work held on rotary tables is limited to the maximum size of the work that can be held on the rotary table.

The radius of the arc cut in work using circular interpolation is only limited by the capability of the control unit to calculate the necessary coordinates. Circular interpolation is the only practical method of generating a radius on a turning centre.

The mill-turning centres must have programmed control of the rotary movement of the main spindle, so that the work can be positioned for cross drilling and milling operations to be performed on the work. The headstock then can be classified as a circular head and since the axis of this rotary movement is on the *Z* axis it is designated as *C* axis control. This rotary movement of the spindle has to be powered by a different motor than the one which drives the spindle for turning operations.

4.3 Machine operating systems

There are a number of different operating systems used for controlling the movement of work or tool tables of numerically controlled machine tools:

(a) Positioning control system
(b) Line motion control system
(c) Contouring control system.

For convenience and for reference purposes each system has been given a symbol.

4.4 Positioning control (symbol P)

This is also known as point-to-point control. As the name implies, the system is mainly used for moving the work or tool from one point (position) to another position. There is no control on the path followed to reach that position. Control ensures that only when the table has come to rest the tool or work is at the desired position. Point-to-point operations can occur on all axes for the rapid positioning of tools or work. Figure 4.12 illustrates point-to-point positioning on the *X* and *Y* axes. If the tool is at A (*X*5, *Y*5) and absolute coordinate (*X*20, *Y*30) are input, the movement will take place simultaneously at an angle of 45 degrees on both the *X* and *Y* axes until the tool reaches B (*X*20, *Y*20), and then will continue on the *Y* axis only until the tool is at C (*X*20, *Y*30). On some machines the movements occur consecutively; on these machines the path followed from D (*X*30, *Y*30) to F (*X*55, *Y*15) will be first on the *X* axis to E at (*X*55, *Y*30) and then on the *Y* axis to F at (*X*55, *Y*15). This method of movement takes longer than when both axes move simultaneously.

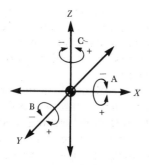

Figure 4.11   Designation of rotary movements

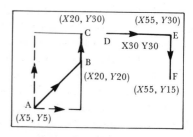

Figure 4.12   Positioning movements

The feed rate for the movement is normally automatically selected by the control unit, and does not have to be specified. However, there are control units where the feed rate has to be programmed; for these machines the speed of positioning selected should be the fastest feed rate available. The feed rate available is normally between 5 and 10 m/min depending on the control system used.

Response time, damping and hunting

A table does not reach the desired feed rate instantaneously. There is a small interval while it accelerates to the programmed feed rate, as shown in Figure 4.13a; this interval is referred to as the *speed response time.* The shorter the response time the more efficient is the drive system. A more critical response is the time for the carriage to come to rest after the feed rate is switched off. The ideal condition is for the carriage to stop instantaneously when it is at the desired position, as shown in Figure 4.13a. However, if the maximum feed rate is used up to the programmed position, there is a high probability of the table overshooting the desired position before coming to rest because of its momentum. The error in the position generates a signal which results in a reversal of movement, and the table will oscillate before coming to rest in the desired position. The oscillation of the table is known as *hunting.* While this type of movement may be acceptable for a positioning system where machining occurs after the carriage has stopped moving, it cannot be used for any machining operations. In order to reduce hunting it is necessary to reduce or *damp* the response of the system to the stop signal. The amount of damping is important. If the system is underdamped, i.e. has not enough damping, hunting occurs as shown in Figure 4.13b. If the system is overdamped, i.e. has too much damping, the stop signal has to be generated at some distance before the carriage has to come to rest, as shown in Figure 4.13c. Manufacturers attempt to provide *critical damping*, which is the term used to designate an acceptable response, as shown in Figure 4.13d.

The drive system illustrated in Figure 4.13d is referred to as progressive. Another type of drive system developed is known as *stepped*, as shown in Figure 4.14. Progressive drives are more expensive than stepped drives. With stepped drives the carriages move at the rapid feed rate of 5 to 10 m/min to within 10 mm of the programmed position. The feed is then automatically reduced to a fine feed rate of the order of 50 mm/min to within 0.5 mm of the programmed position, when the feed rate is finally reduced to creep feed of the order of 5 mm/min. Three stages are found to be the optimum number for a stepped drive, as there is no appreciable increase in accuracy of positioning with more stages. The signals for the change of feed rate are generated automatically by the control system. Stepped drives reduce the problem of the inertia of the carriages tending to overshoot the programmed coordinates. When hydrostatic slideways are used, switching off or reducing the oil pressure helps to reduce any overrun of the programmed position by the carriages using either progressive or stepped drives.

Unidirectional movement

To minimize the possibility of backlash in the screws used for moving the tables, the system may use a unidirectional approach. With this technique the final movement to a programmed position is always in the same direction – usually positive. If successive movements are in opposite directions, the table overshoots in the negative direction and the motion is then reversed so that the position is approached in the required direction. Obviously unidirectional movement can only be used for positioning moves when machining operations are not taking place.

When the positioning movement is completed, machining can occur. On the third axis of some drilling and milling machines the amount of movement may be controlled by preset limit switches. The movement can take place at predefined feed rates, with fast approach of the tools to near the work surface.

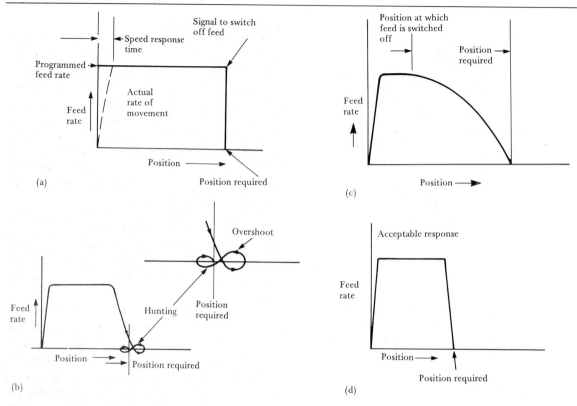

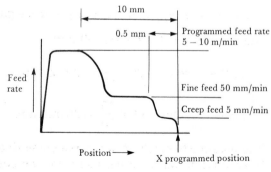

Figure 4.13(a)   Ideal condition for movements; (b) underdamped system; (c) overdamped system; (d) critical damping with progressive drive system

Positioning control is ideally suited for the punching or blanking of shapes during press work; the movement of the punch is activated when the table has come to rest.

When the amount and rate of movement of the Z axis can be controlled, the system is more accurately designated two-axis positioning and one-axis line motion control.

Figure 4.14   Stepped drive system

4.5 Line motion control (symbol L)

This system is also referred to as para-axial or major axis machining. The main difference between this system and a positioning system is that it is possible for the rate of movement to be programmed to suit the machining conditions. This permits machining to be carried out during the movements which take place along the major axes of the machine. Movement can occur along two axes simultaneously, but since the movement is at the same feed rate on both axes the resulting path is at 45 degrees to the axes of the machine.

A line motion control system can be used to produce a wide range of work by the milling of shapes that are composed of lines at 90 degrees and 45 degrees, as shown in Figure 4.10a. The system can also be used on turning centres for the production of stepped shafts, as shown in Figure 4.15b.

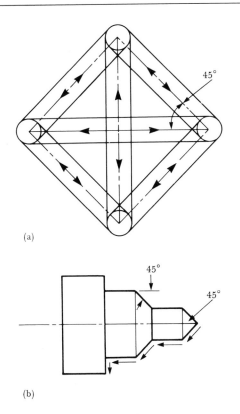

(a)

(b)

Figure 4.15   Line motion control: (a) milling machine; (b) lathe

If a different rate of movement can occur at the same time on different axes, the control system is more accurately designated contouring or continuous path control.

4.6 Contouring control (symbol C)

This is also referred to as *continuous path control*. With this system the movement of the tool or work is under continuous control for position, direction and rate of movement.

There are various control systems of varying complexity used on machine tools. There may be from two to eight controllable axes; movement can take place on a number of axes simultaneously, all at different feed rates if necessary. Care has to be exercised to ensure that no interference occurs when operating on the different axes.

Linear interpolation

Interpolation is a process of connecting up a number of fixed points with the smoothest possible curve or line. The simplest system provides 'linear interpolation', where any two points are joined by a straight line. The geometrical equation of a straight line is $Y = mX$. Therefore linear interpolation may be referred to as first-order interpolation. With a linear interpolation system the movement to programmed positions can occur on a single axis or any two or more axes in a straight line.

The value of the feed rate programmed is that required for the path to be followed. But if that value is sent to all the table driver motors, only movements at 90 and 45 degrees can be made. The system is then identical to line motion control.

When linear interpolation movements have to occur simultaneously on more than one axis, the computer in the control unit calculates the feed rates required on the individual axes to produce the desired shape. The orthogonal (rectangular) components of the feed rate are automatically sent by the control unit to the respective motors.

$$\text{feed rate input to the } X \text{ motor} =$$
$$\text{feed rate } F \times \frac{\text{movement on the } X \text{ axis}}{\text{total length of movement}}$$

With reference to Figure 4.16, if a feed rate of 200 mm/min has been programmed and movement is to take place from point A $(X100.0, Y100.0)$ to point B $(X140.0, Y130.0)$, the movement on the X axis is $(140 - 100) = 40$ and the movement on the Y axis is $(130 - 100) = 30$. Then

$$\text{total length of movement} = \sqrt{(40^2 + 30^2)} = 50$$

The feed rates sent will be as follows:

$$\text{To the } X \text{ motor: } 200 \times \frac{40}{50} = 160 \text{ mm/min}$$

$$\text{To the } Y \text{ motor: } 200 \times \frac{30}{50} = 120 \text{ mm/min}$$

It is possible to machine any curve using linear interpolation, but it is necessary for the computer to calculate sufficient coordinates around

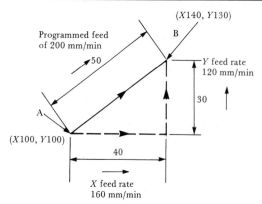

Programmed feed
of 200 mm/min

B (X140, Y130)

50

Y feed rate
120 mm/min

A

30

(X100, Y100)

40

X feed rate
160 mm/min

Figure 4.16   Rectangular components of feed rate

the arc so as to produce the shape within the tolerance band or zone (see Figure 4.17a). This technique is now viable with the use of computer-aided design and graphical techniques. It is necessary to specify the acceptable tolerance band or 'skin thickness', and the computer calculates all the coordinates required; this method may be referred to as *true path*.

Circular interpolation

The more common method of generating an arc is by circular interpolation. The reason why circular curves have been so widely used in the design of a profile of a component is that it is comparatively easy to obtain circular rotations; in addition, designers can draw circular arcs with a pair of compasses.

Circular interpolation may also be referred to as second-order interpolation, because the equation of a circle is $Y^2 = R^2 - X^2$, where R is the radius of the circle, X is the horizontal component of a point on the circle, and Y is the vertical component of the point on the circle.

With computer-controlled machine tools the calculation of the coordinates and feed rates required for circular interpolation is carried out within the control unit of the machine, and sent to the drive units as required. In order that the computer can carry out the necessary calculations it has to be provided with information. The tool will be at the start of the arc at the end of the previous movement sequence. The direction of movement (clockwise or counter-clockwise)

around the arc has to be programmed. The other information required can be provided as:

(a) The rectangular coordinates of the centre of the arc and the finish positions of the arc, as shown in Figure 4.17b.
(b) The rectangular coordinates of the centre of the arc, together with the radius of the arc and the subtended angle
(c) The rectangular coordinates of the end of the arc and the radius of the arc
(d) The polar coordinates of the angle subtended between the start and finish points of the arc and the radius of the arc.

Normally the control system provides both linear and circular interpolation, changing from one to the other as required. This system is expensive but produces more accurate forms.

Parabolic interpolation

The third type of control system which uses parabolic interpolation requires specialized techniques and control equipment; with this system the movement on any arc is along a parabola. Parabolic interpolation may also be referred to as second-order interpolation because the geometrical equation for a parabola is $Y = pX^2$.

An advantage of parabolic interpolation is that it enables the machine to move the tool or work between three non-straight-line positions in a smooth parabolic curve. This is particularly suited for sculpturing work as required in moulds or dies. For parabolic interpolation it is necessary to specify the direction of movement and the rectangular coordinates of the start, intermediate and end points of the curve required, as shown in Figure 4.18.

Cubic interpolation

There are curves which are of third-order or cubic interpolation, and it is possible for computers to calculate the necessary coordinates. For these systems the geometric equation or mathematical law of the curve has to be input together with the acceptable tolerance zone. The computer then calculates all the coordinates and feed rates necessary for the resultant curve

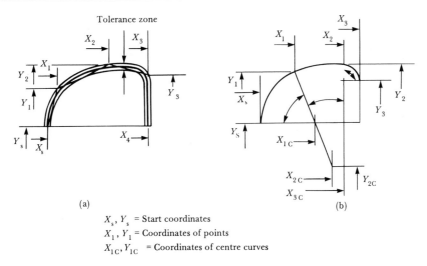

X_s, Y_s = Start coordinates
X_1, Y_1 = Coordinates of points
X_{1C}, Y_{1C} = Coordinates of centre curves

Figure 4.17(a)   Linear interpolation of curves; (b) circular interpolation of arcs

to be smooth and continuous. Control systems are being developed for this type of work, but the majority of the components required can be produced using either linear or circular interpolation.

4.7 Selection of control mode

Continuous path systems can also normally work in both the point-to-point mode and the line motion mode. The mode of operation is designated during programming by using a code which causes the control unit to operate in the mode required. The code is input as a preparatory function employing a word address format; the address used is G followed by two digits. Word address format is explained in Chapter 8. Typical codes are as follows:

G00   results in the machine operating in a point-to-point mode

G01   indicates linear interpolation

G02   results in the machine moving in a circular arc clockwise

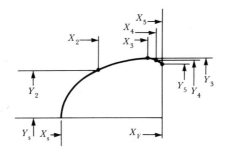

X_s, Y_s = Start coordinates.
X_2, Y_2 = Intermediate coordinates of first parabola.
X_3, Y_3 = End coordinates of first parabola and start coordinates of second parabola.
X_4, Y_4 = Intermediate coordinates of second parabola.
X_5, Y_5 = End coordinates of second parabola.

Figure 4.18   Parabolic interpolation

105

G03   results in the machine moving in a circular arc counter-clockwise

G06   indicates parabolic interpolation

On some machines it is possible to have different modes of control on different axes. Frequently the control system provides circular interpolation on any two axes and linear interpolation on the remaining third axis. This type of control system is referred to as 2CL. Similarly, 2PL indicates point-to-point control on two axes and line motion control on a third axis; 5C is a machine which has contouring control on five axes.

4.8 Adaptive control

With all the above systems, the work/tool positional relationship is controlled and the cutting speed and feed rate are programmed. Some control systems provide manual override controls for programmed speed or feed. With these controls the machine operator can vary the conditions while machining is taking place. This depends on the judgement of the operator; generally the override provided for feed rates is from 20 to 120 per cent of programmed feed. Speeds and feeds can only be automatically varied under program control by the control unit using adaptive control (see Section 2.10).

Questions

4.1 What are the factors considered when identifying the designation of the different addresses used for the axes of movement on NC machines?

4.2 With the aid of line diagrams showing two different types of machine tools, explain how the designation of the axes is modified from the Cartesian system to suit the different types of NC machines.

4.3 What designatory addresses are used for angular rotational movements on an NC machine?

4.4 What type of control exists on a machine that is specified as 2CL?

4.5 Explain why critical damping is important in the functioning of NC machine tools.

4.6 Describe the difference in the movements that occur when a machine is in (a) point-to-point mode or (b) linear interpolation mode.

4.7 Describe the difference between (a) first-order interpolation, (b) second-order interpolation and (c) third-order interpolation.

4.8 How are the different modes of operation selected during part programming?

4.9 Explain how it is possible to machine curves using the true path technique, and specify the information that is required.

Chapter Five

Output transducers

5.1 Transducers

A transducer is a device that converts variations of one physical variable into another physical variable. Transducers that employ different operating principles are used on numerically controlled machine tools for a number of purposes, such as:

(a) Monitoring the position of a carriage on a slideway
(b) Measuring the speed of rotation of a spindle
(c) Measuring the temperature of the tip of a tool
(d) Monitoring the power being transmitted by a shaft (torque measurement)
(e) Measuring the flow of oil or cutting fluid
(f) Measuring the pressure of oil in a hydraulic system.

Except for the transducers used for monitoring position, the transducers used for the other purposes are no different to those used for other applications. Details of the principles of operation of the transducers can be obtained from most books on instrumentation. However, without the position transducers closed loop systems for numerical control could not have been developed.

5.2 Positional transducers

The transducers used on machine tools for monitoring position depend on converting amount of movement into an output signal of some form. This output signal from the transducer does not usually have much strength, and has to be converted and amplified by a signal processing unit into a form which is strong enough and suitable for transmission to the control unit. The signal to the control unit is usually either electrical current or electrical pulses. Output transducers are required on closed loop systems to provide the signals which are fed back to the control unit to be compared with the input signals. (See Chapter 6 for an explanation of closed loop and open loop systems.)

There are different ways in which positional transducers can be classified:

(a) Linear or rotary
(b) Analogue or digital
(c) Absolute, semi-absolute or incremental
(d) Principles of operation.

Accuracy and resolution

There are two terms that are used with transducers: accuracy and resolution. The measuring *accuracy* of the transducer is the smallest unit of movement that it can consistently and repeatedly discriminate. The *resolution* is the smallest unit of length that separates two positions that can be recognized by the transducer. If a rule has 100 divisions in a 100 mm length, the resolution is 1 in 100; however, the divisions between markings may not all be exactly 1 mm, in which case the rule is not accurate.

Positioning of transducers

As explained in Chapter 3, the positioning of the transducers on the machine is of extreme importance and should conform to Abbe's principle of alignment. This requires that the equipment for measuring the movement of the work should be collinear with the work movement and as close to the actual machining area as possible (Figure 5.1). For Abbe's principle to be fully implemented it would be necessary to have the measuring elements or transducers in contact with the face or edge of the work just machined and close to the cutting tool. For milling and turning operations this is not possible for a number of practical reasons, such as:

(a) Availability of space in the cutting area
(b) Stopping of the machine spindle to position transducers when change of direction is required
(c) Interference to the contact of the transducer by swarf.

The use of laser beams reflecting off the work would appear to be viable, but interference with the light beams by swarf and cutting fluid would create problems. The practical solution is to monitor the movement of the carriage on which

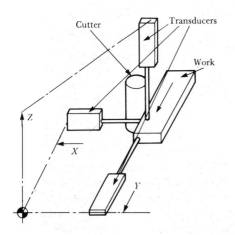

Figure 5.1 Positioning of transducers for Abbe's principle of alignment

the work or tool is mounted (see Chapter 6 for a fuller explanation).

The transducers used for monitoring the position of the carriages should be connected directly to the moving part, rather than be driven by gearing or linkages. Monitoring the carriage movement of CNC machining and turning centres provides satisfactory results and ensures machining accuracy such that these machines can produce components well within economic manufacturing tolerances, and generally better than those produced on other types of milling and turning machines. However, there is a certain category of work which has to be machined to a more precise size and with smoother surface texture than work produced on machining and turning centres. To produce work of this category requires a material removal technique such as grinding.

In-process measurement

On CNC cylindrical grinding machines there is a minimum of two main movements monitored with position transducers; the infeed movement on the X axis of the headstock with the grinding wheel spindle and the transverse movement of the work table on the Z axis. The X axis movement controls the depth of cut and diameter of the work, the Z axis movement controls the length of the work ground or the length of shouldered sections ground. The transducers generally used for position monitoring on grinding machines are the linear type which are described later in this section. One of the problems in producing work of the highest accuracy on cylindrical grinding machines is that although the movement of the head in which the spindle of the grinding wheel is mounted can be monitored and controlled, the diameter of the grinding wheel mounted on the end of the spindle can change. The change in size is due to the grinding wheel wearing, but the wheel can still be an effective cutting tool because of the self sharpening effect associated with correctly selected grinding wheels. In addition, diamond truing or dressing which can occur frequently will also reduce the diameter of a wheel. The control can compensate for diamond dressing but there is also the problem that the diamond itself wears. There are obvious practical difficulties in monitoring the exact diameter of a grinding wheel by direct contact.

The reduction in the diameter of the wheel caused by wear during the grinding of one component is quite small, but the size of the work being ground can be affected when the amount of wear is coupled with the deviation in the work/tool relationship resulting from possible deformation in the linking structure. The deformation can be due to such effects as thermal gradients caused by cutting fluid flowing away from the grinding zone heating a limited section of the machine body, or heat generated in the spindle bearings from the high speed of rotation of grinding wheels affecting the alignment of the spindle. To reduce these effects and achieve the superior accuracy required it is necessary to directly monitor the work size during the actual grinding operation i.e. in-process measurement. One type of instrument used for this purpose is an external diameter calliper gauge as shown in Figure 5.2.

This type of gauge is mounted on the work table in line with the diameter to be gauged and traverses with the work. To use the gauge the gap between the calliper contact points has to be set to a particular size. The callipers can be set to a test piece with a diameter of the required size. Alternatively, the first component can be used to set the callipers; in this case the diameter selected for control is finish ground and its size confirmed by direct measurement.

In both cases the gauge body is moved forward until the calliper contact points are over the diameter to be used for control. The calliper arms are closed under manual control until the contact points touch the work, and then the handle on the top of the gauge body is used to set the callipers in position. The setting of the contact points only takes seconds. The gauge does not generate an absolute value of the diameter being monitored but registers a zero setting i.e. a null signal is generated when the distance between the contact points is the required size (zero deviation from the set diameter). The diameter usually chosen for control

Figure 5.2   Table mounted external diameter gauge (courtesy Marposs Ltd)

is the one with the smallest tolerance, as with this gauge only one diameter of the work can be directly monitored at a time. The difference in the programmed and actual positions of the wheelhead when the wheel is in contact with the finish ground diameter is recorded within the control unit and generates a signal which is fed back to the control unit. The signal is used to offset the wheel head position on the X axis to compensate for wheel wear and other effects. The compensation is applied to the programmed positions for all other diameters of the first component and ensures that the work is produced within tolerance. The callipers need to be set only once for each batch of components and on successive components the diameter gauge is programmed to move forward so that the callipers contact the work when the particular diameter used as the reference is being ground. The signal from the calliper gauge overrides and offsets the programmed position of the wheel and the work is ground to suit the calliper gauge

i.e. correct size. The offset compensation generated can be updated on every component if there is excessive deviation in the wheel head position due to wheel wear etc. With this method the size of only one diameter of the work can be precisely controlled. There are calliper gauges for internal diameters which operate on the zero setting principle.

The transducers normally used in the table mounted diameter calliper are linear variable differential transformers (LVDT). These transducers have a small measuring range and one type works on the principle that when a core moves in a coil a change in voltage occurs. LVDT transducers are capable of discriminating smaller differences in position than the transducers used for monitoring the position of carriages. The small measuring range of LVDT transducers makes them unsuitable for measuring the movement or position of machine carriages. However, the transducers used to monitor the position of the assembly carrying the arms with the calliper

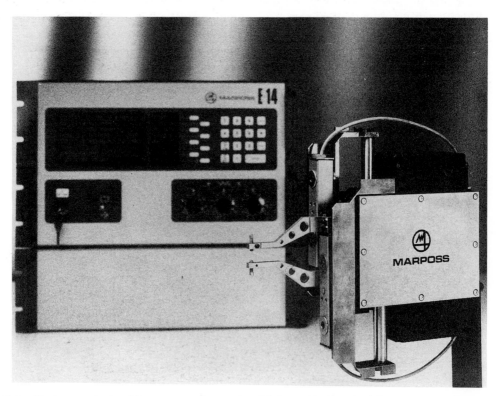

Figure 5.3   Bed mounted calliper gauge (courtesy Marposs Ltd)

contact points on the bed mounted calliper gauges require a greater range than the LVDT transducers.

An alternative technique is used to control the grinding wheel head position for external diameters where the work has a number of diameters or the work has to be produced in small batch quantities. A calliper gauge that can be used for monitoring the diameters is shown in Figure 5.3. The boxed unit in the figure is a control unit for CNC grinding machines.

This gauge is bed mounted clear of the front of the work table and directly opposite the grinding wheel; since the gauge is bed mounted the table carrying the work traverses to and fro on the Z axis in front of the gauge. Therefore, all the diameters on the work can be aligned with the gauge which is capable of measuring any diameter within its range. The gauge has a measuring range of 100 mm and diameters can be ground to a reported accuracy of 1.5 microns. The movement of the arms with the calliper contact points is monitored with its own position transducers. Initially it is necessary to calibrate the gauge for the range required when the grinding machine is being set up. The calibration of the gauge is checked using a test piece with a diameter of suitable size. The test piece is mounted on the machine and the callipers are brought under manual control (jog) to contact the test diameter. If necessary the reading of the gauge can be adjusted to the test diameter; ideally the test diameter should be in the middle of the measuring range.

In operation the gap between the calliper contact points is programmed for the diameter to be ground. When grinding commences on that diameter the gauge body is programmed to move forward so that the diameter of the work can be monitored and controlled while it is being ground. When the gauge is in use the control of the wheel head position is dependent on the signal from the gauge rather than the signal from the position transducers monitoring the wheel head movement i.e. in-process control.

Being in-process the monitoring of the size of the diameter of the work while it is being ground is subject to temperature effects. Therefore, for

Figure 5.4   Shoulder location probe (courtesy Marposs Ltd)

work requiring utmost accuracy the component is removed from the machine when it has been finish ground and measured more precisely under temperature-controlled conditions away from the machine i.e. post-process. If the measured size of any work indicates that corrections are needed a signal is generated by the measuring machine that offsets the signal generated from calliper assembly. This control is post-process on an in-process gauge. In-process measurement is only required when the work tolerance is very small and tool wear is significant although the cutting tool remains efficient, such as in grinding. If the work to be ground has tolerances that do not require the use of either the table or bed-mounted calliper gauges the control of wheel head position is restored to the signal from the transducers monitoring the position of the wheel head.

For certain classes of grinding work it is also necessary to monitor the position of a shoulder in multi-diameter work to ensure that the length ground of each diameter is correct. A shoulder location probe being used to monitor the position of a shoulder is shown in Figure 5.4. The shoulder location gauge is table mounted and the contact stylus is set to a shoulder whose position has been finish ground, and measured and confirmed as correct. The probe itself does not measure the position of the shoulder but

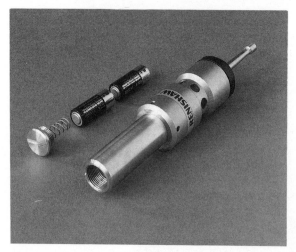

Figure 5.5   Job contact probe (courtesy Renishaw Metrology)

generates a signal at which the control records the position on the linear transducer mounted on the work table (Z axis). The difference between the recorded and programmed position provides a compensation offset value for control of the length of other shoulders being ground. The shoulder location gauge can also be used as a pre-process gauge to locate the work on the machine by determining the position of the shoulder prior to grinding. This nullifies problems created by variation in the depth of centre holes etc.

Probes

Probes are not transducers but types of switching device which are extremely sensitive. There are probes that have a stylus movement which is omnidirectional (XY and Z) – other probes have only two directions of movement. Two types of probes are: job contact probes and touch trigger probes. A job contact probe designed for mounting in a spindle is shown in Figure 5.5 and can only be used with materials that are electrically conductive. When the tip of the stylus of the probe shown in Figure 5.5 touches a surface an electrical circuit is completed. Light-emitting diodes (LED) in the body of the probe light provide a visual indication when the stylus contacts the work. The electrical circuit is from the tip of the stylus of the probe through the spindle and machine body to the machine table and work. The probe is powered by batteries located in the shank of the probe assembly. Although job contact probes can be used on CNC machines for visual indication of contact, the probes primarily used are touch trigger probes.

Touch trigger probes are similar in appearance to job contact probes but are more technically advanced tools and can be used on any material hard enough to deflect the stylus on contact. The probe is interfaced to the machine control unit as shown in Figure 5.6 which also shows the principle of supporting the stylus. When the stylus of the probe is deflected a change in the electrical characteristics of a circuit within the probe triggers a signal which is sent to the CNC control unit. To trigger the signal the deflection of the stylus is of the order

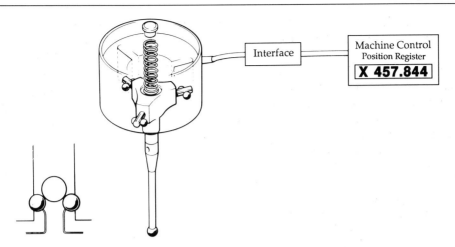

Figure 5.6   Touch trigger probe operating principle (courtesy Renishaw Metrology)

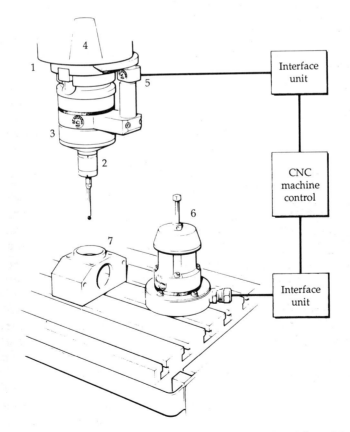

Figure 5.7   Touch trigger probe transmission systems (courtesy Renishaw Metrology)

of 0.005 mm depending on the length of the stylus. The design of the probe permits the stylus to deflect and prevents any damage occurring. The stylus returns to its original position when the cause of the deflection is removed. The maximum amount of overtravel tolerated depends on the length of the stylus. For a stylus of 50 mm length the amount of overtravel is of the order of 15 mm sideways and 5 mm axially with the shank. The relationship between the centre of the stylus contact sphere to the spindle axis in which the probe is mounted has to be determined by the use of a datuming procedure when the probe is first mounted on the machine. There are a number of different size and shape styli available for different applications such as:

(a) Detection of edges as explained above (shoulder location)
(b) Locating tools as explained in Chapter 2 and Chapter 10
(c) Locating work as explained in Chapter 3 and Chapter 10
(d) Tracing the shape of a three-dimensional model as explained in Chapter 13.

In all applications the work or probe is moved until the stylus is deflected and the signal is triggered and sent to the CNC control unit. Software provided in the control results in the power to the drive motor control being switched off and any movement of the carriage or probe which caused the contact to be made is stopped. Additionally, the reading of the position transducers at the contact point is automatically registered within the control. As shown in Figure 5.7, the probes can be hard wired to the control or the connection is via an inductive transmission; there is also an optical transmission system using infra-red light. It is repeated that touch trigger probes do not measure, but provide machines with a sense of touch which enables a surface (edge) to be detected accurately and consistently. All measurements are made with the position transducers fitted to the machine tool. There are two basic forms of transducers used for position monitoring: linear and rotary; the linear transducer provides greater accuracy.

Linear transducers

Linear transducers measure the actual movement of the carriage because they move with the carriage, and only move when the carriage moves. For this reason linear transducers provide more accurate results than rotary transducers, but because they are longer they tend to be more expensive.

For short movements (up to 200 mm) linear transducers can be mounted on the end of the carriage. For longer movements the transducers are usually mounted on the side of the carriage.

The material of the transducer has to be specially selected so that its thermal expansion rate is equal to that of the machine tool. Inaccuracies could arise on long measurements if the thermal expansion rates were different. Another problem with linear transducers is preventing oil fumes, coolant and small particles of swarf contaminating the scale and the reading head. To totally enclose the transducer is difficult, and special concertina-type guards have to be used.

The accuracy with which linear transducers measure movement of a carriage is of the order of 0.005 mm per 300 mm travel, compared with an accuracy of the order of 0.02 mm per 300 mm travel for rotary transducers.

Rotary transducers

It is important to note that the accuracy referred to, for both rotary and linear transducers, is the accuracy of measuring the movement of the carriage, and is not the resolution of the transducers. Rotary and linear transducers using the same principle of operation have the same resolution. In addition, the accuracy quoted is not the accuracy of the work produced, which is influenced by many other factors as explained in Chapter 3.

For rotary transducers to be used for measuring linear movements it is necessary to convert the linear movements to rotary movements. A specific angle of rotation of the transducer is directly related to a defined length of carriage movement.

One convenient technique of measuring a linear movement is to mount the rotary transducer on the end of the leadscrew which is used to move the carriage. The measuring element of the transducer will make one complete revolution when the carriage moves a distance equal to the lead of the screw. When rotary transducers are mounted on a shaft which can be twisted under load, the movement recorded may not be the actual linear movement. An alternative technique to placing the transducer on the end of the driving screw is to use a rack and pinion system. The rack is secured to the side of the moving carriage. The rotary transducer is mounted on the same spindle as the pinion which meshes with the rack. Special gears have to be used which do not have any backlash.

In both rotary and linear transducers there are two forms of output: analogue or digital.

Analogue and digital output

As shown in Figure 5.8a, the strength of an analogue signal changes continuously and gradually with the change in position of the carriage. By contrast, a digital signal changes in steps (Figure 5.8b).

An example of a mechanical analogue device is a barrel or thimble type micrometer that has a measuring screw. As the thimble is rotated, the distance between the anvils gradually changes. The actual reading on the thimble can be open to individual interpretation by different persons. There are now micrometers with a small digital display on the thimble, but it is still possible to have different interpretations of the last digit.

The number display on an electronic calculator is an example of a digital reading. With a digital output the signal is generated in pulse form during a change in position of the carriage. One pulse is generated by a movement corresponding to the smallest unit of length movement for which the machine is designed, i.e. a single pulse could mean a movement of either 0.01 or 0.001 mm depending on the design of the control system of the machine tool.

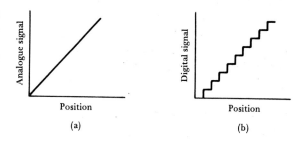

Figure 5.8(a) Analogue; (b) digital signals

Absolute, semi-absolute and incremental

The signal generated by transducers can be related either to a datum position, i.e. absolute, or to the previous position of the carriage, i.e. incremental or relative. Semi-absolute transducers are mainly rotary types which can measure the amount of rotation of a screw, and hence the amount of linear movement of the carriage, but require an additional method of measuring the number of rotations of the drive screw.

With suitable electrical circuitry and a memory store in which the pulses created during movements in different directions can be added or subtracted, the actual position of the carriage at any time can be established and displayed on the control unit. The position will, however, be relative to the position the carriage was at when the machine was set up.

Principles of operation of output transducers

A number of different types of positional transducer were developed during the early years of numerically controlled machine tools; each tool manufacturer produced its own transducer. With the advent of more sophisticated control systems, more efficient transducers have been developed by specialist manufacturers who supply them to the machine tool makers. At one time output transducers that utilized a variety of physical characteristics were available for monitoring the position of the tool or work, but now typical transducers in general use are optical gratings, encoders, Inductosyns and Magnescales. These are described in the rest of this chapter.

5.3 Optical gratings

An optical grating is a strip of glass marked with a series of equally spaced parallel lines; the lines and the spaces are of equal width. There are optical gratings which are known as 'diffraction gratings' that have between 400 and 2000 lines per millimetre; these are used for investigating the properties of light and have applications in testing optical instruments. Optical gratings for use as transducers on machine tools have considerably fewer lines per millimetre, generally between 20 and 200, and are known as 'coarse gratings'. For machine tool control, gratings are used in pairs. One, known as the *scale* grating, can be either metal or glass. It is mounted on the carriage of the machine and moves with the carriage, and has to be as least as long as the maximum movement of the carriage. The other grating, which has to be transparent, is known as the *index* grating and is stationary. The width of the lines and spaces on the index grating are the same as on the scale grating.

There are two techniques of using optical gratings. In one application the lines on the index grating are positioned so that they are angled and cross the lines on the scale grating. In the second technique the lines on both the gratings are parallel.

Crossed optical gratings

These gratings are arranged so that the lines on the index grating are inclined at a very small angle to the lines on the scale grating, as shown in Figure 5.9. Where the lines cross the spaces they create a diamond-shaped obstruction. This causes the intensity of the light passing through to be a maximum at the widest part of the diamond shape formed by the spaces, and a minimum at the apex of the diamond shape. Dark and light bands are created that are known as 'moiré fringes'; the name 'moiré' arises from the French word *moirer*, which means having the appearance of watered silk. The distance between the fringes is dependent on the angle of inclination of the lines. It has been found that a distance of 10 to 15 mm between the fringes is a suitable value.

An arrangement of the reading head is shown in Figure 5.10. A light source is positioned on one side of the gratings. An optical lens causes parallel beams of light (collimated light) to pass through the pair of gratings, and the beam of light falls upon light-sensitive cells (photocells). When the scale grating moves a distance equal to half the pitch of the lines, the widest part of the diamond will have moved to the position previously occupied by the apex on the width of the index grating. The obstruction caused by the lines crossing the spaces therefore moves at right angles to the direction of the movement of the scale grating. The dark and light bands therefore move in a direction at right angles to the movement of the grating, and a varying intensity of light impinges on the photocell. The photocell converts the changes in light energy into electrical energy; the current cycle or waveform is sinusoidal, as shown in Figure 5.11. It is

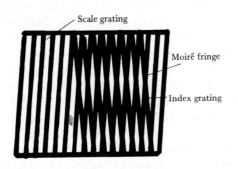

Figure 5.9   Crossed optical gratings

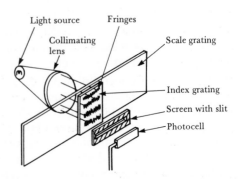

Figure 5.10   Optical reading head for transparent gratings

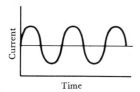

Figure 5.11   Output signal from photocell

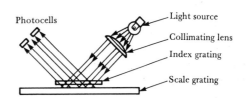

Figure 5.12   Optical reading head for reflective gratings

possible for a particular strength of the electrical signal generated to operate (trigger) an electrical device which sends a pulse of electricity as an output signal. In some systems the signal strength which triggers a pulse occurs once per fringe pass. For these systems the number of pulses generated per unit of length is equal to the number of lines on the grating per unit of length.

There are electrical circuits where in each waveform (one fringe pass) pulses can be triggered at different levels of current strength and transmitted to the control unit. This provides more feedback signals for smaller distances and enables more precise measurements of actual position using gratings with fewer lines per millimetre.

The pulses are generated when the grating is moving either to the left or to the right. In order for the pulses to be added or subtracted to the count there have to be two photocells to provide information on the direction of movement. These photocells are referred to as *direction discriminators*. The two gratings are separated by a distance given by:

separation distance = (pitch of lines on
    gratings × n) + (pitch of lines on gratings/4)

The value of n can be any convenient integer to provide a practical separation between the photocells.

When the scale grating is made of polished metal, the light source is positioned on the same side as the photocell and the light is reflected off the polished metal through the index grating, as shown in Figure 5.12.

Parallel optical gratings
These have a similar arrangement of light source and photocells as the crossed gratings, but the variation in light intensity is created in a different way. With this type it is important that the width of the lines and spaces are equal and are the same on both gratings. The index grating and scale grating are arranged so that the lines on the two gratings are parallel. The changing intensity of light is caused when the scale grating moves. The light passing through the gratings is a minimum when the lines of the index grating cover the spaces of the scale grating, and a maximum when the lines on both gratings are coincident. The electrical signal generated by the photocell has a similar waveform to that generated by the crossed gratings, and pulses are transmitted in the same way. To distinguish the direction of movement, direction discriminators are used as for crossed gratings.

Since pulses are generated only when movement is taking place, the feedback signal is essentially incremental for both the crossed and the parallel techniques of using optical gratings. One advantage of both methods is that small errors in the spacing of individual lines and spaces are not critical; and inaccuracies in the spacing or width of individual lines are averaged out. The optical cell spans a number of lines and the electrical energy is generated over a number of lines and spaces.

Optical gratings using both the crossed and the parallel techniques are available in both rotary and linear forms. Rotary gratings are normally fitted on the end of the leadscrew which moves the carriage. There are systems where the gratings are on the same spindle as a pinion which meshes with a rack clamped to the side of the moving carriage. A typical rotary grating has a band of approximately 20 mm width on a mean track diameter of 150 mm with 2540 lines per revolution. The arrangement of a reading head for rotary gratings is shown in Figure 5.13. Four index gratings are used as direction discriminators; the fifth index grating is used to mark each complete revolution.

5.4. Encoders

An encoder is a transducer that provides a serial or parallel digital value of an angular or linear movement. The value may be absolute or incremental.

There are two types of encoders: absolute encoders and pulse generators.

Absolute encoders

An absolute encoder, used for monitoring the position of a table, is a strip of metal or glass on which is printed a scale calibrated from a machine datum. The markings on the scale are in a binary coded form. One advantage of an absolute encoder is that the reading of the position of the carriage is retained after power switch-off or power failure, unlike an incremental system such as a diffraction grating or a pulse generator.

The markings can alternate in different ways:

(a) Conductive and insulating
(b) Transparent and opaque
(c) Magnetic and non-magnetic.

With the conductive and insulating type there is a separate brush in contact with each track, and an output signal is provided when the brush is in contact with a conductive marking. With the transparent and opaque markings, a light source is positioned on one side of the encoder. On the other side of the encoder there are photocells opposite each track. An output signal is generated when a transparent space is in line with a photocell. Where the code markings are magnetized, magnetic reading heads are used.

The optical and magnetic types are faster in operation than the contact type because there are no mechanical linkages which have to move.

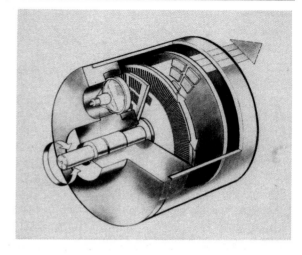

Figure 5.13   Optical reading head for parallel rotary gratings (courtesy Heidenhain GB Ltd)

At fast rates of operation there may be a problem with the contact type caused by the brushes bouncing and creating false readings.

Encoders are available in both linear and rotary forms.

Linear encoders

Shown in Figure 5.14 is a typical linear encoder. The instrument is attached to the side of a moving table, and has to be as long as the carriage movement. The reading head is fixed in position, and has a light source on one side of the scale and a photocell opposite each track on the other side. One problem with linear encoders is that there is difficulty in making the actual reading head physically small enough to be able to achieve a resolution (differences in readings) better than 0.1 mm. In addition, if it is required to position the table to an accuracy of 0.1 mm over a length of 819.2 mm there would have to be 14 tracks and 14 reading heads using the binary number system (8192 in binary is 10 000 000 000 000).

When using a true binary system for the codes, a problem can occur when moving from one set of markings which has three tracks 'live' to the next set which only has one track live. At the intersection of the markings it is possible for the photocells to be energized by all four tracks,

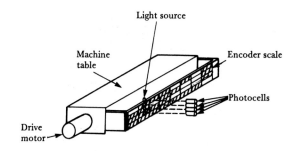

Figure 5.14   *Absolute linear encoder*

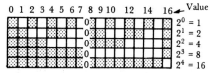

(a) Binary coding

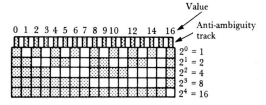

(b) Binary code with anti-ambiguity track

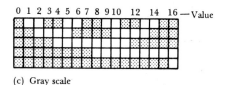

(c) Gray scale

Figure 5.15   *Encoder scales: (a) binary;*
(b) binary with anti-ambiguity track; (c) Gray

which will result in ambiguity errors. Figure 5.15a illustrates this problem. When moving from the set of markings representing 7 to that representing 8, at the intersection of the markings the photocell could be energized by all four tracks, which represents 15. One technique developed to overcome this problem is to provide an additional outermost track with markings which will 'switch on' the reading heads only when they are in the middle of the markings. This track may be referred to as the 'anti-ambiguity' track, as shown in Figure 5.15b.

Another technique developed to reduce the ambiguity problem is to devise other codes, one of which is known as the 'Gray' or 'progressive' code. The code pattern for the Gray code up to 16 is shown in Figure 5.15c. It can be seen that a change in the decimal number by one unit necessitates a change of only one sector in the Gray code.

Rotary encoders

Because of the problem of the number of reading heads required for linear encoders, rotary types are more commonly used. Rotary encoders are generally mounted on the end of the screws which drive the work tables or move the tools, but can be driven by pinions meshing with racks mounted on the side of the carriages.

A typical rotary encoder consists of a disc on which there are a number of annular rings or tracks, as shown in Figure 5.16. Frequently an anti-ambiguity track can be provided on the outermost ring if required. It can be seen that the

rings are divided into a number of sections or sectors; movement from one sector into the next results in the carriage moving the smallest unit of length which the control unit can distinguish.

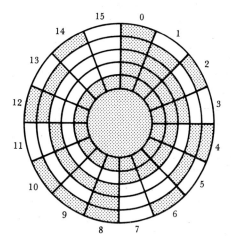

Figure 5.16   *Rotary encoder*

119

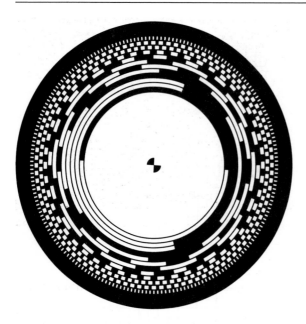

encoder respectively. Figure 5.17 is a rotary encoder with 11 tracks, providing 2048 counts per revolution. Each track is 1 mm wide, and the maximum diameter of the outside track is 68 mm. The anti-ambiguity track can be seen as a series of uniformly spaced small white lines on the outside of the coded tracks.

The rotary encoder is semi-absolute as the scale is only for one revolution of the disc, which occurs when the carriage moves a distance equal to the lead of the driving screw. Therefore either there has to be another position monitoring system with a greater resolution, which will check to determine that the table is moved to the position of the correct lead of the screw, or the number of complete revolutions of the encoder has to be counted. Both systems can be found on machine tools.

Figure 5.17   Eleven-track rotary encoder with anti-ambiguity track (courtesy Graticules)

This is known as the resolution of the encoder. The encoder shown in Figure 5.16 will provide a count of 16 per revolution. Encoders produced by different manufacturers of transducers for machine tools have either 10 or 12 tracks; these have 1024 and 4096 counts per revolution of the

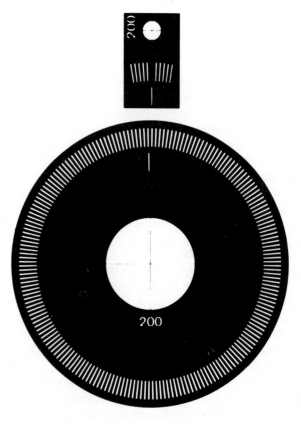

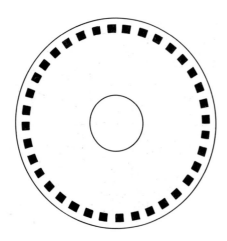

Figure 5.18   Rotary incremental encoder

Figure 5.19   Rotary incremental encoder for 200 pulses per revolution (courtesy Graticules)

Pulse generating encoders

This type of encoder is sometimes referred to as an incremental encoder or digitizer. It has only one ring of markings on the disc, as shown in Figure 5.18; the ring has alternate transparent and opaque segments. When there is an opaque segment between the light source on one side of the disc and the photocell, no power is generated by the photocell. Conversely, power is generated by the photocell when there is a transparent segment in position. The alternate non-generation and generation of power results in pulses being transmitted to the control unit. The required resolution is obtained by dividing the circumference of the disc into the segments required. For example, if the leadscrew has a pitch of 5 mm, and a resolution of 0.01 mm is required, a disc with a mean track of 100 mm diameter would have to be divided into 500 segments; each segment would be 0.63 mm wide at the mean track. Encoders made by different manufacturers have a different number of pulses generated per revolution, varying from 60 to 1270. The incremental encoder shown in Figure 5.19 will generate 200 pulses per revolution; the index head provides one marker pulse for each revolution. The track is 4 mm wide and has a mean diameter of 48 mm.

Pulse generating encoders are available in which the ring has conducting and insulating segments which are read with brush contacts.

As explained previously, incremental transducers can provide a digital readout of position of the carriage. It is necessary for direction discriminators (as explained earlier) to be fitted to determine direction of movement.

It is possible to use a rotary pulse generator for speed measurement. A disc with 60 transparent sections has to be mounted on the spindle. The pulses from the photocell are fed into a digital counter with a 1 s time base; the spindle revolutions per minute can be read directly on the display.

5.5 Inductosyns

Inductosyn is the trade name of a type of transducer used for position monitoring on

Figure 5.20   Inductive and capacitive Inductosyns (courtesy Hightech Components)

machine tools. There are two types of Inductosyn: inductive and capacitive. Each type is available in both linear and rotary forms. Inductosyns are capable of accuracies of 0.0025 mm over 300 mm. The flat tape and flat beam sections shown in Figure 5.20 are inductive Inductosyns; the capacitive Inductosyn is the annular form.

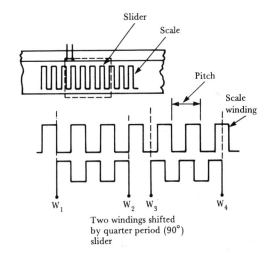

Figure 5.21   Windings of Inductosyns

Linear inductive Inductosyns

The linear type consists of two parts: the scale, which is attached and moves with the work table; and the slider, which is 100 mm long and is fixed in position. There is a small gap of 0.178 mm between the slider and scale. As shown in Figure 5.21, both the scale and the slider are flattened U-shaped windings of copper. The windings are bonded first on to an insulating material and then on to a suitable backing material which provides the necessary strength and rigidity. The backing material can be either spring steel strip, typically 18 mm wide and 0.25 mm thick, or rectangular steel sections of the order of 50 mm wide by 9.3 mm thick.

When the steel strip Inductosyn is mounted on the machine, it has to be specially tensioned to ensure that the copper winding is uniform throughout its length and that the scale does not sag. Alternatively the strip can be supported on a spar section to provide mechanical strength. Scale windings of the steel strip can be up to 40 m long, and can be 'wrapped' around a rotary table to provide position monitoring.

The 9.3 mm thick sections are made in unit lengths of 250 mm, and are joined together to make the required total length. Lengths up to 2 m present little problem, but if additional 2 m lengths have to be joined together special care has to be taken to ensure that the feedback signals generated are correct when passing from one 2 m section to the next.

On the slider, as shown in Figure 5.21, there are two separate short sections of the copper winding of the same form and size as that on the scale. The two sections are displaced axially to each other by a quarter of the pitch of the copper winding.

When an alternating electric current of between 200 and 10 000 cycles per second (H_z) flows through the copper winding on the scale, there is induced in the copper windings of the slider an electric voltage. The relationship between the induced voltage V_a in one of the sliders and the displacement X is:

$$V_a = kV \cos (2 \pi X/P)$$

In the other copper winding of the slider the induced voltage V_b is:

$$V_b = kV \sin(2 \pi X/P)$$

where k is the coupling ratio between scale and slider, V is the voltage input to the scale windings, X is the linear displacement of the slider from the zero datum, and P is the spacing (pitch) of the poles (typically 2 mm).

It can be seen that for a particular linear position there are two specific voltages generated in the slider windings. Control of the position of tables is achieved by applying to the slider windings a constant voltage from the input of the control unit. The voltages applied to the slider are of the same value but opposite polarity to those that would be induced in the slider by the scale at the desired position. This results in the applied voltages and induced voltages cancelling each other in the slider, and a null voltage condition exists when the carriage is at the desired position. At any other position a current will flow in the windings either in one direction or the reverse direction depending on whether the induced or the applied voltage is the greater. While a current flows in the windings, power is supplied to the motor which drives the leadscrews and causes the table to move. When the table reaches the position at which the induced voltage and applied voltage are equal, power to the motor is switched off and the table movement stops.

Rotary inductive Inductosyns

These work on the same principle as the linear type. The equivalent of the scale is normally attached to the end of the screw that drives the carriage, and the equivalent of the slider is fixed to the bearing housing of the screw that moves the carriage.

For a particular diameter the scale can only be of a fixed length, and therefore the position where the voltages in the slider and scale are equal occurs at repeated complete revolutions.

Figure 5.22 Capacitive Inductosyns (courtesy Hightech Components)

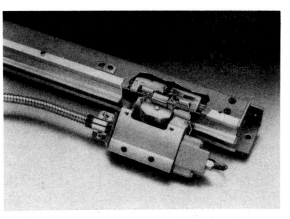

Figure 5.23 Magnescale and reading head (courtesy Stanmatic Precision Ltd)

In order to measure movements greater than the lead of the drive screw, there has to be a separate means of monitoring complete revolutions.

Capacitive Inductosyns

The operation of this type of Inductosyn is based on the capacitive coupling between two closely spaced elements that are directly attached to fixed and moving parts. Interleaved comb-shaped precision conductor patterns, as shown in Figure 5.22, are attached to suitable mechanical supports. The centre-to-centre spacing of the teeth of the comb is called the pitch, which is typically 0.4 mm. A duplicate set of patterns is mounted on both the fixed and moving parts, so that the teeth of the comb patterns pass over each other as the parts move. The capacity between the comb patterns varies cyclically, with one complete cycle for each pitch distance travelled, and an incremental output is generated. If a second set of patterns is added in a specific relationship to the first, it is possible to subdivide the pitch interval with great accuracy and also sense the relative direction of motion.

5.6 Magnescales

A Magnescale is the trade name of a position monitoring system that uses a magnetic element

as a transducer. A magnetic pattern in a sine waveform of constant pitch is recorded on a rod or strip. For lengths of up to 3 m, the measuring element is a rod of 2 mm diameter. For lengths over 3 m the measuring element comprises a magnetic strip which is as long as the maximum movement of the carriages. The construction of a Magnescale unit is shown in Figure 5.23. The rod or strip is fixed in position, and a head is attached to the moving carriage and moves along the rod or strip. As the head moves over the magnetic element a sine wave output of 0.2 mm pitch is generated. The sine wave output is electrically interpolated to obtain an incremental signal in pulse form, which can be used in the same way as the other devices which generate incremental signals.

Questions

5.1 What is a transducer, and what are they used for on NC machine tools?

5.2 Describe four different ways in which transducers can be classified, giving examples of each class.

5.3 Explain the difference between accuracy and resolution of transducers.

123

5.4 Explain why transducers should be positioned according to Abbe's principle.

5.5 Why are linear transducers more accurate than rotary transducers for the measurement of the movement of machine carriages?

5.6 Describe two different techniques of using optical gratings for providing signals.

5.7 What are moiré fringes, and how are they formed by optical gratings?

5.8 Explain the difference between an absolute encoder and an incremental encoder.

5.9 What is an anti-ambiguity track, and why is its use necessary with encoders?

5.10 Describe the principle of operation of (a) Inductosyns and (b) Magnescales.

Chapter Six

Principles of operation of NC machine tools

6.1 Basic principles

On manually operated machine tools, the machining of a component starts with the machine tool operator studying the drawing to determine the design requirements of the workpiece. Those that would affect the machining process would normally be size and tolerances, surface finish and geometric accuracy. After the workpiece has been mounted on the machine and the tools positioned, the operator then uses the trial and error technique of taking a cut and measuring the size of the work produced. This continues with successive changes of the work/tool relationship until it is judged that the component will be produced to the design specification. It may require a number of settings of the work/tool relationship to remove the various inaccuracies that can occur before an acceptable component is produced. When the operator considers that the workpiece is correct, the finished machined workpiece is checked, usually by an inspector, against the desired specification stated on the drawing.

It can be considered that the manufacture of a satisfactory component follows a circle or loop, as illustrated in Figure 6.1:

(a) The operator studies the drawing
(b) The operator sets the machine and the part is machined
(c) The part is checked against the drawing.

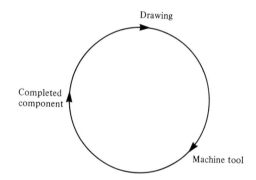

Figure 6.1 *Closed loop in machining of component*

The circle or loop is not complete until the part coincides with the specification. The operator is an essential link in the loop and therefore influences the results, which are subject to human error. One advantage of using skilled machinists to operate machines is that accurate components can be produced on defective machines, the machinist making the necessary allowances for the machine's deficiencies. Numerically controlled machines must be accurate at all times, as there can be no operator intervention to change input or output signals when the machine is working.

Every attempt must be made to produce components that meet the design specifications. If a component has to be remachined, the extra time results in increased costs.

125

One of the reasons for the introduction of numerical control was to reduce the time being taken to position the tools when machining complex components. As stated in Chapter 1, the time taken to position a tool manually is inversely proportional to the degree of accuracy required. The size of the work is dependent on the position of the tool. In addition, on manually controlled machines it is only possible to measure the size of the work when the machining has taken place i.e. post process.

After the first satisfactory component has been produced the difference in size of further components produced on manually operated machine tools is caused by variations or error in the work/tool relationship between different components. There are three basic types of error:

Positional (static) error

This error results from incorrect work/tool positioning before cutting takes place.

In order to reduce this type of error it is necessary to have an efficient method of monitoring the work/tool relationship at all times.

Reference (kinematic) error

This error is caused by the tool or work deflecting due to the forces existing under dynamic conditions.

These errors can be reduced by having tools and tool holders of adequate strength and rigidity with effective movement actuating mechanisms and strong constructional machine members.

Tool wear error

Wear of the cutting tool will cause a change in the size of a component.

If single point lathe tools or milling cutters wear to such an extent that there is a significant change in the size of the work the tools will have large wear lands; with the result that there will almost certainly be a deterioration in their cutting efficiency and the tools will probably have to be changed.

However, on grinding machines, because of the self-sharpening effect of correctly selected grinding wheels, the grinding wheel could re-

duce in diameter but still be an efficient cutting tool. Ground components normally have smaller tolerances than turned or milled work and so the wear of the grinding wheel could result in the work being out of tolerance, although the wheel head is in the same position as when the previous correct component was produced. Therefore, to produce accurate components it is necessary to continuously monitor the size of the work at the same time as it is being ground and to make any necessary compensation in the work/tool relationship.

For machining and turning centres sufficient accuracy can be obtained by monitoring the position of the work table or tool post and correcting any deviations occurring. This is because during milling the work is held in a holder or fixture which is clamped on to a work table. When the table moves the work must also move. The amount of movement of the work table can be automatically measured and controlled. It is reasonable to assume that, during milling, the change in size and shape of the work will depend on the amount that the work table moves. With numerical control the movement of the work table is automatically controlled.

For manually controlled drilling operations the positions of the centres of the different holes has to be marked out, or the work can be held in a jig. When a jig is used, the positions of the

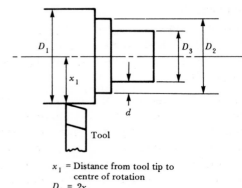

x_1 = Distance from tool tip to centre of rotation
$D_1 = 2x_1$
$D_2 - D_3 = 2d$
d = Change in position of tool required to change from turning diameter D_3 to diameter D_2

Figure 6.2 *Tool relationship during turning*

centres of the holes drilled are controlled by the positions of bushes which guide the drills. Each type of component requires a different jig, which can be very expensive. Jigs are not used on numerically controlled machines. Therefore for drilling operations on numerically controlled machines the work has to be clamped to the work table, and the hole positions will depend on the positions of the work table which is automatically controlled.

To turn work to the desired size on a lathe it is necessary to first establish the distance from the tool point to the centre of rotation of the work; this is shown as x_1 in Figure 6.2. The diameter of the work being machined will be twice this distance. Also as shown in Figure 6.2, any change in work diameter will be twice the change of position (radial movement) of the tip of the tool along the cross-slide. The movement of the tool holder can be controlled accurately with a numerical control system.

There are a number of stages in the production of a component on a numerically controlled machine tool:

First stage   The shape of the component has to be converted from the pictorial representation (the drawing) into a numerical format, and a part program created. The part program contains the sequence of operations and the numerical format which represent the movements that the cutter or work has to follow to produce the component.

Second stage   It is necessary to enter the details of the part program on to a suitable medium such as punched tape which can be 'read' electronically.

Third stage   After the information has been input to the control unit the component is produced by controlling the movements of the cutter. There are two systems used for controlling the movement of the carriages on numerically controlled machine tools: closed loop and open loop.

6.2 Closed loop systems

With these systems, components of correct size and shape are produced by continuously moni-

toring (measuring) the movement and position of the carriages (work tables) on which the tools or work are mounted, while the component is being machined. The amount of movement is controlled to specified requirements; this ensures that the actual size of the work is the desired (designed) size. Any difference between actual and desired size is corrected while the component is being machined, that is the machining 'loop' is closed:

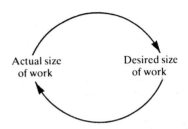

On machine tools which use the closed loop system a series of steps have to be completed during the machining of the component:

Step 1: input signal compared with output signal   An input signal which relates to the desired position of the work or tool is fed into the control unit. It is compared with an output signal which is fed back to the control unit. The output signals are generated by various measuring instruments known as transducers which monitor (measure) the actual position of the tool holders or work tables.

Step 2: error signal sent to power unit   Any difference or error between the input and output signals is transmitted to a power unit.

Step 3: power unit drives servo motors to reduce error to zero   The power unit provides power for the servo motors which cause the movement of carriages. or an action, to reduce or correct that error.

Step 4: when actual position equals desired position   When the actual relationship of the work to tool or other feature coincides with the desired relationship, movement stops.

Step 5: input signal from memory store   The next input signal is fed from the memory store into the control unit and the next action (operation) commences.

These steps are repeated for every operation until the component is completed. It is very important to realize that only when there is a difference (error) between desired and actual relationships that movement occurs; thus the system is known as an error-operated closed loop system.

The input information has to be digital, i.e. discrete numbers, but the output information from the position transducers can be either digital or analogue depending on the type of transducer used. If the output is analogue then the signal may have to be changed in an analogue to digital converter before being compared with the input signal.

Figure 6.3 shows the principle of operation of a closed loop system for controlling the positional relationship on one axis only. The switches and electrical resistors are intended to symbolize the electronic circuitry in a silicon chip. There are two alternatives shown for the first stage of producing the part program:

(a) Manually converting the drawing details into tool movements, and composing the part program; or
(b) Using computers and a graphic screen to produce the part program.

If the part program is produced manually, it is necessary to enter the details manually on to a suitable medium such as punched tape which can be read automatically. If the part program is produced using the graphic screen, the punched tape can be produced by directly linking the computer to the punched tape unit. Both these methods will be detailed in Chapter 13.

It is becoming more common for the part program to be stored on magnetic discs (floppy discs), or retained in the memory of the computer, and transmitted direct to the control unit as required. Further details of the methods of storage are given in Chapter 7.

There has to be a separate feedback system for every feature of the machine which is to be controlled. Every linear or rotary movement has to be measured with an output transducer which monitors the movement. If the tool selec-

tion is to be automatically controlled then each tool has to be identified in some way which can be read automatically by a transducer. Spindle speed and feed rate also have to be monitored and controlled. Different instruments are available for speed measurement.

One instrument used for measurement of rate of movement is a small electrical generator known as a tacho-generator. For measurement of feed rate the tacho-generator is mounted on the end of the leadscrew that moves the work tables or tool holders. For spindle speed measurement the tacho-generator is either mounted on the end of the main spindle, or driven by a non-slip belt by the spindle. The faster the rotational movement, the higher is the output voltage of the electricity generated. This is compared with the 'input' voltage signal which defines the desired rate of movement.

Control also has to be exercised on all operational features, such as:

(a) Selection and turning on or off of cutting fluid
(b) Direction of spindle rotation etc.
(c) Selection of tool.

Machine tools using closed loop systems are more expensive than similar capacity machine tools using open loop systems, because of all the various output transducers and associated control equipment. It is found that closed loop machines produce work consistently more accurate than open loop machines.

6.3 Open loop systems

With open loop systems only the first two stages detailed previously for closed loop systems are necessary. With open loop systems special electrical motors are used, and there are no output transducers for positional control. Consequently there is no monitoring of the work size while it is being machined. The actual size of the component is not known until it is measured after it is completed; that is, the loop is not complete and is said to be 'open'. A block diagram of an open loop system is shown in Figure 6.4.

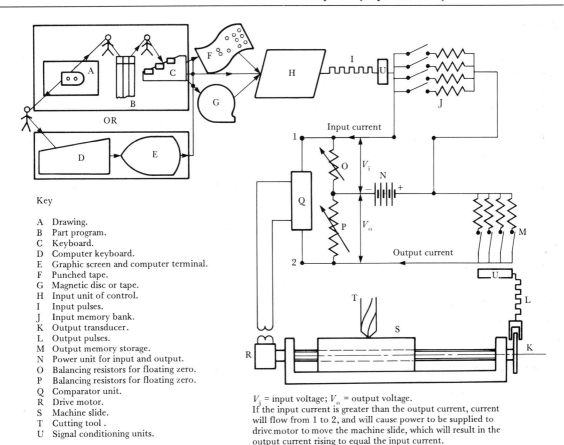

Key

A Drawing.
B Part program.
C Keyboard.
D Computer keyboard.
E Graphic screen and computer terminal.
F Punched tape.
G Magnetic disc or tape.
H Input unit of control.
I Input pulses.
J Input memory bank.
K Output transducer.
L Output pulses.
M Output memory storage.
N Power unit for input and output.
O Balancing resistors for floating zero.
P Balancing resistors for floating zero.
Q Comparator unit.
R Drive motor.
S Machine slide.
T Cutting tool.
U Signal conditioning units.

V_i = input voltage; V_o = output voltage.
If the input current is greater than the output current, current will flow from 1 to 2, and will cause power to be supplied to drive motor to move the machine slide, which will result in the output current rising to equal the input current.

Figure 6.3 *Principle of operation of an error-operated closed loop system*

Open loop systems have only evolved because of the development of special electrical motors known as stepper motors. An explanation of the operation of stepper motors is given in Chapter 3.

The accuracy of components produced on machine tools with open loop systems is well within accepted machining quality levels. Open loop systems are generally better and quicker at manufacturing components than manually operated machine tools. There is no significant difference in the speed of operation of open loop machines compared with closed loop machines.

6.4 Input signals

A variety of different forms have been used for the input signal, such as:

(a) Variations in the strength of electrical current, operating electrical or electronic relays

(b) Differences in pressure, operating valves in hydraulic or pneumatic circuits

(c) Pulses of light, operating light-sensitive electrical units

(d) Pulses of electricity, operating electronic switching circuits (this is now a commonly used method).

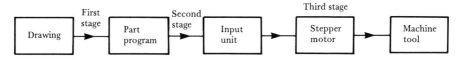

Figure 6.4 *Block diagram of open loop system*

The signal can be in either of two forms: analogue or digital. The difference between analogue and digital signals is explained more fully in Chapter 5. Analogue signals are continuously variable, and provide information which is changed by events. Digital signals are incremental, individual or separate pulses, which give readings that change their value by single numbers of constant amounts. It is this type of signal which is found to be more suitable and is mainly used as input in numerical control.

Methods of input

There are two methods by which the information can be fed into the control unit of the machine:

Directly Using rotary switches, push buttons or keyboards.

Indirectly Using magnetic tape, punched cards, punched tape, or magnetic (floppy) disc, or from the memory of a computer.

The direct method is useful when the components being produced are fairly simple and do not require a lot of input information. The switches, buttons or keyboards are all part of the control unit of the machine. This method is now generally referred to as *manual data input* (MDI). The introduction of computer numerical control (CNC) with visual display units (VDUs) or screens has resulted in an increase in the use of this method.

With some CNC machines, questions are displayed on a screen. The questions, which the operator has to answer through the use of the keyboard, relate to the work size and material and the operations to be performed. In addition, details have to be supplied of the tools required to produce the part. This technique of inputting information is referred to as *conversational*. Conversational programming will be explained further in Chapter 13. When a satisfactory component has been produced the information can be stored for future use, generally using punched tape but also magnetic cassette tapes or magnetic floppy discs.

The indirect method has the advantage of saving of machine time, as the tape, cards or discs containing the information for the manufacture of the part can be produced away from the machine while the machine is working on another part. The information stored on the tape or cards is fed to the machine when required using a special tape or card reader, which is usually part of the control unit of the machine tool. If a computer is linked to the machine it is possible for the part program to be stored on the discs used as memory storage and then loaded into the control unit's memory as required at the setting-up stage. It is possible for a number of programs to be stored on discs, but normally only one program is stored on one roll of punched tape or one cassette of magnetic tape.

If the information for the machining of the component has to be changed (edited) during manufacture, it is possible to use the keyboard or buttons on the control unit to make the necessary alterations. A new tape or disc is produced with the revised information.

Questions

6.1 Explain how NC has helped to reduce the manufacturing time of a component.

6.2 Detail the different stages in the production of a component on CNC machines, explaining why each stage is necessary.

6.3 Describe the principle of operation of CNC machines using (a) closed loop systems and (b) open loop systems.

6.4 Draw the block diagrams of the above systems and specify the different constructional items that would be needed for each block.

6.5 Detail the difference in the operation of (a) DC servo motors and (b) stepper motors.

6.6 Describe the different methods of inputting information into the control unit of a CNC machine, detailing the advantages and limitations of each method.

Chapter Seven

Information storage

7.1 Input signals

The computer in the machine tool control unit has a program specifically written to handle information to operate the machine. The information required is provided in the part program which contains the necessary data on tools, movements of work and tool and mode of operation etc. for each component.

As stated in the previous chapter, the format of the input signal containing the numerical information to control the operation of CNC machines is generally a series of pulses of electricity. The pulses are arranged in different patterns, each pulse pattern representing a particular character. For transfer of data between control equipment it has been standardized that a pattern of seven or eight pulses or no pulses in a specific combination is all that is required to define a character.

Since electrical circuits can be either OFF or ON (two states) it is convenient to use binary arithmetic when allocating numerical values to the pattern of the pulses. This is because in binary arithmetic there are only two symbols '0' (zero) and '1' (one). The symbol 1 being the highest value is used to represent a switch being 'on' i.e. circuit energized. Therefore, a single pulse represents a *BI*nary digi*T* and is known as a *BIT*.

A single set of pulses with different numbers and pulse patterns is known as a '*BYTE*' (*BY* eigh*T*). As well as numbers, it is necessary to represent letters and symbols in a number format, therefore a system of different pulse patterns representing different characters has had to be developed. These different pulse patterns are known as *codes*.

At one time each manufacturer of numerical control equipment devised and used their own coding system to define the different characters, which caused confusion and incompatibility. It also meant that separate machines had to be available for preparing the input signal. Fortunately there are now only two systems in general use: the ISO (International Standards Organization), and the EIA (Electronic Industries Association). The ISO codes are similar to the ASCII (American Standard Code for Information Interchange) codes which are used for transferring data between computers and peripheral equipment, printers etc.

For numerical control purposes only seven of the eight pulses are used for designating a binary value. Therefore, it can be seen that the highest binary number that can be represented by seven pulses is 1111111 which in denary (conventional decimal) notation is: 64 + 32 + 16 + 8 + 4 + 2 + 1 = 127. Since it is necessary for one combination of pulses to represent one code there are 127 possible codes. It has been found that 127 codes are more than sufficient for the purpose of numerical control, as only about 52 codes are actually used for numerical control of machines.

ISO codes

Table 7.1 shows the majority of the ISO codes used for numerical control purposes. Other codes may be used by manufacturers of control system for special applications.

Table 7.1 *ISO codes used for numerical control*

Code	Character		Purpose
9	Tab		Used in block formats
10	Line feed		End of block
13	Carriage return		End of block
32	Space		
35	Hash sign	#	Macro indicator
36	Dollar sign	$	
37	Per cent sign	%	Program start and tape rewind stop
40	Left bracket	(	Control out for operator intervention
41	Right bracket	)	Control in after operator intervention
42	Asterisk	*	End of block marker
43	Plus sign	+	
45	Minus sign	−	Dimension indication
46	Decimal point	.	Dimension indication
47	Right slash	/	Block delete
48	Digit	0	Number
49	Digit	1	Number
50	Digit	2	Number
51	Digit	3	Number
52	Digit	4	Number
53	Digit	5	Number
54	Digit	6	Number
55	Digit	7	Number
56	Digit	8	Number
57	Digit	9	Number
58	Colon	:	
61	Equal sign	=	
65	Letter	A	Rotary movement around X axis
66	Letter	B	Rotary movement around Y axis
67	Letter	C	Rotary movement around Z axis
68	Letter	D	Third feed function
69	Letter	E	Secondary feed function
70	Letter	F	Primary feed function
71	Letter	G	Preparatory function
72	Letter	H	
73	Letter	I	Arc centre offset on X axis
74	Letter	J	Arc centre offset on Y axis
75	Letter	K	Arc centre offset on Z axis
76	Letter	L	
77	Letter	M	Miscellaneous function
78	Letter	N	Sequence number
79	Letter	O	
80	Letter	P	Tertiary axis parallel to X axis
81	Letter	Q	Tertiary axis parallel to Y axis
82	Letter	R	Tertiary axis parallel to Z axis
83	Letter	S	Speed function
84	Letter	T	Tool function
85	Letter	U	Secondary axis parallel to X axis
86	Letter	V	Secondary axis parallel to Y axis
87	Letter	W	Secondary axis parallel to Z axis
88	Letter	X	Primary axis
89	Letter	Y	Primary axis
90	Letter	Z	Primary axis

Codes 0 to 47 inclusive are used for control purposes and are referred to as control codes or characters.

The characters represented by codes between 0 and 32 are for the operation of the keyboard or transfer of information and do not appear on any printout.

48 to 57 inclusive represent numbers 0 to 9 and are known as 'numer' characters.

65 to 90 inclusive represent upper case letters A to Z and are referred to as 'alpha' characters.

Binary coded decimal (BCD)

The size of the workpiece will determine the magnitude of the dimensions, and dimensional numbers with up to six digits are common. However, although the highest denary number that can be represented in one seven-bit byte is 127, as shown in Table 7.1 some of the numbers have to be used as codes for letters and symbols. Therefore, it is necessary to represent dimensional information using a system known as binary coded decimal (BCD). With this technique the highest decimal number that is represented in a single byte is nine. In the ISO coding this has been allocated code 57 which in binary = 0111001 (bits numbered from the right to the left). In BCD the value for each digit is in binary along the byte using codes 48 to 57 inclusive but successive bytes form the denary (decimal) digits. The magnitude of an individual digit is determined by its position in the number. Some control systems do not recognize the code for the decimal point. For these systems the number in BCD has to be in the smallest unit for which the machine control system is designed.

Parity

As stated previously, the signals now used have a maximum of eight pulses to define the characters, but only seven of the pulses have a binary value – the remaining pulse is known as a '*parity pulse*' or '*parity bit*'. The parity pulse was introduced as a safety measure to prevent errors occurring due to a pulse not being detected for some reason. It was considered that the probability of two pulses being missed in one byte simultaneously was so small that it did not warrant the extra cost involved in providing for the eventuality. As the reliability of the control equipment has improved the use of 'parity checking' has tended to be discontinued, although for punch tape parity checking is still quite common. The ISO system is referred to as an '*even parity system*', because the receiving device will perform a parity check and the input signal will only be accepted by the control unit when the number of pulses sensed in reading the code of the input signal is an even number. Therefore, a pulse may be required in the eighth position (the parity bit) to make the number of pulses in a character an even number. For example, if it is required to input the code 50 which has a binary value of 0110010, there will be pulses in two, five and six bit positions (bits numbered from the right to the left) resulting in three pulses in the byte – an odd number. Therefore, to make the number of pulses up to four, an even number, an additional bit is required in the eighth bit position. The pulse in the eighth position should not be considered when calculating the binary value being input.

The EIA system uses '*Odd parity*' i.e. the number of pulses for one character has to be an odd number. The fifth pulse is used as the parity pulse.

7.2 Indirect methods of input

As stated in the previous chapter, it is possible to store the part program information using the following media:

(a) Punched cards
(b) Magnetic tape
(c) Magnetic discs
(d) Punched tape.

A means of reading these media has to be provided at the control unit. Most computer-controlled systems are capable of using more than one method. There is a portable unit to which the part program can be transferred from a desk-top computer. The program is stored on either magnetic discs or magnetic tape. The unit is carried to the machine and the program transferred into the control unit. A portable unit is shown in Figure 13.10. These portable units can save a firm having tape readers or disc drives fitted to every machine.

Punched cards were never widely used in numerical control work except for inputting information into large mainframe computers. Magnetic tape was used with quite a few of the early controls, especially for continuous path machines, but has gradually dropped out of use. The magnetic disc is proving to be a very convenient medium because of the ability to store a large number of part programs on one disc. At one time punched tape was very widely used and it is still used on a number of controls. For shop floor use punch tape has a number of advantages.

7.3 Punched cards

These are specially designed cards approximately 185 mm wide by 80 mm long. There is provision for there to be up to 80 columns of holes across the card. Each column has a number of positions in which holes can be punched;

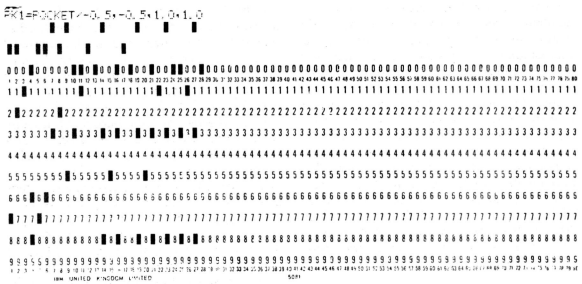

Figure 7.1 Punched card APT part programming statement

the positions of the holes indicate a number or a letter according to some predefined code. Special machines are used for punching the cards. The reading of the cards requires an electrical contact or photo-electric cell to detect the presence of absence of a hole in the card. All the information for one operation only is contained on one card. Figure 7.1 shows a card used in preparing a part program.

As stated previously, cards were never as widely used as punched tape or magnetic tape for storage of information for numerically controlled machines. Their main application was in conjunction with part programming languages, where they were used in the initial stages of inputting information to the computers. It is now very unusual to find them used in computer control systems for machine tools. One of the problems is that cards are more bulky than tape or discs. In addition, under workshop conditions a pack of punched cards can become greasy and dirty. Moreover, it is difficult to handle a large pack and the sequence can easily be altered. To a certain extent there is protection against this problem in that the operation number or card number is the part of the input information which is read first. If a card is out of sequence its operation number will also be out of sequence; the control unit will not accept the

information, and a warning light will be displayed. The cards then have to be passed through a special machine that rearranges them in the correct sequence.

7.4 Magnetic tape

This method of storage works on the principle that when a magnet is moved relative to a wire an electric current is induced in the wire. Magnetic tape is a plastic tape on to which magnetic oxide has been sprayed. The magnetization of the tape can be either in pulses or in density, i.e. digital or analogue. Generally the magnetic signal is in a type of pulse form. The pulses alter or change in frequency or intensity. When the tape passes near a coil the magnetized pulses induce pulses of electric current in the coil; these form a code that will operate relays or switches. The tapes are made in widths varying between 6 and 40 mm, and there can be a number of parallel tracks on which it is possible to record information. Each track requires its own reading/writing head.

The tape generally used now is 6 mm wide with a double track, only one track being read from each edge of the tape. The tape can be held in a cassette very similar to audio tapes; frequently much smaller cassettes known as mini-cassettes are used. It is essential for the tape to maintain a constant speed as it passes the reading heads. It is possible to store a lot of information on a very short length of magnetic tape – densities of 60 to 250 bytes per mm are possible.

Great care has to be taken when handling magnetic tape. It must not be touched with the fingers, and must be kept clean. In the machining environment ferrous metal swarf is a continuous hazard, which can cause problems. One danger with magnetic tape is that if it is affected by an external electric or magnetic field its information may be corrupted, i.e. distorted and changed. It is good practice to make back-up copies of important programs on separate tapes which should be stored in another location.

7.5 Magnetic discs

Magnetic discs are available in a number of sizes and magnetic storage capacities. The discs generally used for the storage of part programs for computer numerically controlled machine tools are of two physical sizes: one is 133 mm (5.25 in) in diameter, and the other is smaller and is 90 mm (3.5 in) in diameter. The magnetic storage capacity of both sizes of discs can be classified as single, double or high density which relates to the maximum number of bytes the discs are capable of storing: either 360 kilo bytes; 720 kilo bytes (KB) or 1.4 mega bytes (MB), respectively. (*Note*: although kilo = 1 thousand and mega = 1 million, a kilo byte is actually 1024 bytes and a mega byte is 1 048 576 bytes.)

Although it is possible to configure high density disc drives to read and write in single or double densities, there can be problems trying to use low density drives to read discs of any density which have been formatted and written on high density drives.

The discs are made of thin plastic (Mylar is frequently used) which is coated on both sides with a magnetic oxide. There is a central hole in the disc through which a drive spindle in the unit locates the disc in the drive unit so that the disc is correctly aligned to the write/read head.

The disc is rotated by pressure on the hub of the disc from flanges on the spindle. When the disc rotates, data can be transferred by the write/read head. The head contacts the disc surface with a precise pressure. If the surface of the disc becomes contaminated with dust or other particles, the contact between the head and the disc may be affected, and the particles can cause wear and cut grooves in the head contact surface. This may cause problems in data transfer.

The discs should never be placed near a magnetic field, because the magnetized regions on the disc can be affected by the external field. If the magnetized regions are changed by magnetic fields or the disc surface is contaminated and the program altered, the disc is said to be *corrupted*. There is no visual indication that a disc is corrupted; this only becomes known by reading the disc into a computer. A corrupted disc can create production problems and even danger to personnel if it is tried out on a machine tool. It is recommended that if the discs are likely to be placed near electrical transformers or other electrical equipment creating magnetic fields, precautions should be taken such as storing the discs in a metal box which will provide a shield.

The larger discs (5.25 in diameter) are held permanently in a black cardboard case which must never be removed. There is a window in the case through which the write/read head can contact the magnetic oxide on the disc. The discs are very flexible and are commonly known as *floppy discs*. Extreme care has to be exercised in the handling of the discs, and they should never be bent or creased. To store and carry the disc, it should be placed in a protective envelope with the window in the black case on the inside of the envelope to prevent dust, smoke particles etc. becoming attached to the magnetic oxide on the surface of the disc. The surface of the disc exposed through the window must never be touched with the fingers, because the oils or salts on human skin can be deposited on the surface of the disc. If it is necessary to write on the case of the disc for identification purposes, a soft felt tip pen should be used with light pressure.

The smaller discs are held in a fairly rigid plastic case which gives more physical protection. There is a window in the case that is covered by a shutter. The shutter is moved aside automatically when the disc is loaded into the disc drive unit so that the write/read head can contact the disc.

Before the disc is capable of storing information the magnetic coating on the surface of both sides of the discs has to be arranged in a special form; this is referred to as *formatting*. During formatting a series of concentric rings known as tracks are formed in the magnetic coating. Each track is subdivided into sectors, and it is on the tracks in each sector that the information is stored. It is possible to store a number of part programs on both sides of the disc.

To store a program on disc it is necessary to create the entire part program in the memory of an external desk-top computer, and then transfer it to the disc. The use of a computer for the creation of part programs will be explained in Chapter 13, and see Appendix C for an explanation of computers. When a part program has been saved on a disc it is advisable to obtain a copy of the program on a printer connected to the computer.

To use discs as a means of storing programs, there has to be a program in the computer known as a *disc filing system* (DFS) or *disc operating system* (DOS). One of the designed uses of the DOS is specifically for transferring information to and from discs. When the part program is first stored on the disc it has to be given a name, and the location of the part program on the disc is identified in a special part of the disc known as the *catalogue* or more commonly *directory*. To locate a particular part program the write/read head reads the directory for details of where the program required is stored on the disc. The head moves radially across the disc to the track or tracks on which the information is stored, and as the relevant sector moves under the head the information is transferred. A program may be stored on a number of different tracks.

When the disc is formatted, as well as arranging the magnetic layer in a series of concentric rings the magnetic field strength is set so that it induces in the write/read head a constant signal equal to that induced by an 'on' bit. An 'on' bit is stored on a magnetic disc as a magnetized region that will induce a pulse in the write/read head that will be amplified and cause a 5 V electrical pulse to be sent to the control of the computer.

On magnetic discs each bit in the byte is arranged consecutively i.e. in series. The bytes also follow one another in series. Within the computer there is a clock which generates time signals, and the pulses (bits) are sent at regular periods. Each period is a constant number of clock intervals. When the bit indicating the start of the program is received, a change in the signal voltage level is sensed and the computer expects to receive the next bit at the next period; the signal at the write/read head is sampled at regular periods. After the start bit is detected the data bits (normally up to seven) are received; and finally two stop bits and on some systems a pause, which is the byte finish signal. Some systems require a parity bit to be sent before the

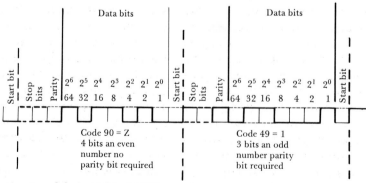

Figure 7.2   Magnetic signal format for ASCII

stop bits. This procedure is followed for each byte until the byte signalling the end of the program is received.

Figure 7.2 shows some examples of the form of the signals for typical characters (read the figure from the right). There are two methods by which a part program stored on a disc can be loaded into the machine tools control unit's memory.

The first method is to read the program from the disc into the memory of an external computer and then to send the program to the computer in the machine tool control unit. The external computer and the machine tool control computer are usually linked via a serial or RS232 interface. Full details of this interface can be obtained from technical literature on computer communications. Essentially the interface consists of electronic circuitry in both the external computer and machine tool control unit connected via cables with plugs and sockets. It is necessary to ensure that the speed that the external computer sends data is the same as that which the machine tool control unit can receive the data. The speed with which the program is sent is known as the *baud rate*; it varies from 110 to 9600 bytes/s. As well as the baud rate it is also essential to set the same values for the number of start bits, data bits, parity, stop bits and pause on both the sending and receiving computers.

In the second method the program is transferred directly from the floppy disc. For the machine tool control computer to be able to accept data it would have to have the disc drive units and the necessary software to interpret the signal from the discs. The disc rotates at speeds of the order of 300 rev/min. This results in the data being written or read at speeds approaching 100 000 bits/s. If the equipment to which the data is to be sent cannot work at that speed the information is stored in a buffer memory until required. As stated previously, each bit in the byte is arranged consecutively i.e. in series on the magnetic coating. The bytes also follow one another in series. However, one of the functions of the disc drive electronics is that when the stop bits at the end of each byte are read the bytes are converted from the series format so that all the bits in a byte are transmitted simultaneously i.e. in parallel.

Part programs created on discs by one type of computer may not be able to be read by another type of computer.

7.6 Punched tape

At one time punched tape, which is also known as perforated tape, was the main method of storing information for inputting into the control unit of the machine tool. Although magnetic tape and magnetic discs (floppy discs etc.) can store a much greater amount of information within a smaller space, great care has to be taken in their handling and storage. On punched paper tape about 400 rows of holes can be punched in a metre length of tape. A high density magnetic disc of 90 mm diameter can store up to 1.4 M bytes, this would be capable of storing more than 500 typical part programs. Provided punched tape is kept dry there is very little possibility of the information stored on the tape becoming faulty. Punched tape is therefore a very safe medium for the storage of data, also torn tape can be repaired if required, although it is fairly easy to punch a new tape.

Punched tape materials

The tape can be made of various materials such as paper, plastic (Mylar), laminated paper and plastic, and foil. Each material has particular advantages. Paper tape is the cheapest material and is generally the material used for trials, or if there are only a small number of components to be produced. It can easily be damaged and is subject to very rapid deterioration if it comes into contact with oil. There are no problems in punching paper tape. Paper tape is weakened each time it is fed through the tape reader; as a result of the weakening it is only advisable for a particular paper tape to be used about 100 times, after which it may tear along the sprocket holes. Mylar is very durable but can stretch under load, which may cause reading errors. Laminated tape is best for long service, as there is virtually no limit to the number of times it can be used. Foil tape is the most expensive and normally is only

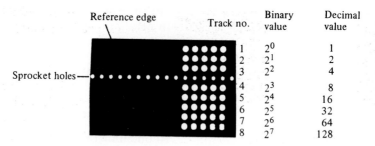

Figure 7.3   Details of punched tape

used as a master from which a working copy of paper is made as required. The paper tape copy is destroyed after it has been used. Tools used for punching foil tape must be kept sharp; the material tends to blunt the punches used, and the blunt tools may not punch correctly.

It is possible to obtain tapes in different colours. The colours can be used as a means of identifying tapes that are to be used on different machines, or for indicating whether the program has been proved (one colour for untried programs and another for programs which have been successful). Consideration must be given to the colour of tapes which are to be used on photo-electric tape readers. Some colours are more capable of reflecting light, or may allow light to pass through and cause errors when the tape is being read.

Dimensions of punched tape

A full specification of punched tape is given in BSS 3880:1973. Various widths of tape have been used, ranging from 12 to 200 mm, but the width is now standardized at 25.4 mm (1 inch). During the development of the first British numerically controlled machines that used punched tape, a tape used had a maximum of five holes across the tape. However, the tape now used has provision for a maximum of eight holes to be punched across the tape, as shown in Figure 7.3. The eight holes are known as data holes and are 1.83 mm diameter and at a pitch of 2.54 mm. Each hole position is known as a *channel* or *track*. As stated previously, tapes usually have an additional continuous line of holes along the length of the tape. These holes are known as feed or sprocket holes and are 1.17 mm in diameter and at the same pitch as the data

holes. A sprocket wheel engages in the feed holes, which are used to drive the tape through the reading heads. The sprocket holes are not in the centre of the tape, but 9.96 mm from one edge, and there are positions for three tracks of holes to be punched between the sprocket holes and one edge of the tape. This edge is known as the tape *reference edge*; the track nearest to the reference edge is track number 1.

At the start and end of the tape there are sections of tape which only have the line of small sprocket holes punched. This section at the start of the tape is known as the *leader*. The section at the end is called the *tail* or *footer*. The lead and tail sections are used when loading the tape into the reading head and connecting it on to the spool spindle. The length of the lead and tail are dependent on the design of the spooling system of the tape reader, but a length of 400 mm will generally be found convenient for handling.

One advantage of punched tape is that, when the codes for various characters are known, it is possible to read the tape visually to check sections of the program.

Another advantage of punched tapes over magnetic tapes is that it is possible to insert amended sections into a punched tape by splicing the tape. For short tapes which can be copied quickly this technique is more trouble than it is worth, but for long tapes it is a practical solution. The amended section of tape can be joined to the original tape using adhesive patches. Figure 7.4 shows the two types of patches used: (a) splicing patch and (b) correction patch. A splicing patch has all the holes including the sprocket holes punched, and is used to join punched sections of tape. The correction patch

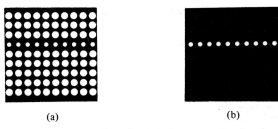

(a)   (b)

Figure 7.4   Patches for joining punched tape: (a) splicing; (b) correction

only has the sprocket holes punched. This can be used to join non-punched sections of amended tapes together, or can be overlaid on a section of the tape that has only a few rows or tracks to be amended; the new holes are then punched with a hand held punching unit. Damaged or torn tapes can be repaired with splicing patches. It is recommended that the manually operated splicing units which are available should be used to ensure the correct alignment of the sprocket holes in the tape. The adhesive patches should be placed on both sides of the tape.

Punched tape codes

The tape is read by mechanical probes (fingers) or light passing through the holes punched in the tape which activate (energize) electrical circuits. A hole in the tape is the equivalent of a magnetized region on a floppy disc or magnetic tape, therefore a single hole represents a *BI*nary digi*T* (*BIT*), and each row of holes across the tape is a byte.

The information is stored on punched tape using the same ISO or ASCII codes as on magnetic discs. Figure 7.5 shows the ASCII codes for numbers and letters, with and without parity. The EIA system uses odd parity i.e. the number of holes punched across the tape has to be an odd number. The fifth track is used as a parity track.

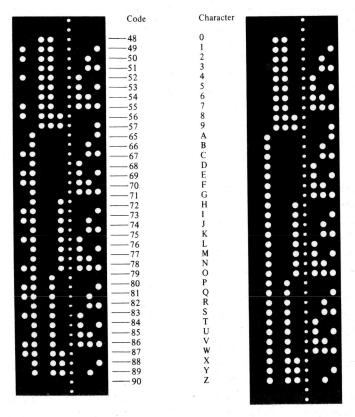

Figure 7.5   ASCII punched tape codes: (a) with parity; (b) without parity input

As explained previously, in order to represent dimensional information a technique known as binary coded decimal (BCD) is used. With this technique each digit in the number appears on successive rows on the tape. A value of 325.78 would thus appear on the tape as shown in Figure 7.6. As mentioned previously, some control systems do not recognize the code for the decimal point; the number on the tape in BCD has to be in the smallest unit for which the machine control system is designed. If the smallest unit for which the machine control responds is 0.01 mm then a value of 41.00 would be punched as shown in Figure 7.7. From the above it can be seen that the value for each digit is in binary across the tape but in denary (decimal) along the tape. The magnitude of an individual digit is determined by its position in the number.

Production of punched tape

Punched tape can be produced using:

(a) Teletypewriter
(b) Punch unit linked to a computer
(c) Punch unit linked to the machine control unit.

The most frequently used method of producing punched tape has been the teletypewriter. This machine has a normal QWERTY typewriter keyboard, together with a number of extra keys which operate an attached unit on which the tapes can be punched or read. (The name QWERTY comes from the sequence of the lettered keys, starting at the top left-hand side of the keyboard.) The teletypewriter is electrically powered. When a key is depressed the corres-

ponding letter or number will appear on a sheet of paper as on a normal typewriter, but in addition a number of holes are punched across the tape for each character typed. Therefore as the sheet of paper is typed, a length of paper tape is punched. As explained previously in this chapter that each number, letter or symbol on the typewriter keyboard has its own combination of holes across the tape. On some machines there is a switch with which it is possible to change the particular combination of holes (codes) which represent different characters according to the requirements of the machine tool control unit.

If the wrong keys are pressed and the error is noticed immediately, it is possible to take corrective action. There is a special button known as the 'backspacing' button on the tape punch unit. Pressing the button the number of times the errors have occurred will cause the tape to backspace the same number of rows. Only the tape moves; the printout paper remains stationary. The incorrect codes are deleted with use of an additional key known as a 'delete' or 'rub out' key on the keyboard. This key will cause a complete line of holes (eight holes) to be punched across the tape. The delete key must be pressed the number of times the tape has been backspaced. The correct keys can then be pressed and the corresponding code will be punched on the tape. When the tape is being read the control unit will ignore the signal generated by the complete rows of holes. The backspacing facility is only practicable for correcting errors that are noticed within the last few characters printed.

There are some teletypewriters that have

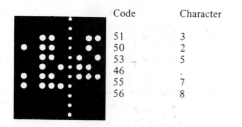

Code	Character
51	3
50	2
53	5
46	.
55	7
56	8

Figure 7.6 Binary coded decimal value of 325.78 with parity and decimal

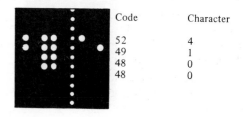

Code	Character
52	4
49	1
48	0
48	0

Figure 7.7 Binary coded decimal value of 4100 with parity

facilities for storing a limited amount of information, such as a single line, which can be checked for accuracy by reading the printed copy; when the line is seen to be correct, it can be sent to the punch unit and the corresponding tape produced. The next line is entered, and when it is seen to be correct it is sent to the punch unit; and so on. One technique that has been proved beneficial is that when the part program is created, each operation is written on alternate lines on the program sheet. The program sheet is loaded into the teletypewriter and the information is typed on the blank lines below each handwritten line. A visual check to compare each line can quickly reveal any mistakes.

If the tape has to be edited, or if errors are found after the complete tape has been produced, a correct section of tape can be inserted by a splicing technique. It is more practical to produce a new tape with the use of the tape read unit on the side of the teletype. On this unit there are three buttons which control the feeding of the tape through the read unit:

(a) One button will cause the tape to feed through the tape read unit continuously and automatically.

(b) A second button will switch off the read unit.

(c) A third button will cause the tape to be fed forward and be read one line at a time.

While the tape is being fed forward, either automatically or one line at a time, a new tape can be punched. When the section of tape which has to be amended is reached the following actions, which can be very time consuming, will produce a new tape with the amendments:

1 The tape read unit is switched off, but the tape punch unit is left on.
2 The correct keys on the keyboard are pressed and the new codes punched on the new tape.
3 The tape punch unit is switched off.
4 The tape read unit is switched on.
5 All the incorrect portion of the original tape is fed through the tape read unit using the single-line read button.
6 The tape punch unit is switched back on and

the remainder of the tape is fed through the read unit; the tape is produced as before.

The introduction of the computer has enabled a convenient method of producing punched tape to be developed, using tape punch units connected to the computer. The computer is used in a text editing mode (see Chapter 13 for more detailed information).

On some computer numerically controlled machine tools it is possible to enter the information by manual data input or to edit the information using a keyboard on the machine. To save the information for the next time the component has to be produced, it is convenient and common practice to have a tape punch unit connected to the machine which can be used to produce a tape for storage.

Punched tape readers

Arrows are usually printed on the tapes which indicate the direction in which the tapes should travel through the tape reader. The tape is fed through the reader by a sprocket wheel which engages with a line of holes. These holes are off centre on the tape, and care has to be taken when initially loading the tape into the reader.

A tape should not be rolled up into a very tight coil, as it may then become tangled as it is being fed through the tape reader. Tangling is particularly troublesome when 'tumble boxes' are used to hold the tape. These are containers into which the tape is allowed to fall freely after it has passed through the reader; they are used for tapes 3 to 10 m long. It is more convenient to hold longer tapes on a spool or reel and feed the tape on to another servo-driven spool. Care has to be taken to ensure that the tape is threaded correctly; rollers on some form of spring tensioning device press on the tape to ensure that it is correctly tensioned and not overloaded. If the tape is less than 3 m long and it is required to read the tape for each component, such as on non-CNC controls, the ends of the tape can be joined together to form a continuous loop. This will save rethreading the tape through the reader each time it has to be used.

It must be remembered that all the pulses for

one character are read simultaneously i.e. in parallel. There are different types of punched tape reader: pneumatic, electromechanical and photo-electric.

Pneumatic tape readers

There are not many pneumatic types in use. One application was on the control unit of one of the early models of milling and drilling machines by a company named Moog. The tape was 'read' by a jet of air being blown on to it. If a hole was present in the tape, the jet of air passed through the hole and actuated a form of pressure-sensitive switch.

The Moog reader is a 'block' reader where all the holes in 20 rows (one block) are read simultaneously.

Electromechanical tape readers

The tape readers generally used on control systems for machine tools are of the sequential type. These units simultaneously read all the holes in a single row across the tape at a time. The next line of holes is then read, and so on until all the holes in one block (operation) are read in sequence. It is the sequential readers which were mainly used on control units for the second-generation numerical control machines. Typical sequential readers read the tape at speeds of 30 rows of holes per second, which is found to be satisfactory. This is because in many cases the information for one operation usually requires less than 30 rows of holes. The information for an operation is fed in, and the tool or work movements or other actions are carried out; then the details for the next operation are read. This results in the tape being subject to many starts and stops.

On numerically controlled machines without computers in their control unit, a buffer store may be used (see Chapter 2). A buffer store is a memory store from which the information can be more quickly transmitted to the control unit than when read directly from the tape. The information for an operation is read from the tape into the buffer store while the machine is completing the preceding operation. As soon as the buffer store is empty, the information for the next operation is read off the tape and stored. A

buffer store saves the time of waiting for the information to be read from the tape.

There are available electromechanical readers which operate at much faster speeds than 30 rows per second. The problem with a fast electromechanical reader is the number of moving parts in its construction. There are different designs of mechanisms used, but they have either a row of eight spring-loaded pins or eight star wheels. A pin will pass through the tape if there is a hole in the tape at its position, and will cause a switch to be activated to permit a pulse of electric current to flow. It is important that the tape is fed squarely through the reading head so that each of the row of holes is in alignment with the row of pins, and the presence or otherwise of all the holes in a row is detected simultaneously.

Electromechanical tape readers are slower in operation than the photo-electric types.

Photo-electric tape readers

The photo-electric readers are more expensive than the electromechanical, and operate at speeds of at least 150 rows of holes per second. Light is directed on to the tape from a source on one side of the tape, as shown in Figure 7.8. If a hole is present in the tape the light passes through and energizes a light-sensitive diode (photocell). The energy generated is used to send a signal to the control unit. There is one diode for each track across the tape. Care and attention must be given to the reading head, with a regular programme of maintenance to prevent bulb failures.

Care must also be taken when selecting the colour of the tape to be used with photo-electric

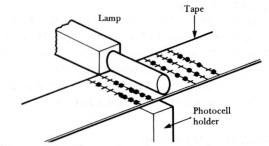

Figure 7.8   Photo-electric punched tape reader

readers. It is essential that the tape is opaque; some colours permit light to pass through, and faulty signals can be generated.

Tape for use on photo-electric readers may not need sprocket holes, but may instead be fed forward by a powered roller known as a capstan. On computer-controlled machines it is normal practice for the complete tape to be read in one continuous pass of the tape. The tape is automatically reversed in direction of movement at the end of the program by a command in the program; the information is then verified as the tape moves in the reverse direction. All the information is stored in memory, from where it is sent to the control unit as required.

Behind the tape reader (BTR)

This does not read punched tapes but is a printed circuit card with electronic circuitry interfaced into the control unit to which a floppy disc drive or desk-top computer can be connected. Information can be input from discs or the computer which will simulate the input from punched tape and provide an alternative method of inputting the information. One particular application of a BTR is to update an older NC machine and extend its useful life.

7.7 Block formats

The part program is stored using the same codes in the same form on punched tape, magnetic tape and discs. The information necessary to control the machine tool has to be input to the control unit in the form of words. A *word* is a group of rows or bytes which define an element of information, such as a dimension, a spindle speed or a feed rate.

The number of words necessary to define all the controlled functions for one position or operation is known as a *block*. There are different systems for arranging the blocks: the fixed sequential system, the tab sequential system and the word addressed system.

Fixed sequential system

The fixed sequential system was mainly used with the first numerically controlled machines, and is rarely used with computer-controlled machines. In this system each block of information contains exactly the same number of words, and all words follow in the same sequence in each block. Every word has to be represented even if only by a series of zero characters or its previous value. No letters are input; the machine therefore recognizes the value of a character by its position in the block. This system results in needless information being supplied and very long programs being produced.

Tab sequential system

In the tab sequential system the words must occur in a predetermined sequence, but they are preceded by a tab character. The name 'tab' comes from tabulation, which is used to arrange the words applying to the same item underneath one another. The ASCII code used for tab is code number 9, but other codes can be used on some control systems.

A tab character indicates the beginning of a new word but does not identify it, for the same tab code is used for each word. If a particular word is unchanged in consecutive blocks it may be omitted, but a tab character must always be inserted. Generally the same number of tabs must always be present in every block even if one is not followed by any data. An exception to this rule is that it is possible to reduce the number of tabs by the entering of a block marker. There are no letters (addresses) in the blocks; the number of the tab in the sequence of tabs enables the control to recognize what the data following a tab is to be applied to. The first character entered at the start of the program must be a block marker, and no tab precedes the sequence number. The sequence number always follows a block marker. The second item in a block is usually a preparatory code (G code); therefore the data following the first tab is always a preparatory code. The data following the second tab is the X coordinate etc. A full line could consist of:

a sequence number TAB a preparatory function (G code) TAB X TAB Y TAB Z TAB feed TAB speed TAB tool number TAB miscellaneous function (M code) EOB

To illustrate this, the information to be input for the line

N020 X150.0 Y56.0 F200 M08 EOB

would be

020 TAB TAB 150.0 TAB 56.0 TAB TAB 200 TAB TAB 08 EOB

This line does not contain any G code, Z coordinate, speed or tool number, but all the tabs have to be entered.

Word addressed system

The word addressed system is the method generally used today for the numerical control of a wide variety of machines and other equipment such as plotters. In a word addressed program there must be an identifying letter (an alpha character) preceding each word. This character, which is called an *address*, differs for each machine function. The machine control therefore recognizes a word by its address rather than by its position in a particular sequence. The words need not be programmed in any particular order, although for the sake of convenience they usually are. A number of words make up a block, and adjacent blocks are separated by a special character known as a *block marker*. Further information on word addressed format is given in Chapter 8.

To provide the information to cause a table to move from one position to another requires first an address letter, i.e. the byte which represents the alpha code for the machine feature concerned. It must be remembered that the input is in number form, and therefore the control recognizes the address not as a letter but as a number. Its number in binary activates a particular switching circuit and allows the information

that follows to be directed to a particular memory. The address can be considered to be a key which unlocks a particular door and allows the information to pass through and be stored.

Questions

7.1 What are the advantages and limitations of using the following as a means of storing part programs: (a) punched cards (b) magnetic tape (c) punched tape (d) magnetic discs?

7.2 Describe three different techniques of producing punched tape.

7.3 Describe two methods of feeding punched tape through tape readers.

7.4 Describe three different methods of reading punched tape automatically.

7.5 Which method of reading punched tape is most likely to be used with CNC machines? Give reasons why that method is used.

7.6 Describe the use of (a) tumble boxes (b) reels and (c) loops as techniques of supporting punched tape before and after it has passed through the tape reader.

7.7 Explain how it is possible to identify the track numbers on punched tape.

7.8 What do the letters ASCII represent?

7.9 How would the number $X - 127.45$ appear on punched tape using ASCII?

7.10 What is a parity track, and why is it necessary?

7.11 What are the following as related to ASCII: (a) control characters (b) numer characters (c) alpha characters?

7.12 Explain how the above characters could be stored on a magnetic disc.

7.13 Detail the different systems of arranging the blocks of information in a part program.

Chapter Eight

Structure of part programs

8.1 Part programs

A part program contains all the information for the machining of a component which is input to the control unit. The control unit provides the control signals at the correct time and in the correct sequence to the various drive units of the machine.

The input information required is in a series of blocks; one operation requires one block. Within each block there may be different types of data.

8.2 Types of control information

There are three basic types of input information in a part program for control of the machine:

(a) Information on desired work/tool relationship (position, direction and amount of movement)
(b) Information on rate of change of work/tool relationship (feeds and speeds)
(c) Information on auxiliary functions (mode of operation, selection of metric or imperial units, tool selection, use of cutting fluids, work or tool loading etc.).

Information on the work/tool positional relationship is referred to as *dimensional information*, and is input as a series of coordinates for all the axes such as X, Y and Z. Information on the

rate of movement is input as values of feed rate and speed, and is referred to as *non-dimensional information*. Information on auxiliary functions is input in special codes; this is also referred to as non-dimensional information.

The information for control of rate of change of work or tool and for auxiliary functions was referred to at one time as 'management information', because these functions are normally part of the operator's handling (managing) duties on manually operated machines. Today *management information* refers to data used in assessing the operation of the business, such as the time the machine is actually removing material, and how long a particular tool is efficient in cutting material. This type of information is not included in the part program.

With some control systems, as each operation takes place the details of the operation are shown on the screen of the control unit. There are occasions during machining that require operator intervention, and the information for this purpose can be provided in the part program. To provide the details which will be displayed on the screen, they have to be inserted between parentheses (brackets). The right-facing parenthesis (is referred to as 'control out' and will cause the control unit to ignore the characters following the parentheses. The left-facing parenthesis) is referred to as 'control in' and in effect will switch the control back on.

145

The different methods of creating part programs will be described in later chapters, but it is important to remember that the information actually input into the control unit is in word addressed format. There will be occasions when it is necessary to edit the part program manually; in addition, a knowledge of programming will provide operators with an additional interest to ensure that the machine is working at maximum efficiency. For these purposes alone it is essential to understand the structure of the part program.

8.3 Word addressed format

As described in Chapter 7, a word consists of a letter followed by a number. The letter is referred to as the address, and between the letter and the number there may be an algebraic sign; normally only the negative (minus) sign has to be entered if required. Different words can have a different number of digits; some words have one digit and other words up to seven digits. A word may also contain characters such as / or # for special purposes.

The manufacturer of the control system decides on what letter is to be used as the address for each of the words; this information will be found in the manual for the control system. For convenience certain letters have become standardized for particular words.

An operation which is defined in a block can contain a single word or a number of words. If there is to be no change in a word it is normally not necessary to enter it in the succeeding blocks. However, there are certain cases when it is essential to repeat details. When the function remains active and it is not necessary to repeat words, the words are termed *modal*; when it is necessary to repeat words, they are termed *non-modal*. A typical block has the following format:

/N05G02X+or−042Y+or−042Z+or
−042F04S04T03M02EOB

The letters are the address of the word, and the following zero indicates that any leading zeros in the word may be suppressed; the next numbers give the total number of digits that can appear in the word before and after the decimal point. There are some words which do not have a decimal point, and for these words only one number is stated. However, it is important to be aware that there are some control systems which permit the omission of trailing zeros. The block format for suppression of trailing zeros is indicated as:

/N50G20X+or−420Y+or−420Z+or
−420F40S40T30M20EOB

It is not necessary for spaces to be left between words, but for easy reading of the program the words may be spaced. When the program is read by the control unit the space will be ignored. If the program is edited when in the control unit, and a new program is printed, the revised program will not contain any spaces.

When creating the part program, it is convenient to compose all the information first in a tabulated form. The manufacturers of the machine tool usually provide a form which has a layout suitable for the control system fitted to the machine tool. Different control units require part programs in particular formats; a part program written for one machine may not be suitable for another machine.

As explained in Chapter 7, the part program can be stored on punched tape, magnetic tape or floppy discs, or may be stored in the memory of a computer. When punched tape is used it is necessary for the program to contain additional characters which are required for the feeding of the tape through the reading head.

There are certain conventions or rules which should be followed. A program start character must appear at the beginning of the program before any other information. The percentage sign % is the recommended symbol used for the program start, and for punched tapes it is also used as a rewind stop character. At the end of the program another character has to be entered to indicate the end of the program. For punched tapes this character may also be the rewind character, which causes the tape to automatically rewind back to the start. For some control systems, such as the Bridgeport BOSS 5 or 6 control, an additional character may have to be entered at the end of the tape; the character used by these control systems is upper case E. An alignment character can be programmed after the %, and the character used is the colon :. It is possible to use the alignment character as an intermediate rewind stop.

8.4 Optional block skip character (/)

There are certain types of components which may be produced from the same blank or casting but have different combinations of holes, or holes of different size which require different operations. For this type of component it is possible to program all the operations required for all the components, and then the operator can omit certain operations for particular components. This is possible by the use of an optional block skip function. The control unit has to have this facility for it to be used.

There are two techniques of applying this facility. On one control system the function is activated by the insertion of a / (slash) before the sequence number. When the program is read by the control unit any blocks which are preceded by the character are omitted when a switch on the control unit is activated (switch is on). This method is illustrated in program 1 below. An alternative method of utilizing this facility is that a / is entered after the sequence number in the first block that has to be omitted, and at the end of the last block in the sequence that has to be omitted. The option switch on the control unit has to be on for the blocks to be omitted. This method is illustrated in program 2 below:

Program 1	Program 2
/N100X15.0Y32.0	N100/X15.0Y32.0
/N105X30.0	N105X30.0
/N110Y64.0	N110Y64.0/

The block skip character may be referred to as the 'block delete character' by some manufacturers. It is useful in reducing the number of programs required for the machining of suitable components. The operator does have to be aware of the possible omission of particular

blocks, and has to ensure that the option switch is activated when required.

8.5 Sequence or block number (N)

This is the sequence number or operation number, and is usually the first word which appears in any block. The letter N is generally used as the address, but other letters such as H or P have been used by different manufacturers of control systems. The system manual issued by the manufacturers will provide details of which letter is to be used as the address.

For those systems that permit leading zeros to be suppressed, i.e. omitted, instead of N00150 the sequence number is programmed as N150. For those systems that permit trailing zeros to be suppressed and will accept up to five-figure sequence numbers, the sequence number N00150 could be entered as N0015. It is essential to write the part program to satisfy the requirements of the control system by consulting the programming manual for the machine. If leading zeros are suppressed, trailing zeros must be programmed in order to establish the magnitude (size) of the number, and vice versa. Suppressing the zeros is not compulsory but can be helpful in reducing the length of the program.

The number of digits permitted in the sequence number depends upon the control system; some control systems permit a maximum of five, others may only permit a maximum of three. The sequence numbers need not be consecutive, and are mainly used for the convenience of the operator in identifying the different operations. The sequence number may also be used as a reference point in the program to which the process can advance without having to repeat machining operations that have been completed. This may be required, for example, after a tool replacement due to breakage. The sequence numbers can also be used as check points for repetitive programming in computer controlled machine tools. It is convenient for the sequence numbers in the original program to advance in multiples of 5 so that additional lines can be inserted during any editing that may be required, e.g. N5, N10, N15 etc.

The control system of computer controlled machines will *not* arrange the sequence numbers in numerical order but will complete the operations in the sequence that they have been programmed regardless of the sequence numbers. Therefore if blocks have to be added to a program, they *must* be inserted in the program at the desired place. It is not possible to add the blocks at the end of the program, give them the sequence numbers where they should be inserted, and expect the computer in the control unit to rearrange the order.

8.6 Preparatory function (G)

The preparatory function command consists of the letter G followed by two digits; the digits are a code for different actions or operations. Although suppression of either leading or trailing zeros is permitted, it is recommended that the preparatory function is always programmed as a two digit number; this avoids ambiguity. For convenience, preparatory functions are frequently referred to as G codes. The majority of the G codes are modal.

A preparatory function is an instruction for the control unit to perform an operation which is non-dimensional in nature. The functions are used for various purposes, such as:

(a) To select the movement system (point-to-point, contouring etc.)
(b) To select either metric or imperial units for input of dimensions
(c) To make compensation for variation in tool sizes
(d) To select a preset sequence of events (canned cycles).

The majority of control systems permit a number of G codes to be programmed in one line, provided that they do not conflict in their purpose.

There have been attempts to standardize on particular digits for the specific functions commonly used. It is unlikely that all the G codes

available will be used in one system on one machine tool; reference will have to be made to the manual published by the manufacturer of the control system for the G codes available for a particular machine tool. There are a number of codes which have not been assigned functions and are provided for the manufacturer to designate for its particular use; therefore it is possible for G codes with the same number to be used for different functions on different manufacturers' systems.

The most frequently used standard G codes are as follows:

G00   point-to-point positioning
G01   linear interpolation
G02   clockwise circular interpolation
G03   counter-clockwise circular interpolation
G70   imperial units
G71   metric units
G90   absolute dimensions
G91   incremental dimensions

Appendix A gives a fuller explanation of preparatory functions.

8.7 Dimensional information words (X, Y, Z etc.)

The desired work or tool position is programmed using the address for the particular axis. If the machine tool has axes other than X, Y and Z, then the letter used for the address for programming those axes has to be the letter assigned to the axes by the manufacturer.

With dimension words it is essential to enter a negative sign $(-)$ if required; if there is no sign the number is assumed by the control unit to be positive. If the part program for milling and drilling machines has a negative sign on an absolute dimension word for the X axis, movement to the left of the X zero would result. In incremental programming the zero position changes, and an incremental dimension is related to the immediately preceding position. This means that if the position to be moved to on the X axis is to the left of the existing position, the coordinate has to have a negative sign even though the work could still remain to the right of the X zero (absolute).

If the format of the word is shown as X+or−042, leading zeros can be suppressed, and there can be four digits before the decimal point and a maximum of two digits after the decimal point. However, if the format of the word is shown as X+or−330, trailing zeros can be suppressed, and there can be three digits before the decimal point and a maximum of three digits after the decimal point. These methods are illustrated in Table 8.1. Note that 5264.0 could not be programmed in trailing zero suppression because it has four digits before the decimal point.

Table 8.1 *Dimension word formats*

Dimensions	Word format X+/−042 Leading zero suppression	Word format X+/−330 Trailing zero suppression
0.25	X0.25	X000.025
34.25	X34.25	X034.25
150.00	X150.00	X150
5264.00	X5264.00	

It is normally only necessary to enter any changes in the value of a coordinate. If there is to be no change in the value then the word may be omitted from following blocks until its value has to be changed.

8.8 Decimal point

On some numerically controlled machines the numbers programmed have to be in the units of the smallest value that the control system can discriminate. It is with these control systems that the suppression of leading or trailing zeros has the greatest significance. This is shown in Table 8.2, where it is assumed that the system can work to 0.01 mm and that there can be four digits before the decimal point.

On the majority of machines it is possible to program the decimal point. For these machines a whole number must have a decimal point programmed. If the decimal point is omitted the

Table 8.2 *Units of smallest value 0.01 mm*

Dimension to be programmed	Leading zero suppression	Trailing zero suppression
0.25	X25	X000025
34.25	X3425	X003425
150.00	X15000	X015
5264.00	X526400	X5264

control will assume that the number is in units of the smallest value. If the number entered was 1234 and the system can work to 0.01 mm, the control will assume the number has a value of 12.34; if the system can work to 0.001 mm, the control will assume the number has a value of 1.234.

Where a decimal point is programmed, if trailing zeros do not influence the magnitude of the number they can be suppressed at the programmer's discretion, i.e. 10 mm can be programmed as 10. or 10.0 or 10.00.

8.9 Feed rate (F)

There are a number of methods of designating the feed rate, such as:

(a) Millimetres per minute
(b) Millimetres per revolution
(c) Feed rate number
(d) Inverse-time feed rate number.

The most common method of programming feed rate for machining operations is in millimetres per minute. There are no recommended standard preparatory functions for designating whether the feed rate is in millimetres per minute or millimetres per revolution. Some control systems use the following preparatory functions for designating the units:

G94 feed in millimetres per minute
G94 feed in millimetres per revolution.

At one time some machines had a range of fixed feed rates, and a particular feed was specified with a number such as F5. The 5 was the number of the feed to be selected from the range of feeds provided on the machine. The block format would require the feed to be entered using this type of number.

There are control systems that require the feed rate for certain operations, such as slopes or arcs, to be programmed as an inverse-time feed rate number. These feed rates are numbers which are proportional to the reciprocal of the time taken to traverse a slope or arc (see Chapter 9 for further details).

The control unit can store only one feed rate at a time. However, it is not necessary to program a feed rate on every line, as once a feed rate has been programmed it applies to all table drive motors. Obviously a new feed rate may be programmed at any time, and that rate would then apply until changed.

8.10 Spindle feed (S)

A spindle speed is normally programmed at a tool change. The speed would then remain at that value until changed, but does not have to be programmed on every line.

Spindle speeds can be programmed in a number of ways:

(a) Actual spindle speed in revolutions per minute
(b) Cutting speeds in metres per minute
(c) Constant cutting speeds in metres per minute
(d) A number selected from a table provided for the machine.

For the majority of the control systems now in use it is the actual spindle speed in revolutions per minute which is programmed. A constant cutting speed in metres per minute is more beneficial for lathes. With those systems where a constant cutting speed can be programmed, the spindle speed in revolutions per minute will automatically change as the tool moves on the X axis.

There are no recommended standard preparatory functions for designating spindle speeds. Some control systems use:

G96   constant cutting speed in metres per minute

G97   spindle speed in revolutions per minute.

8.11 Tool number (T)

For a large percentage of work it is unlikely that more than 20 tools would be required for a particular component, but there are tool storage facilities on some large machines which can hold more than 200 tools. The storage facilities are referred to as reservoirs or magazines (see Chapter 3). Each tool in these magazines may be individually numbered, or may be identified by the number of its location in the tool storage system. If the latter arrangement applies then, at the setting-up stage, a particular tool would have to be placed in the position in the tool storage unit selected for it by the programmer. (See Chapter 3 for an explanation of a method of identifying individual tools.)

When the control system provides for tool diameter or tool length compensation, the tool number can be used for inputting the diameter or the length of the tool as an extension of the tool number at the start of the program. For example, a diameter of 10 mm for tool number 1 could be shown as T1/10.0; a tool length offset of 80 mm for tool number 2 could be shown as T2//80.0. Some control systems for turning centres use the letter D to indicate tool offset values.

Tool offset values

When tools are mounted on a turning centre (lathe), the amount of extension of the different tools from the tool holder varies according to their size and length. It is possible to create only one X or Z datum zero on the machine with one tool. Tool offset values are the differences in the lengths of the extension of the other tools from the tool which was used to set the zero datum. A tool can have two offset values, one applying to the X axis and the other to the Z axis. Tool offsets can also be used to edit the sizes of work being produced. The same principle applies to the length of milling cutters and drills.

8.12 Miscellaneous function (M)

The miscellaneous function command consists

of the address M followed by two digits. Appendix B contains details of all the commands that can be used.

Miscellaneous functions are similar to preparatory functions in that they are encoded instructions which cause machine features to be activated, such as turning coolant on or causing a tool change to occur. The main difference between preparatory and miscellaneous functions is that the event or purpose for which the preparatory functions are activated occurs before the start of the movements in the block in which they are programmed, whereas the event for which the miscellaneous functions are activated has to occur at the end of the operation (block) in which they are programmed.

Generally only one miscellaneous function is coded in one block. The majority of M codes are modal. There tends to be fewer miscellaneous functions than preparatory functions provided in the control systems; the manual issued by the manufacturer of the control system will give details of the codes available. The most common codes provided are:

M00 program stop
M01 optional program stop
M02 end of program
M03 spindle on clockwise
M04 spindle on counter-clockwise
M05 spindle stop
M06 tool change
M07 mist coolant on
M08 flood coolant on
M09 coolant off
M30 end of tape

Although, depending on the control system, either leading or trailing zeros can be suppressed, to avoid ambiguity it is recommended that an M code should always have two digits.

8.13 End of block (EOB)

On the part program sheet the end of the block may be indicated by the asterisk symbol *. However, it will be found that different symbols are used to indicate the end of a block. At the end of a line (block) a carriage return key has to be pressed, followed by the new line key to commence the next block. It is therefore convenient to use the carriage return or new line character as an end of block marker. These characters will not appear in the printed version as they are control characters (see Chapter 7 for further details). On some keyboards of tape punches or computers the line feed occurs automatically when a carriage return is entered.

The part program structure explained above is sometimes referred to as ISO format or G and M programming. This format has been widely adopted by manufacturers of control systems. However, the actual letters and digits used for the address and the codes assigned for activating specific functions varies between control systems.

For example, to designate inch or metric units the Fanuc controls use G20 or G21 not G70 or G71 used by the majority of the controls.

The Sinumerik control uses the address L and not G when activating canned cycles. Sinumerik refer to drilling cycles as 'protected subroutines'.

For efficient part programming it is essential to use the address and code assigned by the manufacturer of the control when creating a part program for their particular control.

There are controls that can be programmed in two styles such as certain of the Heidenhain controls which uses either the ISO format (preparatory and miscellaneous functions) or the Heidenhain plain-language format.

8.14 Heidenhain plain-language programming

The Heidenhain plain-language programming format is an example of a word address language developed by a manufacturer of a control system. This language was primarily developed for creating the part program at the machine tool control unit, using what is referred to as dialogue technique. The creation of the program is assisted by the keyboard having a number of keys marked with symbols dedicated for inputting specific commands and coordinate values. However, part programs can be written using a text editor as the language is essentially word address. G codes (preparatory functions) are not used but M codes which are referred to as auxiliary functions are used. For the most widely used functions the same numbers are used as those listed in the ISO format detailed in Section 8.12 to activate functions; such as M00 program stop and M06 tool change etc.

The block numbers do not have an address and the first block is designated 0 (zero). This first block contains the term BEGIN followed by PGM (Program) and the program number of up to eight digits such as PGM 1231234. In this block the dimensional units (mm or inch) are also specified. The last block in the program contains the block number and the term END followed by the program number PGM 1231234; and the dimensional units (mm or inch) are confirmed. The program is written between these blocks.

The tools to be used are entered in a tool definition sequence early in the program before any blocks containing movement statements. The tool definition sequence contains the tool number, length and radius. The tool is not selected until a block is entered which contains the words TOOL CALL with the tool number, direction of rotation and the spindle speed.

Linear interpolation is activated by entering the address L followed by the address of the axis (X Y Z etc.) on which movement is to take place and the coordinates of the point to move to. Machining of circular curves or arcs can be activated in one of four ways; one way uses the address C followed by the address of the axes and the coordinates of the point to move to; the direction of movement is designated with the address DR− for clockwise and DR+ for counterclockwise.

Both rectangular and polar coordinates can be input in absolute and incremental units. Rectangular absolute coordinates are entered as X+/−123.45, incremental coordinates are entered as IX+/−123.45. Polar coordinates in absolute units are entered as PR+/−123.45 (radius) and PA+/−30.0 (angle) and in incremental units IPR+/−321.50 (radius) and IPA+/−30.0 (angle). The reference centre for polar coordinates, the pole position, has to be

entered in rectangular coordinates before any polar coordinates.

Heidenhain controls have a number of canned cycles available which are selected by first entering a cycle definition code and entering the necessary details in a specific order and format. The cycle is activated with a CYCLe CALL command.

A part program in the Heidenhain-plain language format can be written as outlined above but can be effectively created as part of dialogue programming, which will be detailed in Chapter 13.

8.15 Word addressed programming of plotters

Plotters for draughting commonly use stepper motors to move the pens or the paper. The information that is input to the stepper motors' control unit usually has to be in word addressed format. It can be extremely helpful to use the programming of plotters for teaching, and learning, the principles of word addressed programming. There are a number of companies that supply plotters, but unfortunately they use different symbols or characters to define lines, arcs etc. for the production of drawings. However, they all use a type of word addressed program format for the input of information. One widely used programming language is the Hewlett-Packard Graphic Language (GL). The word commands in this language are made up of a two-letter mnemonic (the address) followed by a number of permitted parameters (a mnemonic is the name given to letters or characters that are used as an aid to memory).

Questions

8.1 What is a part program?
8.2 Explain the structure of a part program, with reference to (a) characters (b) words and (c) blocks.
8.3 Explain what is meant by word addressed format?
8.4 Specify the three types of information in a

part program required to control a machine.

8.5 What character appears at the start of a part program?

8.6 Detail all the words that can appear in a block.

8.7 What is the purpose of the sequence numbers?

8.8 Explain how it is possible to incorporate information for the operator within a block.

8.9 What is a block skip character, and when is it required?

8.10 What is meant by the terms 'leading zero suppression' and 'trailing zero suppression'? What effect do they have on the numbers?

8.11 What are (a) preparatory functions and (b) miscellaneous functions?

8.12 For what purposes are preparatory functions and miscellaneous functions used?

8.13 What is the main difference between preparatory functions and miscellaneous functions?

8.14 Detail eight frequently used preparatory functions, explaining the reason for their use.

8.15 Detail eight frequently used miscellaneous functions, explaining the reason for their use.

8.16 What is meant by the terms 'modal' and 'non-modal'?

8.17 What is the purpose of an end of block character?

8.18 What character is generally used as an end of block character? What other character may be used?

Chapter Nine

Writing part programs

9.1 Creating part programs

There are a number of different methods of creating part programs:

(a) Manually writing the part program directly in word addressed format
(b) Using part programming languages such as APT
(c) Using a digitizing technique
(d) Using graphical techniques with computer-aided machining (CAM) programs
(e) Using conversational programming techniques directly at the machine tool
(f) Verbal programming using computers with voice recognition facilities.

For some components it is easier and faster to manually create the part program directly in word addressed format. If there are only a few numerically controlled machines in a firm it may not be economic to have expensive computer facilities and the necessary skilled labour required.

When using the digitizing technique, a stylus or probe or optical device is traced over a model, and blocks are created either automatically when the stylus moves a significant amount or by the operator pressing a button to store the coordinates of the position of the stylus. The model can be mounted on the machine, and it is the position of the slides which provides the coordinates. Additional blocks will have to be added through the keyboard to complete the program.

The use of computers with part programming languages such as APT, or graphical techniques for the production of part programs, considerably reduces the labour involved in producing part programs for complicated components. These techniques are also viable if the machine shop labour is not qualified and there are a number of different types of machine tools. Programming languages and graphical techniques require skilled labour and expensive equipment. The use of these techniques will be explained in Chapter 13.

Conversational programming is a very effective method, but requires skilled and well trained labour. This technique of creating part programs is carried out at the machine tool, and can only be used on machine tools that have the required facilities. Generally the machine tools and associated control units are very expensive. With the development of graphical methods using computers, it may not be economical to have all the individual machine tools equipped with the essential computer facilities required for conversational programming. A problem is that different control systems require different operational procedures for the inputting of the necessary information, and a setter skilled on one machine may not be capable of operating another machine with a different control system. One big advantage of conversational programming is that it restores to the machine setter an important responsibility for the running of the machine. With suitable skilled labour the machine can be very productive, because the setter will realize what material removal rates a particular machine is capable of achieving with different tools. Although conversational programming can be used on any machine tool, it is probably more effective on turning centres because there is less variety in work holding techniques than on milling and drilling machines.

If the part programs are created away from the machine, it is essential that the part programmer and the setter work together as a team. They both have an extremely important role to play in the efficient operation of the machine and the economic production of components.

Developments are taking place which enable a programmer to input information to a computer by speaking into a microphone. The range of words is not large, but is sufficient for the creation of part programs. The computer has to be 'taught' to recognize the words spoken; to do this, the programmer repeats a series of words and numbers in a fixed order. This method will probably be the quickest available when it is fully developed.

9.2 What the programmer has to know

The writing of a part program is more than stating the geometry of the component or deciding the path to be followed by the tool. It requires skill and judgement in selecting the tools to use, the methods of locating and clamping the part, the cutting speeds, the feed rate, the depth of cut and so on. To write efficient (for economic manufacture) part programs requires an extensive background knowledge of machining practice and techniques.

It is extremely important that the programmer always keeps in mind that the part program contains the instructions for a particular machine tool. The instructions must comply with the requirements of the particular control system and machine tool; these requirements will be found by studying the manuals published by the manufacturers. As stated previously, a program acceptable on one machine tool may not be acceptable on another machine tool of the same make but with a different control system.

When the machine and control system are switched on there are certain features such as mode of operation (G00) or dimensional units (G71) which are automatically selected; these are referred to as *default values*. Default values do not need to be entered in the part program, however, it is recommended that all details be entered including default values.

Before actually beginning to write a part program, it is essential that a number of factors are considered relating to the drawing, the component, the machine tool and the control system.

9.3 Drawings for numerically controlled machines

A typical view of an engineering drawing is shown in Figure 9.1. This figure is intended to show the standard practice of using centre lines; these are drawn first on an engineering drawing in order to position or locate the representation of the component on the paper. It is convenient to use the centre lines as a suitable starting point for the drawing; it is also practice to use them as reference points for dimensioning the component.

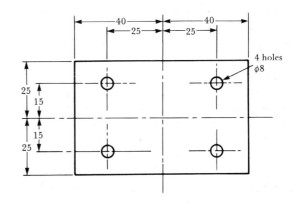

Figure 9.1 Centre line dimensioning

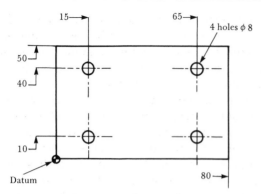

Figure 9.2 Coordinate or base line dimensioning

two edges meet is referred to as the *datum point*.

It is convenient and practical to locate the majority of components from their outside edges, and time can be saved if the features of the component (hole centres etc.) are dimensioned from these datums or reference edges (as shown in Figure 9.2) instead of the centre lines. The method of dimensioning from the datum edges, which is referred to as coordinate or base line dimensioning, enables easy transcription of the coordinates of the various features of the component. The use of only one arrow head and a short dimension line makes the drawing easier to read. It is the part programmer who specifies which feature of the component should be selected as the datum edges, consideration being given to such factors as:

(a) The type of component
(b) The sequence of machining operations
(c) The method of work holding
(d) The direction the cutting forces will act.

One method of finding the centre of an actual component is first to determine the overall size of the component by measurement. Then measure half the width and mark a line, and measure half the length and mark a line. Obviously where the lines cross is the centre. The centre is thus located from two adjacent outside edges. To determine the position of the centre accurately, it is essential that these edges are at a known relationship. The position where these

If the dimensions on the drawing issued by the drawing office or designer are not from a suit-

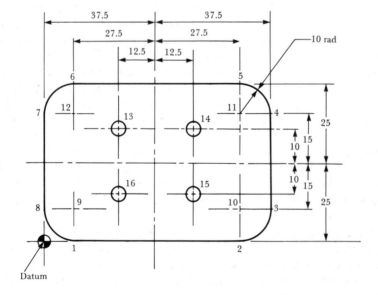

Label	X	Y
1	10	0
2	65	0
3	75	10
4	75	40
5	65	50
6	10	50
7	0	40
8	0	10
9	10	10
10	65	10
11	65	40
12	10	40
13	25	35
14	50	35
15	50	15
16	25	15

Figure 9.3 Labelling of positions

able datum, the programmer has to convert the dimensions on the drawing. This does not necessarily mean redrawing the component. An alternative method of dimensioning which is convenient for ease of programming is to label with letters or numbers all the relevant features of the component, such as:

(a) Centres of all the holes
(b) Start and end points of lines and curves
(c) Centres of slots and arcs.

The labels and dimensions can be referred to in a tabular format (see Figure 9.3). The table shows only the X and Y coordinates.

It is necessary to specify whether the dimensions are in imperial (inch) or metric (millimetre) units. This was done on some machines with a switch on the control panel, but now it is more usual to include the selection of units with a preparatory function in the part program. In general:

G70   designates imperial units
G71   designates metric units

On some control systems the default value will be imperial units while on others it will be metric units. It is advisable always to program the desired units. If it is possible to designate the units by a preparatory function, the part program can contain both inch and millimetre units. However, within a particular block there can be only one type of unit. On those machines that have a switch or push button on the control unit with which to set the particular units, the units have to be the same throughout the program.

Absolute and incremental dimensions
These are two methods of dimensioning components to be machined on numerically controlled machine tools; these methods are related to different programming techniques. When all the dimensions stated in the program are taken from the datum point, the programming is referred to as *absolute* programming. The units may be referred to as absolute units. One advantage of this method is that it is fairly easy to check and correct a part program written for components dimensioned using this method. If a mistake is made in the value of a dimension entered in a block of a part program, it will only affect the movement in that block or operation. If the rest of the programming is correct, the movements will take place at the correct location on the component. It is comparatively easy to relate the dimensions quoted in the part program with dimensions on the drawing of the actual component. Once the error has been corrected there should be no further problems.

With the *incremental* method the separate features of the component are dimensioned from each other in the order in which movements are made in the machining sequence. Each dimension is relative to the previous dimension. In effect the zero changes from one position to the next. If a mistake is made in the value of a dimension, all the following movements will be out of place by the incorrect value. It is difficult to check through an incremental part program unless the amount and direction of each movement is followed exactly. Another method of checking the correctness of the coordinates in a part program written completely in incremental format is to program for the tool to return to the point from which it started. If all the X coordinates are algebraically summed, the total should be zero. The coordinates of the other axes can be checked in the same way.

For certain types of component such as those containing a number of equally spaced holes of the same diameter, the incremental technique can reduce programming by the use of a looping technique. Incremental programming is also suitable where scaling or mirror imaging is to be used. However, as explained above, a mistake in one dimension can affect all the others that follow.

Figure 9.4 shows the difference between absolute and incremental programming. Assuming that the first move is from the start point, and it is intended to move from the start to each point on the drawing in the order ABC etc., the movements in absolute and incremental programming are as shown in Table 9.1. It can be seen that with absolute programming the values of the coordinates are the dimensions on the

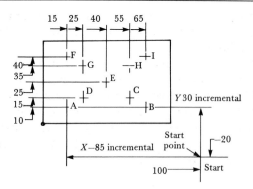

Figure 9.4 *Absolute and incremental movements*

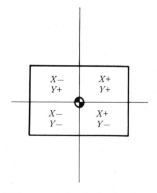

Figure 9.5 *Effect on 'sign' of coordinates of position of datum*

drawing. With the incremental method the zero point is the point being moved from.

The part programmer specifies the method of programming by the use of a preparatory function. The recommended standard preparatory functions are as follows:

G90 absolute programming
G91 incremental programming

It is important to check the programming manual issued by the manufacturer of the control system to ascertain the G code to be used.

Position of datum

To a certain extent the choice of the actual position of the zero datum is influenced by the

Table 9.1 *Movements in Figure 9.4 in absolute and incremental programming*

Position	Absolute		Incremental	
Start point	X100	Y−20	X0	Y0
Move to A	X15	Y10	X−85	Y30
A to B	X65	Y10	X50	Y0
B to C	X55	Y15	X−10	Y5
C to D	X25	Y15	X−30	Y0
D to E	X40	Y25	X15	Y10
E to F	X15	Y40	X−25	Y15
F to G	X25	Y35	X10	Y−5
G to H	X55	Y35	X30	Y0
H to I	X65	Y40	X10	Y5
I to start	X100	Y−20	X35	Y−60
Check sum			+150	+65−65
			−150	

shape of the component. The position of the datum can also be affected by the method of work holding and the actual machine tool to be used to produce the component.

If the component is of the type that its shape can be contained within a cuboid, a convenient position for the datum selected for X, Y and Z zero is the top front left-hand corner of the cuboid. This will mean that, when part programming in absolute coordinates, all the X and Y values will be positive. If any other position for the zero datum is selected, some of the X and Y coordinates will be negative; this is illustrated in Figure 9.5. However, it is important to note that some machine tools only accept positive dimensions, and may also require that any movement from the zero datum has to be in a specific direction. The requirements of the control system must be ascertained and complied with. There are machines that have a machine fixed zero feature; for these machines all the coordinates must be related to the machine datum. There is usually a particular position on the machine table where the corner of the work has to be placed. The distance from this position to the machine datum is known, and its values have to be added on to the component coordinate dimensions.

It is becoming more common for a machine to have a floating zero facility. This allows the zero datum to be located at any position on the machine table.

For dimensions in the third direction (usually the Z axis), the finished top surface of the workpiece can conveniently be taken to be the Z zero datum. This will result in dimensions into the work on the Z axis having negative values. This can be helpful because the negative sign can indicate the work changing size. The main reason for selecting the top surface of the work to be the Z zero datum is that it is the most convenient surface to which to set the tools.

If a hole, such as a bore in a casting, exists in the component before it is loaded on to the machine tool, it may be necessary to locate from the inside of the bore. It would then be desirable for the centre of the bore to be the datum for the X and Y axes. For this type of component it may be convenient to use polar coordinates instead of rectangular coordinates (see Section 9.10).

For components that are to be produced on a turning centre, as shown in Figure 9.6, the centre line of the work (the axis of rotation) is usually taken as the X zero datum. It is the convention that the absolute X coordinates are entered as positive values even on those turning centres where the tools are at the rear of the machine. There are control systems which require the X coordinate to be diameters; on other control systems the radiuses have to be entered. It is convenient to take the end face of the work that is furthest from the chuck or work holder to be the Z zero datum. There are control systems for lathes that require that some feature of the machine, such as the end of the spindle nose, to be taken as the Z zero datum. For these systems it is necessary to know the depth or width of the

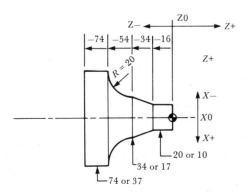

Figure 9.6 Base line dimensioning of turned components

work holder (chuck etc.) and the amount of protrusion of the workpiece from the face of the work holder.

It is essential that the programmer provides the setter of the machine tool with full details of which feature or part of the component has been selected as the zero datums. This information must be provided on the setting-up detail sheet, as it will be required when setting the component on the machine.

9.4 The component

The writing of the part program is influenced by the following factors related to the component:

(a) Size of blank
(b) Material of workpiece
(c) Amount of material to be removed
(d) Surface texture required
(e) Dimensional tolerance required
(f) Rigidity of component
(g) Work holding positions.

Size of blank

It is essential that the work to be machined on a numerically controlled machine for milling and drilling operations has a face and two edges prepared that can be used for accurately locating the component in the work holder. The size of the blanks should be to known dimensional tolerances. If the size of the work cannot be maintained within the tolerances then it is necessary for the part program to contain additional operations to cater for the largest possible component.

For milling and drilling work it is possible to accommodate components that are different in size from those for which the part program was written, by using probes fitted in the same type of holder as the tools in the main spindle.

For certain components it is possible for the operator to prepare the blanks on a manually controlled machine, which must be located near the numerically controlled machine. The numerically controlled machine would continue machining other components. This method of working can be favoured by the

159

operators, because it gives them some responsibility for the machining rather than just the tasks of loading the work and pressing the start button. This method of using the machine operators is only feasible if the machining time of the programmed component is of sufficient duration. It is essential that the numerically controlled machine should be kept fully occupied, and not be stopped while waiting for the operator to carry out tasks such as the loading of a workpiece.

For work on turning centres the component is usually held in a collet or the jaws of a chuck. If there is sufficient length of material which can be held securely, the work blanks can be cut off to the required length beforehand, or prepared by preliminary machining of the section that will be gripped by the collet or chuck jaws. Alternatively, the component could be machined from bar stock and 'parted off'. The decision on which method is to be used for a particular component will be influenced by the following:

(a) The stock size of the component may pass through the bore of the machine spindle, and there may be a bar feed unit on the machine.

(b) The number of components to be produced at one set-up of the machine will alter the economics.

(c) The parting-off operation can be one of the longest operations, and the cutting off of blanks by a power saw can be more economical.

(d) It may be more convenient to have cut off blanks if further machining (second operations) are to be carried out on that part of the material which was held by the work holder in the first sequence of operations.

(e) The use of work blanks may be preferred if the work is to be loaded by robots or automatic work handling devices.

(f) If the machining operations are of sufficient duration, the premachining of the blanks on a lathe close to the NC machine by the machine operator will provide the operator with some responsibility.

Material of the workpiece

The machineability of the material of the workpiece together with the cutting tool material will mainly affect various aspects of machining, such as:

(a) The cutting speeds that can economically be used

(b) The selection and type of coolant

(c) The feed rate

(d) The tool geometry.

A miscellaneous function can be used to program the selection of coolant and also to cause the coolant to be turned on and off. The recommended M codes are:

M07   selects and turns on the coolant in a flood, i.e. continuous flow

M08   selects and turns on the coolant as a mist

M09   turns off all types of coolant

M50   selects and turns on coolant number 3 (if provided)

M51   selects and turns on coolant number 4 (if provided)

Obviously the machine tool must have provision for the different coolants and types of coolant delivery for these functions to be selected.

Amount of material to be removed

For economic reasons, the minimum number of cuts should be taken to remove the material. Ideally it should be necessary to take only two cuts on a component, one to remove the bulk of the material and one to finish. The maximum amount of material removed per cut will be dependent on the strength of the cutting tool and the power of the drive motors. Control systems that have adaptive control can be programmed so that optimum metal removal conditions exist.

Turning operations

For turning operations when metal removal is of more significance than accuracy of size or surface finish, the amount of movement f of the tool per revolution (mm/rev) of the work is dependent on the type of material being cut. The

Table 9.2 *Feed rates*

Work material	Feed rate (mm/rev)
Cast iron	0.1 to 0.3
Alloy steel	0.06 to 0.2
Mild steel	0.15 to 0.6
Copper alloys	0.2 to 0.8
Aluminium alloys	0.3 to 1.0

stronger or harder the material, the lower has to be the feed rate. For minimum power consumption it is better to have large depths of cut and small feed rates, rather than large feed rates and light depths of cut. The most reliable values for feed rates can be obtained by having a test program that can be used for the cutting of the particular materials used in a firm. Table 9.2 is given as a guide to the feed rates that could be used for straight turning operations with high-speed steel turning tools. The lower values are for less rigid components. For parting-off or groove forming operations the feed should be no more than half of the lowest value. The feed rates used with cemented carbide or ceramic tools would probably be in the middle of the range.

The feed rate F for turning operations is calculated using the formula

$$F = fN$$

where F is the feed rate in millimetres per minute, f is the feed rate in millimetres per revolution of the work, and N is the spindle speed in revolutions per minute. The work spindle speed would have to be determined before the feed rate can be calculated (see Section 9.6 for calculation of spindle speed).

For a work spindle speed of 500 rev/min, the feed rate F programmed for a feed f of 0.8 mm/rev would be

$$F = 0.8 \times 500 = 400 \text{ mm/min}$$

With a spindle speed of 750 rev/min, a feed of 0.05 mm/rev results in a feed rate given by

$$F = 0.05 \times 750 = 37.5 \text{ mm/min}$$

This would be programmed as either F36 or F38, because feed rates in mm/min are not programmed with a decimal point, and the increase in feed rate values is usually in steps of 2 mm/min. However, feed rates in mm/rev are programmed with a decimal point.

Milling operations

For milling machines the feed rate F programmed is calculated using the number of teeth in the cutter and its rotational speed. The rotational speed would have to be determined before the feed rate can be calculated (see Section 9.6 for speed calculation). The relationship is

$$F = fTN$$

where F is the feed rate in millimetres per minute, f is the feed per tooth on the cutter (which is dependent on work material), T is the number of teeth on the cutter, and N is the spindle speed of cutter in revolutions per minute.

Manufacturers of milling cutters publish recommended feed rates per tooth for their tools when cutting different materials. These should be considered only as guide, as it is unlikely that the material and conditions when the feed rates were determined would be identical to the material etc. of the actual workpiece. Practical tests using special programs prepared for the purpose will provide information on suitable rates for actual machine tools. Many firms have their own feed rate tables which their experience over the years has proved suitable for their products. Table 9.3 gives a guide to typical values of feed rates to be used with high-speed steel end mills. The lower values are to be used with the smaller-diameter end mills, and the top

Table 9.3 *Feed per cutting edge*

Work material	Feed per cutting edge (mm)
Cast iron	0.05 to 0.25
Alloy steel	0.04 to 0.2
Mild steel	0.05 to 0.3
Copper alloys	0.1 to 0.35
Aluminium alloys	0.1 to 0.4

values are when face cutting. Cemented carbide tools would use the middle to top of the range.

The values of feed stated are when cutting on the periphery of the cutter; when feeding the cutter vertically into the work the feed rates should be halved. It must be remembered that only slot drills should be used for feeding directly into the work, as the teeth on end mills do not extend to the centre of the cutter. When a cutter is cutting a slot of width equal to its full diameter it is advised that, as a trial, the depth of penetration of a cutter should not be more than half its diameter, i.e. a cutter 10 mm in diameter should not penetrate more than 5 mm. Tests should be carried out to determine the optimum depth of penetration for different materials. One problem with increasing the depth of penetration using the larger-sized end mills and face mills is that the power required to remove material is proportional to the depth of penetration.

Drilling operations

For drilling operations, the feed rate is more dependent on the diameter of the drill than on other factors. This is because the high thrust loads generated during drilling tend to cause the drills to bend. Typical values of feed rates for drilling are given in Table 9.4. The lower values of feed rate should be used with materials that are harder and more difficult to cut.

Further remarks on feed rates

A tool library which contains all the details of the tools available is very helpful and time saving. Details of the information a tool library could contain are given in Section 9.6.

The feed rate can also be programmed using the 'magic three' method (see Section 9.6). A feed rate of 50 mm-min is programmed as 550. If imperial units are being used, a feed rate of 0.50 in/min is programmed as 350.

As explained in Chapter 8, there are no recommended standard preparatory functions for designating whether the feed rate is in units of millimetres per minute or millimetres per revolution. Some control systems use the

Table 9.4 *Feed rates for drilling*

Diameter of drill (mm)	Feed (mm per rev)
1.0 to 2.5	0.02 to 0.06
2.6 to 4.0	0.04 to 0.10
4.1 to 6.5	0.08 to 0.16
6.6 to 11.0	0.10 to 0.24
11.1 to 17.0	0.18 to 0.32
17.1 to 25.0	0.20 to 0.36

following preparatory functions for designating the units:

G94   feed per minute
G95   feed per revolution.

At one time some machines had a range of fixed feed rates, and a particular feed was specified as a number such as F5. The 5 was the number of the feed to be selected from the range of feeds provided on the machine. The block format of the program would require the feed to be entered in this type of number.

There are control systems that require the feed rate for certain operations, such as slopes or arcs, to be programmed as an *inverse-time mode*. For slopes the feed rate number (FRN) is given by

$$FRN = \frac{K \times \text{feed rate in mm/min}}{\text{distance to be travelled}}$$

For arcs, the FRN is given by

$$FRN = \frac{K \times \text{feed rate in mm/min}}{\text{radius of the arc}}$$

The value of K is dependent on the length of the slope or radius of the arc:

K	For slopes of length (mm)	For arcs of radius (mm)
10	0 to 99.98	0 to 99.98
100	100 to max.	100 to max.

Surface texture

Smooth surface texture will usually require small feed rates, and attention has to be paid to the shape of the profile of cutting tool, as shown in Chapter 1.

For turning operations the value of the feed rate will be mainly dependent on the type of cutting tool used, the work material, the spindle speed and the surface finish required. The profile of the cutting tool is important for producing a smooth surface texture, because a tool with a nose radius will produce a smoother texture than a sharp pointed tool. Too large a nose radius will cause the tool to tend to rub against the work instead of cutting.

Except when using turning tools made of diamond, the amount of movement f of the tool per revolution of the work for finish turning operations is normally of the order of 0.05 mm. Diamond turning tools can only be used on lathes which are very rigid and have special spindle drive arrangements and high-quality bearings.

Dimensional tolerance
Small tolerances may require fine finishing cuts to obtain the required values. The tolerance may also be required for the computer in the control system to determine the coordinates for the interpolation of curves. The value of the coordinates will influence the deviation of the actual programmed path from the desired curve (see Section 4.6).

Where the designer has specified a particular dimension on the drawing, such as 25.0 with limits −0.2, −0.4, the value programmed would be in the middle of the tolerance zone, here 24.7.

Rigidity of the component
This has to be considered when deciding the method of work holding and positions of support. Care must be taken with components that are very flexible. Support must be provided under thin components when they are being drilled.

9.5 Work holding positions

This has to be decided when planning the tool movements so that the clamps etc. do not interfere with the programmed movements of the work or tool.

Where automatic clamping devices are provided on the machine tool, a miscellaneous function can be programmed to activate and release clamps. The recommended miscellaneous functions are:

M10 clamps on
M11 clamps off.

If the machine has automatic work changing facilities the recommended miscellaneous function to be used are:

M60 for work change to take place
M68 to activate clamps
M69 to release clamps.

9.6 The machine tool

The machine tools suitable and available for producing the component must be known. It is no use writing a program for a particular machine tool if it is overloaded with work. Other factors to be considered are the machine running costs and the capacity of the machines in relation to the work size. The capacity of the machine tool has to be known in terms of factors such as:

(a) Power of drive motors, which will decide material removal rates and depths of cut
(b) Spindle speeds available
(c) Amount of movement possible on all axes.

The format of the machine has to be known. The designation of machine format is explained in Chapter 4:

2CL contouring control on two axes and line motion control on the third axis
3D positioning, line motion and contouring control on three axes.

The types of tool and the tool holding facilities available must be considered. Manual or automatic tool changing will influence the program. If automatic tool changing facilities are available, it is important that the tools normally held in the tool magazine should be known, together with the capacity of the tool magazine. Usually the tools are changed, either manually or automatically, when the spindle is fully retracted; this position is frequently referred to as the *home position*. With automatic tool changing, the spindle nose may have to be programmed to be at a particular location so that the tool transfer mechanism can work correctly. There are control systems where the spindle nose is positioned automatically when the tool change code is programmed. The recommended M code for designating a tool change operation is M06.

Selection of cutting tools
The selection of the most suitable cutting tool requires very careful judgement, and has to be based upon the choice of tool material and tool form. There are three main tool materials used on numerically controlled machine tools: high-speed steel, cemented carbides and ceramics.

High-speed steel
There are now many different compositions of high-speed steel. These form a group which possesses the toughest properties of all the cutting tool materials used on machine tools. They also have the added advantage that they can easily be reground when required. The most common composition is with 0.8 per cent carbon (C) and additional alloying elements of 18 per cent tungsten (W), 4 per cent chromium (Cr) and 1 per cent vanadium (V). It is extremely important for maximum cutting ability that the correct tool geometry is ground on the tool. These steels are excellent cutting tool materials and are used for the widest variety of machining operations.

Cemented carbides
These tools are used as tips. At first they were brazed on to steel shanks, but they are more commonly clamped on to steel holders. There are many different grades of cemented carbide. The ISO method of classifying the carbides is based upon the applications of the tools, and has three ranges:

P range   For long chipping steel.
K range   For cast iron.
M range   For materials whose machining properties are between the above.

In each application range there are subdivisions indicated by numbers, e.g. P01, P10, P40 etc.

Every manufacturer of cemented carbide tools issues a catalogue of its tools with their recommended applications. To obtain full use of these tools, the following recommendations should be observed:

(a) The cutting speed should be such that sufficient heat is generated to cause the material immediately in front of the tool to plasticize and hence be easier to remove. If the speeds used are too low the cemented carbide is liable to chip and fracture.
(b) Care should be taken not to abuse the tip by accidentally hitting it.
(c) The tool overhang should be kept to a minimum as far as operational conditions will permit.
(d) When using cutting fluids, it is essential that a continuous heavy flow be provided to prevent alternate heating and cooling of the tool.
(e) The machine tool should be very rigid and strong to obtain maximum utilization of the cemented carbides.
(f) The grade of cemented carbide must be very carefully selected for the particular application.

In 1970 a cemented carbide tip was developed which had a surface coating of 0.005 mm of complex carbides. These tips are referred to as laminated, and it is reported that these tools are 400 per cent more wear resistant than uncoated cemented carbide tips.

The development of tool holders that permit the tip to be indexed has reduced tool resetting times to a minimum. In many cases the method of manufacture enables the tips to be supplied with the chip breakers already formed, but these tips should only be used with the recommended feed rates.

Ceramics

Ceramics are also termed 'sintered oxides', and consist of almost pure aluminium oxide with a very small amount (0.25 per cent) of other additives such as magnesium oxide. Ceramics are extremely hard and have a very high abrasion and erosion resistance; they are particularly suited for the turning of such materials as cast iron to very smooth surface finishes (1 micron R_a being possible). These materials can only be used as tips on clamp-type tool holders, as it is not practical to braze the tips on to steel shanks.

For efficient use of ceramics the machine tools used must be very rigid and powerful, with high spindle speeds to give cutting speeds of up to 25 m/s (1500 m/min). At these speeds tool life can be six times that of carbides, but if the speed is too slow the reverse applies. Ceramics are not suitable for high feeds or large depths of cut because their heat conductivity is low and unequal heating causes cracking and shattering of the tip. This creates problems when applying cutting fluids. Plan approach angles should be fairly large – of the order of 45 degrees.

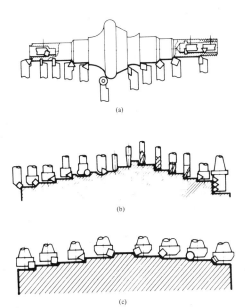

Figure 9.7 Shapes of cutting tools: (a) turning; (b) end milling; (c) face mills (only one cutting shown for (b) and (c)) (courtesy Sumiden Hardmetal)

Shapes of cutting tools

Cutting tools have a variety of shapes, each shape having a particular use. Figures 9.7a–c shows a selection of the uses of the various tool shapes. Figures 9.7b and c, of end mills and face mills, show only one cutting edge for clarity of drawing. The number of cutting edges on a tool depends mainly on the diameter of the tool.

Tool library

A tool library can be set up which contains all the details of the tools available. This will be very helpful and time saving. The library should consist of information such as:

(a) Tool materials
(b) Tool shapes
(c) Tool size (diameter or shank size)
(d) Cutting lengths of all the available tools
(e) Set lengths of all the available tools
(f) Economic speeds and feeds when machining different work materials.

It is an advantage to create a coding system for these details which can be used to identify a particular tool. A very comprehensive system of specifying the designation of indexable hardmetal inserts is provided in BS 4193:Part 1:1980. The standard has seven compulsory symbols and three optional symbols; each symbol identifies the dimensions and other characteristics of indexable inserts.

Spindle speeds

The spindle speed selected will be dependent on the cutting speed (peripheral speed in metres per minute) and the diameter of the work or tool. There is no one cutting speed for a particular work material; the speed used will be affected mainly by the tool material and the desired life of the tool.

The life of a tool can be measured using different criteria, but it is usually taken as the time it is cutting efficiently. From the moment a tool start to cut, it starts to wear. With reasonable cutting speeds the wear is progressive. If the tool is allowed to continue cutting beyond a certain amount of wear, it will finally break down completely, causing considerable problems. If

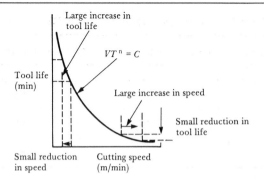

Figure 9.8 Relationship between cutting speed and tool life

the cutting speed is too fast the heat created will cause a very high temperature at the tip of the tool. The tool tip will become softened, and the edge will deteriorate very rapidly.

There are many factors which can influence the life of a tool, but cutting speed has the greatest effect. The relationship between tool life and cutting speed is shown in Figure 9.8. It can be seen that a small reduction in speed at low speeds will result in a relatively large increase in tool life. Conversely, at high speeds a relatively large increase in speed results in a small reduction in tool life. For a particular work and tool material, if the feed rate and depth of cut are kept constant there is a mathematical relationship between cutting speed and tool life. This is known as the Taylor tool life formula:

$$V T^n = C$$

where V is the cutting speed in metres per minute, T is the tool life in minutes, and n and C are constants dependent on the work and tool

materials. Accurate values of n and C for particular work and tool materials can only be obtained by experimental tests, but most textbooks on machining technology will provide information on these values. The following are given as an indication:

Work material	High-speed steel		Cemented carbide	
	n	*C*	*n*	*C*
Mild steel	0.1 to 0.2	80	0.2 to 0.3	240

The optimum life of a tool for any machining operation is a choice between the fastest rate of production and the time spent in regrinding the cutting tool and resetting the tool in position. There is no advantage in producing components in 1 minute if a greater time is spent on changing and resetting the tool. By slightly reducing the spindle speed it will be possible to have the tool cutting for a much longer period without unduly reducing the rate of production.

There are so many different work and tool materials available that it is impractical to try to provide precise information on the cutting speeds for all possible combinations. The most efficient method is to have a test program which will provide information on the machineability of all the materials used by a firm. Tool material manufacturers will usually be willing to provide samples of tools for test purposes, in anticipation of future orders. The cutting speeds in Table 9.5 have been found to be useful for calculating spindle speeds. The lower speeds are used for turning operations when material removal is the

Table 9.5 *Cutting speeds*

Work material	High-speed steel (18%W, 4%Cr, 1%V) cutting speed (m/min)	Other tool materials
Cast iron	20 to 30	For sintered carbide,
Alloy steel	25 to 50	multiply all speeds by 2
Mild steel	40 to 80	
Copper alloys	90 to 120	For ceramic tools, multiply
Light alloys	140 to 200	all speeds by 3

prime concern, and the higher speeds for finishing operations. The lower speeds are also more applicable for milling and drilling operations. It is emphasized that the cutting speeds are only intended as a guide.

As explained in Chapter 3, the speeds generally provided on numerically controlled machines range from 30 to at least 4000 rev/min and are steplessly variable. Developments in bearing design have resulted in increased spindle speeds being provided. It is common to find a limited number of speeds in a geometric progression on manually controlled machine tools.

The spindle speed is calculated using the formula

$$N = \frac{V \times 1000}{\pi D}$$

where N is the spindle speed in revolutions per minute, V is the cutting speed in metres per minute, and D is the diameter of the work or tool in millimetres.

Milling and drilling operations

For these operations, each work material has a spindle speed at which each diameter of tool is at optimum efficiency for material removal and duration of cutting (tool life). The same spindle speed set at the start of the cutting operation can be maintained throughout the operation without detriment to the efficiency of the cutting tool, as there is no possibility of the tool changing diameter in the middle of a machining operation and requiring a different speed. For drills and end mills of less than 3 mm in diameter, it is unlikely that the spindle speeds required to give the cutting speeds indicated in Table 9.5 would be available on most machines. For this size of tool the fastest speed available has to be selected. There are certain specialist machines that may have the high range of speeds required.

Turning centres

On these machines the selection of spindle

Table 9.6 *Cutting speed at various RPM*

Spindle speed selected (rev/min)	Actual cutting speed (m/min)		
	10 mm dia.	15 mm dia.	30 mm dia.
955	30	45	90
319	10	15	30
637	20	30	60

speed can be a problem because, as the diameter of the work changes during machining, the spindle speed in revolutions per minute should change to maintain the actual cutting speed (peripheral speed). It is important that the material is being removed at optimum cutting efficiency. If the spindle speed has been calculated for the smallest diameter of a component with large and small diameters, problems can arise when the tool is turning the larger diameter. This is because the actual peripheral speed (metres per minute) will be too high, resulting in reduced tool life. If the spindle speed is selected which is suitable for the larger diameters, the peripheral speed will be too slow when machining the smaller diameters; this will result in poor cutting conditions. The effect on the cutting speed when turning different diameters at a fixed spindle speed is shown in Table 9.6.

The difficulty in selecting the spindle speed is not of significance if the diameters being machined do not change more than 20 per cent. However, there is a particular problem on turning centres when facing large-diameter workpieces. To overcome this problem there are control systems which provide constant cutting speed at the point of cutting. With these systems the actual peripheral cutting speed (metres per minute) required is programmed. As the radius being machined changes, the spindle speed (revolutions per minute) automatically changes so that the peripheral speed at the tool point is constant. These systems ensure that the tools are cutting at optimum efficiency and that the desired tool life is obtained.

Different control systems use alternative programming inputs to obtain the constant cutting speed, such as:

(a) The cutting speed required in revolutions per minute for a particular diameter of work.
(b) The required cutting speed in metres per minute together with the diameter of the work or tool.
(c) The tool life required, with the mathematical relationship of tool life and cutting speed; from this information the spindle speed is selected to suit the conditions.

There are no recommended preparatory functions for designating which speed is being specified, but it is practice with some control systems to use:

G96 constant cutting speed in metres per minute
G97 spindle speed in revolutions per minute.

The specifying of tool life as a means of selecting spindle speed is only possible on machines that have adaptive control (see Section 2.10).

Spindle speed using magic three method

On some of the older control systems the spindle speed required is programmed in what is referred to as the 'magic three' method. In this method the speed in revolutions per minute is programmed as a three-digit number. The first digit of the number programmed is the sum of the digits in the number of the speed required, plus three. The next two digits of the number programmed are the first three digits of the speed required, rounded up or down to a two-digit number. For example, suppose the speed is 1560 rev/min. There are four digits in the number 1560, and so $4 + 3 = 7$ for the first digit. The first three digits in the number are 156, which is rounded up to 160; this gives 16 for the second and third digits. Hence the speed programmed is 716. Similarly a speed of 790 rev/min would be programmed as 679: three digits gives $3 + 3 = 6$, and the first three digits of the spindle speed are taken as 79.

At first sight the magic three method appears to be confusing, but it must be remembered that the control system is designed to accept numbers in this way. If the programmer is acquainted with the procedure then there should be no problem.

Capacity of machine tools (amount of movement)

On milling and drilling machines that have manual tool changing and work loading, it is necessary to know the maximum table movements in order to program the X and Y coordinates to which the work table can be positioned so that the tools or work can be changed in safety. For this it is necessary to determine the machine setting point. Additionally the machine tool setter has to be provided with details of the machine setting point for the positioning of the work on the machine table during the setting-up stage.

The *machine setting point* gives the values of the machine coordinates when the programmed datum of the work is directly under the centre of the spindle. The machine coordinates referred to are the readings of the scales that most machine tools have fixed to the sides of the X and Y tables. The scales can be used for positioning the tables during setting up. On most computer-controlled machines the control system provides a digital readout (DRO) of table position. Manufacturers may provide the values of the machine coordinates when the centre point of the work table is under the centre of the spindle, or they may be stated on the machine.

So that the weight of the work and the work table is evenly balanced on the machine, the most logical position to place the workpiece is with the centre of the workpiece over the centre of the work table.

Determination of machine setting point

A typical machine could have maximum movements of the machine table of 500 and 300 mm on the X and Y axes respectively. When the centre of the machine table is directly under the centre line of the spindle (Z axis), there should be possible movements of 250 and 150 mm either way on the X and Y axes respectively. If a

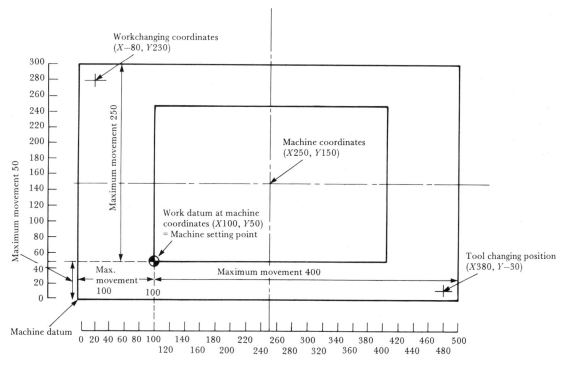

Figure 9.9 Determination of machine setting point

workpiece of 300 by 200 mm is positioned so that its centre is at the centre of the work table, the lower left-hand corner of the work will be at the machine coordinates of 100 on the X axis and 50 on the Y axis (Figure 9.9 shows this position). There will then be maximum movements on the X axis of 400 mm in a positive direction and 100 mm in a negative direction. On the Y axis the table can move 250 mm in a positive direction and 50 mm in a negative direction. It will then be possible to move the work table to program coordinates (X+380, Y−30) for a safe position for tool changing. For work loading the table should move to program coordinates (X−80, Y+230). It is advisable not to use the maximum movements possible, but to allow some latitude for easing any setting-up problems that may occur. It is also advisable to restrict the movement so that there is some clearance allowed to avoid closing the override safety microswitches that limit table movement.

This method of determining the machine setting point is only necessary on large components. For relatively small components held in a stepped jaw vice, it is more convenient to keep the vice in a fixed position on the machine table. It is not necessary to move the vice so as to position the centre of small work over the centre of the machine table. Under these conditions the machine tool setter should provide details of where the machine setting point is located. The coordinates of the tool changing point and work changing point only need to be those that will place the centre of the tool spindle about 60 mm clear of the relevant edges of the vice or work.

At one time the distances between the machine datum and the machine setting point on the X, Y and Z axes were referred to as DAX, DAY and DAZ (datum X, datum Y and datum Z) respectively.

9.7 The control system

As stated previously, it is essential that the manuals provided by the manufacturers of the control system and the machine tool are studied

169

so that the programmer knows what is available for preparatory and miscellaneous functions (G and M codes) and other programming aids.

The first consideration is the modes of operation available, as upon these will depend whether curves can be machined. The mode of operation is generally selected by entry of a G code, but there are control systems such as that provided by Heidenhain which use other commands. Probably the most common modes of operation are positioning (point-to-point) and linear interpolation.

Positioning (point-to-point)

This mode is selected when the tool or work has to be positioned. No machining occurs during the time the movements are taking place. Normally a feed rate does not have to be programmed for a positioning operation; the control unit automatically selects the fastest feed rate available on the machine. However, on some machines it may be necessary to enter a feed rate, and to save time the fastest feed available should be selected for these machines.

On milling and drilling machines, if X, Y and Z coordinates are programmed in the same block for a positioning operation (G00), movement does not normally take place on all three axes simultaneously. If the previous operation has resulted in the tool spindle being in a fully retracted position and clear of the work, movement first occurs on the X and Y axes; when both these slides have come to rest, movement takes place on the Z axis. The reverse sequence of movements takes place if the previous operation has resulted in the tool being in the work. Under this condition, movement first takes place on the Z axis, so that the tool is raised clear of the work to the programmed Z coordinate, and then on the X and Y axes. This method of operating provides a means of ensuring that the tool does not strike the work or clamps when moving at a fast feed rate.

When the machine is in the point-to-point mode, the same feed rate is sent to the drive motors of the carriages on both the X and Y axes. As a result the path of the movement will be at an angle of 45 degrees. If a shorter movement is

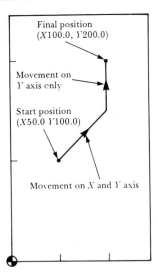

Figure 9.10 *Point-to-point operation*

required on the X axis, that movement will be completed first; then movement continues on the Y axis.

The movements that would take place as a result of the following blocks are shown in Figure 9.10:

N120G00X50.0Y100.0* starting position
N125X100.0Y200.0*   final position

It is not necessary to enter G00 in block N125 since there is to be no change in mode of operation. Additional examples of point-to-point operations are given in Chapter 4.

Linear interpolation

Linear interpolation (G01) is used if machining has to take place:

(a) In a straight line along one axis
(b) In a straight line at an angle on any two axes
(c) In a straight line simultaneously on any number of axes.

It is essential that a feed rate is programmed when operating in the linear interpolation mode. However, once a feed rate has been programmed it is stored in the memory of the control unit, and will remain active until changed.

When operating in the linear interpolation mode, if movement has to occur simultaneously on more than one axis the control unit automatically sends the orthogonal (rectangular) components of the feed rate to the respective motors (see Chapter 4 for further explanation).

If movement is to take place from coordinates (*X20.0, Y30.0, Z−6.0*) *to X80.0, Y70.0, Z−6.0*), as in Figure 9.11, the operations would be programmed as:

N150G00X20.0Y30.0Z−6.0*

    start position, point-to-point operation
N160G01X80.0Y70.0F200*

    finish position, linear interpolation, at a feed rate of 200 mm/min

The block numbers have no significance,and are only inserted in this example so that they can be referred to in the following notes. Block number N150 would result in the work table moving, from the previous position, at a fast feed rate to X20.0 and Y30.0; then the tool spindle would move at the same feed rate to Z−6.0. As shown in Figure 9.11, in block number N160 the work table would move to X80.0 and Y70.0. The feed rate sent to the X and Y axes motors would be different, but their vector sum would result in the programmed feed rate of 200 mm/min along the cutter path. The computer in the control unit would do all the calculations necessary.

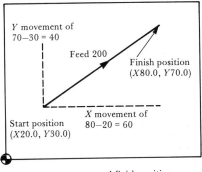

Distance between start and finish position

$$= \sqrt{60^2 + 40^2} \qquad = \sqrt{5200} = 72.11$$

Feed rate sent to X motor $= 200 \times \dfrac{60}{72.11}$

$= 166.41$ mm/min

Feed rate sent to Y motor $= 200 \times \dfrac{40}{72.11}$

$= 110.94$ mm/min

Figure 9.11 *Linear interpolation movement*

9.8 Circular interpolation

In order to machine circular curves, the control system has to be in one of the circular interpolation modes. The standard preparatory functions (G code) used are:

G02   clockwise circular arc
G03   counter- (anti-) clockwise arc

For a number of control systems for milling and drilling machines, it is possible to machine circular arcs in only one of the planes at a time. The plane of operation is selected by the following G codes:

G17   The XY plane
G18   the XZ plane
G19   the YZ plane.

The XY plane is usually automatically selected as the default value. The maximum radius will be limited by the capacity of the machine tool.

When it is possible to provide simultaneous control for circular interpolation on two axes and linear interpolation on the third axis, spirals can be machined and the operation may be called 'helical interpolation'.

Single-quadrant circular interpolation

This mode of operation is used for any arc that is complete in one quadrant, i.e. up to a maximum of 90 degrees. Arcs up to 90 degrees but that lie in more than one quadrant require two blocks of data. The restriction to one quadrant is not a big limitation, as most work will only require a maximum of 90 degrees for the rounding of corners.

Before a circular interpolation move, the tool would have to be positioned at the start of the arc. To program a circular interpolation move from this position the following are required:

(a)   The direction of movement (G02 or G03)
(b)   The coordinates of the end of the arc
(c)   The coordinates of the centre of the arc.

The relevant G codes (G02 or G03) must be programmed for every block in which circular interpolation is required, even if the moves are consecutive.

The block format for a G02 or G03 move is:

Nnnn G02 or G03 XxxxYyyyIiiiJjjjFfff*

where

Nnnn	sequence number
Xxxx	X coordinate of end of arc
Yyyy	Y coordinate of end of arc
Iiii	distance from start point of arc to centre of arc along X axis
Jjjj	distance from start of arc to centre of arc along Y axis
Ffff	feed rate
*	end of block.

Only one of G02 and G03 is to be programmed. Note that the X and Y coordinates would be absolute values if G90 had previously been programmed or incremental if G91 had been entered. The lower case letters, i.e. xxx etc., are intended to represent numbers.

The I and J coordinates are known as the arc centre offset values, and apply when the arc is in the XY plane. If the arc is in the XZ plane the arc centre offset values would be I and K, and if the arc is in the YZ plane they would be J and K.

For the single-quadrant circular interpolation mode, the coordinates of the arc centre offset values I, J and K are generally taken to be the distances *from* the start point of the arc *to* the arc centre. The values of the I, J and K coordinates are not signed (i.e. do not have positive or negative signs), but I0, J0 and K0 values must be programmed when they occur.

Important note   Different control systems may have different requirements to those specified here in selecting the reference point from which the I, J and K coordinates are specified. It is essential to use the correct designation.

The computer in the control unit uses the I, J or K values to calculate the values of the coordinates of the required cutter path to send to the drive units. In Figure 9.12 the curve extends through more than one quadrant; therefore in single-quadrant circular interpolation mode it is necessary to program to move first from A to B and then from B to C. When moving from A to B the arc centre offset values I and J are the distances

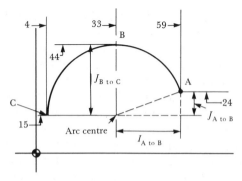

Figure 9.12   Arc centre offset for single-quadrant circular interpolation

from A to the arc centre on the X and Y axes respectively:

I = X coordinate of A − X coordinate of the centre of the arc
= 59 − 33 = 26
J = Y coordinate of A − Y coordinate of the centre of the arc
= 24 − 15 = 9

The program blocks would be:

N50G00X59.0Y24.0*
   start point
N55G03X33.0Y44.0I26.0J9.0F200*
   finish point

When moving from B to C, the I and J values would be:

I = X coordinate of B − X coordinate of the centre of the arc
= 33 − 33 = 0
J = Y coordinate of B − Y coordinate of the centre of the arc
= 44 − 15 = 29

The program blocks would be:

N55G03X33.0Y44.0I26.0J9.0F200*
   start point
N60G03X4.0Y15.0I0J29.0*
   finish point

The computer also generates the value of the feed rates required for the drive units so that the movement along the circular curve is at the required feed rate. The feed rate does not need to be programmed if a suitable feed has been previously programmed. However it is important to remember that, when the machine is cutting circular arcs, the feed rate programmed will be the feed rate at which the centre of the cutter moves, and not the feed rate with which the edge of the cutter moves along the surface. When the cutter is moving round the outside of an arc the centre of the cutter will be travelling faster than the edge of the cutter along the surface of the arc, and the feed rate should be increased. When the cutter is moving round the inside of an arc the centre of the cutter will be travelling slower than the edge of the cutter along the surface of the arc, and the feed rate should be reduced.

In Figure 9.13, while the *outside* of a 20 mm diameter cutter is travelling from A to B, the centre of the cutter is travelling through the arc C to D at a programmed feed rate of 200 mm/min. The arc A to B is clearly shorter than the arc C to D, although the time taken to travel both the arcs is the same:

time taken to travel arc CD

$$= \frac{\text{length to be travelled (mm)}}{\text{feed rate (mm/min)}} = \frac{2 \times \pi \times 30}{4 \times 200} = 0.236 \text{ min}$$

feed rate through arc AB

$$= \frac{\text{length to be travelled (mm)}}{\text{time (min)}} = \frac{2 \times \pi \times 20}{4 \times 0.236} = 133 \text{ mm/min}$$

This feed rate is too slow. For the feed rate through the arc A to B to be 200 mm/min, the

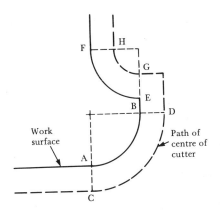

Work surface

Path of centre of cutter

Figure 9.13 Feed rates for curves

feed rate programmed for the centre of the cutter spindle while the cutter moves through the arc C to D should be increased:

time for cutter edge to travel arc AB at 200 mm/min feed rate

$$= \frac{\text{length to be travelled (mm)}}{\text{feed rate (mm/min)}} = \frac{2 \times \pi \times 20}{4 \times 200} = 0.157 \text{ min}$$

feed rate through arc CD

$$= \frac{\text{length to be travelled (mm)}}{\text{time (min)}} = \frac{2 \times \pi \times 30}{4 \times 0.157} = 300 \text{ mm/min}$$

As can be seen in Figure 9.13, when the machine is cutting from E to F on the *inside* of an arc, the length of the arc travelled by the centre of the cutter from G to H is smaller than the arc of the work, and the feed rate should be reduced. The feed rate F programmed when cutting arcs should be as follows:

For cutting around the outside of an arc:

$$\text{feed } F = \text{normal (linear) feed} \times \frac{r_w + r_c}{r_w}$$

For cutting on the inside of an arc:

$$\text{feed } F = \text{normal (linear) feed} \times \frac{r_w - r_c}{r_w}$$

where r_w is the work radius and r_c is the cutter radius.

Multiquadrant circular interpolation

This mode of operation is not available on all numerically controlled machines. For those machines that have this facility, it is selected with a G code; the code used on some control systems is G75. Arcs of up to 360 degrees can be generated in a single block of data. The block format is similar to that used for single-quadrant circular interpolation, i.e.

Nnnn G02 or G03 XxxxYyyyIiiiJjjjFfff*

It is normally an essential requirement that all X, Y, I and J coordinates are entered even if their values are zero or unchanged. Note that the Xxxx and Yyyy values are the coordinates of the end point of the arc; these coordinates would be absolute values if G90 had been previously programmed or incremental if G91 had been programmed.

As before, it is important that the arc centre offset values are specified from the correct reference position. Different control systems may use different reference positions.

It is common practice, when programming in *absolute* units, that the I and J values are the signed absolute rectangular coordinates of the arc centre from the chosen absolute zero (X and Y zero datums) along the X and Y axes respectively. In Figure 9.14 it is intended to move to A and from A to B. The relevant blocks are:

N65G01G75G90X80.0Y50.0F200*
   from start to point A
N70G03X50.0Y20.0I50.0J50.0*
   from A to finish point

It is necessary to enter the G code (G75) before the block in which the multiquadrant move is to be activated. Once it is entered it remains active every time G02 or G03 are programmed. If multiquadrant circular interpolation is to be used in a part program it is normal to enter the G code (G75) in the first line of the program. If absolute programming (G90) had been programmed in an earlier block, and was still

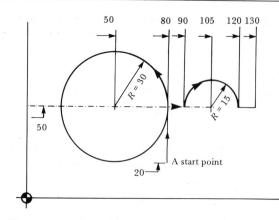

Figure 9.15 *Circular interpolation*

active, there would be no need to enter G90 in block N65.

When programming in *incremental* units, the I and J values are the signed distances from the arc centre to the start point of the arc on the X and Y axes respectively. For Figure 9.14 the relevant blocks in incremental programming to move from A to B would be:

N60G01G75G90X80.0Y50.0F200*
   from start to point A
N70G03G91X−30.0Y−30.0I30.0J0*
   from A to finish point

Note that the centre of the arc and the start point of the arc have the same absolute Y coordinate; therefore the J value is zero (J0), which must be entered in the program. Code G91 has been entered in block N70 to activate incremental programming.

To illustrate the difference between single-quadrant and multiquadrant circular interpolation, an example is given in Figure 9.15 and the relevant blocks are shown in Table 9.7. The start point is at A. It can be seen that with single-quadrant interpolation a lot more blocks have to be programmed.

9.9 Parabolic interpolation

The standard preparatory function for activating parabolic interpolation is G06, but some control systems use G06 for other functions. Parabolic interpolation is not as generally available as

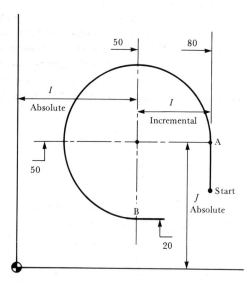

Figure 9.14 *Arc centre offsets with multiquadrant circular interpolation*

Table 9.7 *Programming blocks for interpolation in Figure 9.15*

Absolute programming, single quadrant	Absolute programming, multiquadrant	Incremental programming, multiquadrant
N80G01G74G90X80.0Y50.0*	N80G01G75G90X80.0Y50.0*	N80G01G75G91X0Y30.0*
N85G03X50.0Y80.0I30.0J0*	N85G03X80.0Y50.0I50.0J50.0*	N85G03X0Y0I30.0J0*
N90G03X20.0Y50.0I0J30.0*	N90G01X90.0Y50.0*	N90G01X10.0Y0*
N95G03X50.0Y20.0I30.0J0*	N95G02X120.0Y50.0I105.0J50.0*	N95G02X30.0Y0I−15.0J0*
N100G03X80.0Y50.0I0J30.0*	N100G01X130.0Y50.0*	N100G01X10.0Y0*
N105G01X90.0Y50.0*		
N110G02X105.0Y65.0I15.0J0*		
N115G02X120.0Y50.0I0J15.0*		
N120G01X130.0Y50.0*		

circular interpolation because it requires specialized control equipment. Where parabolic interpolation is provided by the control system, it is available for curves in the three planes X, Y and Z. Circular interpolation requires only two points on the curve (the end point and the centre) for the computer to calculate the cutter path coordinates. Parabolic interpolation requires three points: the start point, an intermediate point and the end point.

There are two methods of programming the required coordinates:

(a) The intermediate point and the end point can be programmed as X, Y or Z coordinates in consecutive blocks.
(b) Alternatively the end point can be programmed as X, Y or Z coordinates, and in the same block the intermediate block is programmed as I, J and K coordinates.

The calculation of the coordinates of the intermediate point can be involved, and it is extremely helpful if computer assistance is available. Using graphical techniques, the creation of the part program is considerably simplified. Graphical numerical control is explained in Chapter 13.

9.10 Polar coordinates

With some control systems it is possible to program a position using polar coordinates. They are used as an alternative method to rectangular coordinates for specifying a position in any of the planes (*XY, XZ* or *YZ*). Where rectangular coordinates use addresses X, Y and Z, addresses of polar coordinates are R and A. The letter R is followed by the value of a radius in the same units used for rectangular coordinates. The letter A is followed by the value of an angle to an accuracy of 0.01 degrees.

A position specified in polar coordinates is by reference to a radius and an angle from an axis through the centre of a circle. As shown in Figure 9.16, the centre of the circle is termed the *pole*, and a horizontal line through the nine and three o'clock positions is referred to as the *polar axis*. Zero angle is at the three o'clock position, and angles measured counter-clockwise are termed positive. The radius only has positive values.

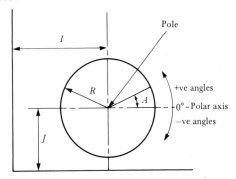

Figure 9.16 Polar coordinates

If the component zero datum is at the centre of the circle (the pole) a position can be specified by a radius and an angle. If the datum point is at any other position is it necessary to specify the position of the pole using coordinates with I, J and K addresses. In the XY plane the coordinates of the centre of the pole would be defined with addresses I and J, the distance along the X axis would be defined as an I value and as a J value on the Y axis.

Point-to-point and both linear and circular interpolation movements can be programmed with polar coordinates using the same G codes as rectangular coordinates, i.e. G00, G01, G02 and G03, to designate the movements required as shown in Figure 9.17.

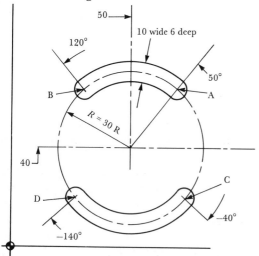

Figure 9.17 *Polar coordinate programming*

N120G00G90R30.0A50.0I50.0J40.0*
    positioning operation to A
N125Z2.0*
    positioning operation
N130G01Z−6.0F100*
    linear interpolation down feed
N135G03R30.0A120.0I50.0J40.0F200*
    counter-clockwise circular interpolation to B
N140G00Z2.0*
    positioning operation
N145R30.0A−40.0I50.0J40.0*
    positioning operation to C
N150G01Z−6.0F100*
    linear interpolation down feed
N155G02R30.0A−140.0I50.0J40.0F200*

    clockwise circular interpolation to D
N160G00Z10.0*
    positioning operation

9.11 Cutter diameter compensation

Cutter diameter compensation can only be used when milling around the outside of the work or inside a pocket in the work; it cannot be used for drilling operations or when milling slots with cutters of the desired size. When writing a part program to control the path of the centre of the cutter for milling operations, it is assumed that the diameter of the milling cutter used will be a particular size. However, it is possible that, owing to regrinding, the diameter of the tool will be different from the assumed size, and consequently the work produced will be larger than that required. Instead of having to reprogram the complete path of the centre of the cutter to suit the operation, it is possible to adjust the relative position of the cutter using cutter diameter compensation.

The information on the diameter of the tool, which the control system has to have to calculate the required compensation, must be input into the control unit's memory before the operation (block) in which the compensation is to be applied. The input of tool diameter can be entered as part of the program at the beginning of the program, or can be input during the setting-up by the operator using manual data input (MDI) through the control unit's keyboard. When the diameter is entered as part of the program, an amended value of the diameter of the tool can only be entered by editing the program. When the diameter is entered by MDI any corrections necessary can be by MDI via the keyboard. The latter method has the advantage of being very convenient when tools may have to be changed.

Tool diameter compensation is activated by the relevant preparatory functions (G codes):

G41   compensation, cutter at left
G42   compensation, cutter at right
G40   cancel compensation

Code G41 is used when the centre of the cutter is to the left of the edge of the workpiece being cut

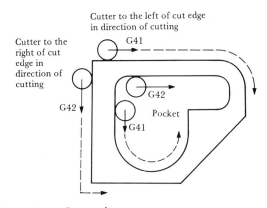

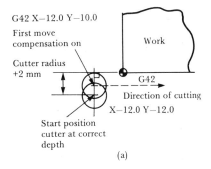

Figure 9.18 Cutter diameter compensation

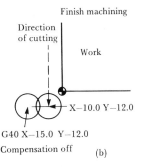

Figure 9.19(a) Applying cutter compensation;
(b) cancelling cutter compensation

when looking in the direction that the tool is travelling, as shown in Figure 9.18. Code G42 activates compensation when the centre of the tool is to the right of the edge of the workpiece being cut when looking in the direction that the tool is travelling, as also shown in Figure 9.18.

Before activating compensation the cutter should be at the programmed depth (Z coordinate). It is also important that the cutter should be positioned clear of the edge of the work, as shown in Figure 9.19a, and at least the radius of the cutter plus 2 mm at right angles away from the edge to be machined. The next move should activate the compensation (G41 or G42), and this programmed move should result in the cutter moving at right angles towards the edge to be cut.

When the cutter has completed the machining, the compensation must be cancelled using code G40. The block containing the G40 should also be programmed so that the cutter moves at right angles away from the machined edge, as shown in Figure 9.19b.

There are three different applications of cutter diameter compensation:

(a) To compensate for variation between actual tool diameter and assumed diameter when programming.
(b) To program the actual size of the work and apply tool compensation to cause the centre of the tool to be automatically offset the required amount. This application can save the programmer a lot of calculations.

(c) To use different tool numbers for the same tool and, by applying different values of compensation, to use the tool to take a number of cuts.

Tool nose radius compensation

Tool nose radius compensation is only applicable to tools used on turning centres. Practically all turning tools have a small radius on the tip of the tool. When the tool is set up on the machine, the X datum is established using the front of the tip, and Z datums are established from the side of the tool tip radius, as shown in Figure 9.20a. There are no problems in producing correct sized parallel diameters or machining straight shoulders and faces. Obviously the radius of the tip of the tool would be left in any inside corners. However, where movement has to take place simultaneously on both the X and Z axes, such as when machining chamfers, angles or turning curves, it is necessary to make allowance for the tip radius. The allowance is referred

177

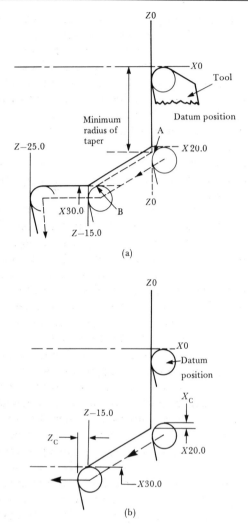

Figure 9.20 Tool nose radius: (a) compensation not applied; (b) compensation applied

automatic adjustment of the tool position, at both the start and the end of the cut, it is essential that the size of the radius at the tip of the tool is entered in the program. In Figure 9.20b the distance X_c is the adjustment necessary at the start of the cut, and the distance Z_c is the adjustment necessary at the end of the cut. When the nose radius is fairly small, such as of the order of 1 or 2 mm, the actual amount of compensation is very small and may not be of practical significance; however, for very accurate work and for tools with a large nose radius it is extremely important. The G codes used for nose radius compensation on a number of control systems are frequently the same as for cutter diameter compensation, i.e. G41 and G42. The codes follow the same rules for direction of movement; the viewing direction is directly above the Z axis, looking in the direction of cutting.

9.12 Screw thread cutting

The recommended standard G codes for cutting screw threads on turning centres are G33, G34 and G35 depending on the type of lead of thread being cut (G33 is for constant lead). The above three G codes are non-modal, and it is necessary to enter them with a separate depth of cut (X dia.) in each block. The pitch of the thread must be programmed, and the control then ensures that the axial movement of the tool is synchronized with the rotational movement of the work.

There are control systems that use G84 or other G codes, and with these codes provision is commonly made to produce progressively reducing depths of cut as the tool advances into the thread form. There are various arrangements for this purpose such as:

to as nose radius compensation. As shown in Figure 9.20a, the edges of the tool tip radius would be positioned at the programmed X and Z coordinates, but as the tool is fed along the slope at the correct angle the tip of the tool would follow the path A to B shown. The diameter of the taper produced would be incorrect. In order that tapers of the correct diameter are produced, the tool position has to be adjusted so that the tool cuts the true size.

So that the computer in the control unit can calculate the necessary coordinates to make an

(a) An initial depth of cut is entered together with the full depth of thread, and the computer in the control unit applies a percentage reduction to each successive depth of cut per pass of the tool. The number of passes is not restricted.

(b) The total number of cuts are programmed and the computer calculates an initial depth of cut so that the fixed number of passes is not exceeded. The same volume of material is removed on each pass, so that the depth gets progressively smaller.

The control unit frequently provides a number of clean-up passes where there is no increase in the depth of cut.

9.13 Dwell

A dwell is a period when movement during a machining operation stops to allow the cutting tool to finish cutting. The duration of dwell is defined as part of the G04 code; for example, G04/5 would result in a dwell of 5 seconds being applied when G04 is entered in the program. A dwell can be programmed at any time in a program, prior to it being required. Some control systems have a default value of 4 seconds for a dwell.

The duration of a dwell can be calculated using the spindle speed and the number of revolutions that the work or tool will make to finish cutting. The number of revolutions required will be dependent on work material and sharpness of tool; a typical value is 20. If the spindle speed is 1200 rev/min, the time per revolution would be 0.05 seconds. The dwell would then be $20 \times 0.05 = 1$ second.

It is possible on some controls to program the duration of the dwell directly as a number of revolutions.

9.14 Programming aids

Programming aids known as canned cycles and subroutines are commonly available on a number of control systems. It is essential to study the programming manual for the control system to determine what programming facilities are available, and the rules to be applied for the efficient implementation of the facilities.

Manufacturers' canned cycles

Canned cycles are sometimes referred to as *fixed cycles*. They are predefined sequences of events (movements) stored in the memory of the con-

trol unit; the sequences are usually repetitive until cancelled. Canned cycles are activated by the insertion of the relevant preparatory function (G code) in the operational block. It is necessary to enter dimensional information with the required G codes for machining to take place. The use of canned cycles reduces the programming required to perform certain operations. Manufacturers of control systems frequently provide cycles for their systems on G codes which have not been assigned standard functions.

It is impossible to provide examples of all the canned cycles provided by different manufacturers. The following are typical examples of manufacturers' canned cycles. It is repeated again that it is essential to study the manufacturers' manuals to determine what cycles are available, and how the information required has to be entered; the only standard that must be followed is that specified by the manufacturer.

In a canned cycle it is usual for the tool or work to move the same amount on the same axis at different parts of the cycle, and therefore the coordinates entered are usually unsigned. Generally, where applicable, either absolute or incremental programming can be used in canned cycles if G90 or G91 has been previously entered. For certain of the canned cycles some of the dimensional values must be in incremental units, but where they are part of the canned cycle it is not necessary to enter G91. The cycles save the entering of at least four blocks in the program, and can form part of a loop or subroutine.

Milling and drilling canned cycles

The following cycles are available on Bridgeport Series 1 CNC machines with BOSS 5 or 6 control (BOSS is an acronym of Bridgeport Operating System Software), and the details are based on information in the machine programming manuals (acknowledgement is made to Bridgeport Ltd for the initial details on the cycles).

Face milling cycle (G77) This cycle can considerably reduce the number of blocks required

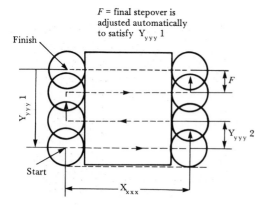

Figure 9.21 Face milling cycle G77

to face mill a surface. The path followed by the centre of the cutter is shown in Figure 9.21. The cycle requires the following block format:

$$NnnnG77XxxxYyyy1Yyyy2Ffff*$$

where

Nnnn block number

G77 activates cycle

Xxxx incremental distance to be milled along X axis, cutter centre to cutter centre

Yyyy1 incremental distance to be milled along Y axis, cutter centre to cutter centre

Yyyy2 Y axis 'stepover' value. Maximum stepover value is diameter of cutter; for efficient cutting a more practical value is 70 to 80 per cent of cutter diameter. Last stepover is automatically adjusted by control to satisfy Yyyy1

Ffff feed rate

* end of block

As an example, to face mill the block whose details are shown in Figure 9.22 with a shell mill of 60 mm diameter, assuming the cutter has been placed at the start position and at the correct Z coordinate, the format is:

$$N105G77X260.0Y240.0Y50.0F200*$$

Pocket milling cycle (G78) This cycle can be used to mill a rectangular pocket. The path followed by the centre of the cutter is shown in

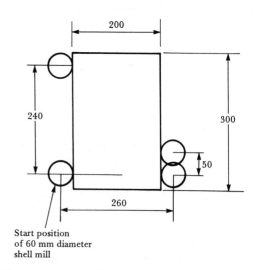

Figure 9.22 Details of component for face milling

Figure 9.23. The cutter has to be positioned at the centre of the pocket and at the desired depth before the cycle is activated. Consideration has to be given to the diameter of the cutter used; if there is a small radius in each corner of the pocket, it will be more practical to use a larger-diameter cutter which can withstand larger cuts and to finish the corners with a cutter of suitable

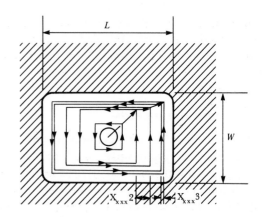

$$\frac{L}{2} - \frac{D}{2} = X_{xxx}1$$

$$\frac{W}{2} - \frac{D}{2} = Y_{yyy}1$$

D = diameter of cutter

Figure 9.23 Path followed by centre of cutter during pocket milling G78

diameter. The cycle requires the following format:

$$NnnnG78Xxxx1Xxxx2Xxxx3$$
$$Yyyy1Yyyy2Ffff1Ffff2*$$

where

Nnnn	block number
G78	activates cycle
Xxxx1	distance from centre of pocket to wall along X axis, less cutter radius
Xxxx2	stepover value on X axis. Stepover is amount cutter moves for each cut. Maximum stepover is diameter of cutter
Xxxx3	stepover for final boundary cut. If it is not programmed, default stepover value for final cut will be 0.5 mm
Yyyy1	distance from centre of pocket to wall along Y axis, less cutter radius
Yyyy2	stepover on Y axis. If not programmed, stepover on Y axis will be same as X axis stepover
Ffff1	feed rate for clearing pocket
Ffff2	feed rate for final cut. If not programmed, default feed rate will be 1.5 times Ffff1 feed rate
*	end of block

As an example, to mill the rectangular pocket whose details are shown in Figure 9.24, 10 mm deep, assuming that a cutter of 20 mm diameter has been placed in the centre of the pocket and at the desired depth, with a 10 mm stepover and a 1 mm finishing pass, the format is:

N115G78X40.0X10.0X1.0Y30.0F200F300*

Internal hole milling cycle (G79) This cycle can be used for milling a circular hole to size. It is more suitable for holes which are too large to be drilled, and which may be difficult to bore with a single-point boring bar. It can be used as a roughing operation followed by a boring bar for very accurate work. The depth of hole that can be milled will depend on the length of the cutting edge of the milling cutter. If the diameter of the cutter is less than half the radius of the hole to be milled, then at least two cycles will

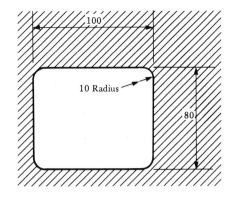

Figure 9.24 Pocket details

$$J_{jjj} = R - \frac{D}{2}$$
$$D = \text{Diameter of cutter}$$

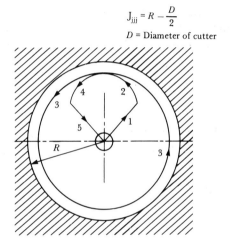

Figure 9.25 Internal hole milling cycle G79

have to be programmed. The cutter has to be placed at the desired depth before the cycle is activated. The path followed by the centre of the cutter is shown in Figure 9.25; the path starts at the centre of the hole and follows the numbered sequence 1 to 5, where it returns to the centre of the hole. The cycle requires the following format:

NnnnG79JjjjFffff*

where

Nnnn	block number
G79	activates cycle
Jjjj	radius of hole to be milled minus cutter radius
Ffff	feed rate
*	end of block

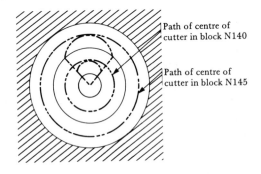

Figure 9.26 Details of hole to be milled

As an example, to mill the hole of 100 mm diameter, shown in Figure 9.26, assuming that the cutter of 20 mm diameter has been placed at the centre of the hole and at the desired depth of 10 mm, there will have to be two blocks:

$$N140G79J20.0F200*$$
$$N145G79J40.0*$$

Turning canned cycles

The following cycles are available on turning centres fitted with Fanuc controllers (acknowledgement is made to Fanuc for the information supplied).

Straight cutting cycle (G77) This cycle can be used for either straight or taper turning; the movements in the cycle followed by the tool tip are shown in Figures 9.27a and b. The block format for both cycles is similar; the taper turning cycle requires the amount of taper as an additional word with the address of I. The tool has to be placed at the start position before the cycle is entered. The tool will advance from

whatever position it is placed in to the coordinate X and then feed along to the coordinate Z. The cycle requires the following format:

For straight turning:

$$NnnnG77XxxxZzzzFfff$$

For taper turning:

$$NnnnG77XxxxZzzzIiiiFfff$$

where

Nnnn	block number
G77	activates cycle
Xxxx	coordinate on X axis
Zzzz	coordinate on Z axis
Iiii	incremental distance in X axis of taper required
Ffff	feed rate
*	end of block

As an example, the component shown in Figure 9.28 is to be turned, assuming that the tool has been placed at the start position and the spindle is rotating. The control recognizes a taper turning cycle from a straight turning cycle by the insertion of the I coordinate in the block:

$$N110G77X70.0Z140.0I36.0F200*$$
$$N115G77X60.0Z140.0I36.0*$$
$$N120G77X50.0Z140.0I36.0*$$

Thread cutting cycle (G78) This cycle can be used for cutting either straight or tapered threads. The tool has to be placed at the start position before the cycle is entered. The movements in the cycle shown in Figure 9.29 require the following format:

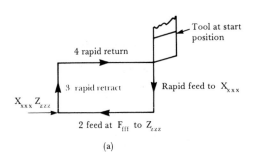

(a)

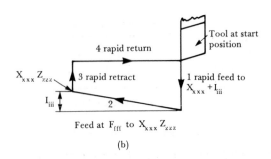

(b)

Figure 9.27(a) Movements in straight turning cycle G77; (b) movements in taper turning cycle G77

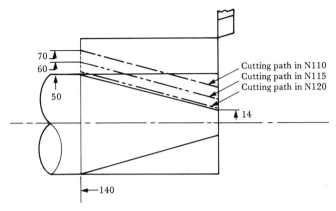

Figure 9.28 Application of taper turning cycle G77

For cutting parallel threads:

NnnnG78XxxxZzzzFfff

For cutting tapered threads:

NnnnG78XxxxZzzzIiiiFfff

where

Nnnn   block number
 G78   activates cycle
Xxxx   coordinate on *X* axis
Zzzz   coordinate on *Z* axis
 Iiii   incremental distance in *X* axis of taper
        required
 Ffff   threading feed rate
   *   end of block

The threading feed rate would be the lead of the thread to be cut per revolution of the work. The programmed block for thread cutting would

contain the relevant coordinates, similar to the taper turning cycle.

Facing cycle (G79)   This cycle can be used for cutting either straight or tapered faces. The tool has to be placed at the start position before the cycle is entered. The cycle shown in Figure 9.30 requires the following format:

For straight facing:

NnnnG79XxxxZzzzFfff

For tapered facing:

NnnnG79XxxxZzzzKkkkFfff

where

Nnnn   block number
 G79   activates cycle
Xxxx   coordinate on *X* axis
Zzzz   coordinate on *Z* axis
Kkkk   incremental distance in Z axis of taper
       required

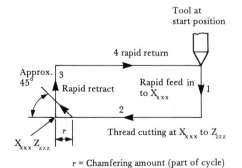

Figure 9.29 Movements in thread cutting cycle G78

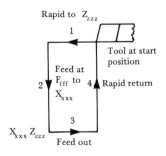

Figure 9.30 Movements in facing cycle G79

Ffff   feed rate
 *   end of block

The programmed block for a cycle for tapered faces would contain the relevant coordinates, similar to the taper turning cycle.

There are other canned cycles provided by both the above and other manufacturers of control equipment, such as area clearances for turning operations.

Standardized canned cycles
The following is not intended to be a definitive description of standard canned cycles available from all manufacturers, but a guide as to how they can be used.

Cycles G81 to G89 are standardized canned cycles. They are mainly used for drilling, tapping and boring operations, but turning centres may use the same number G codes to perform canned cycles which have essentially the same movements but are different in context. For machining centres all the standard canned cycles have similar fundamental movements:

(a) The tool is positioned at the start of the first sequence of movements.
(b) The tool advances on the Z axis at a programmed feed rate.
(c) The tool retracts on the Z axis.
(d) The work moves on the X and Y axes at maximum feed rate to the start position for the next cycle of movements.
(e) The cycle repeats.

Certain of the canned cycles require a particular arrangement of the dimensional information. The Z coordinate is always programmed as an incremental unit, but the coordinates of the next location can be input as absolute units, or incremental units, depending upon whether G90 or G91 has previously been programmed and is still active.

Repeated drilling or turning (G81)   This cycle is possibly the most frequently used. After the tool is positioned at the start, it is advanced on the Z axis at a cutting feed rate to a programmed

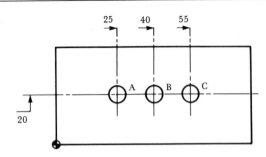

Figure 9.31 *Example component for drilling G81*

position, returned to the start position at maximum feed rate, and repositioned at the next machining position at maximum feed rate; the cycle then recommences. This cycle can be suitable for either drilling or turning.

A typical program for drilling the three holes A, B and C in Figure 9.31, 10 mm diameter and 10 mm deep, without G81 would be:

N100G00X25.0Y20.0Z2.0M03*
    position of centre of hole A. Tool point 2 mm above surface of work, spindle on clockwise (M03)
N105G01Z−12.89F200*
    drills hole A. Moves to Z−12.89 at feed rate of 200 mm/min. Note that 2.89 mm is point angle allowance
N110G00X40.0Z2.0*
    drill retracts to Z2.0 and moves to centre of hole B
N115G01Z−12.89*
    drills hole B
N120G00X55.0Z2.0*
    drill retracts to Z2.0 and moves to centre of hole C
N125G01Z−12.89*
    drills hole C
N130G00X100.0Y−80.0Z10.0M00*
    drill retracts to Z10.0 and moves to safe position. Program stops (M00)

The same operations using G81 would be:

N100G00X25.0Y20.0Z2.0M03*
    position of centre of hole A. Tool point 2 mm above surface of work. Spindle on clockwise
N105G81X25.0Y20.0Z14.89F200*
    confirms X and Y coordinates, moves

downwards total of 14.89 mm, then retracts
14.89 mm

N110X40.0Y20.0*

moves to hole B and repeats Z movements

N115X55.0Y20.0*

moves to hole C and repeats Z movements

N120G00X100.0Y−80.0Z10.0M00*

drill retracts to Z10.0 and moves to safe
position. Program stops (M00)

It is advisable to confirm (enter in the program)
the X and Y coordinates of each hole position.
The Z value is incremental and unsigned.

The following G81 canned cycle is used for
turning on an Audit turning centre fitted with
ANC control (acknowledgement is made to the
manufacturer J. C. Holt for the information
obtained from the machine manual):

G81;Xxxx1;Zzzz1;Xxxx2;Zzzz2;Zzzz3;F*

where

G81   activates the cycle
Xxxx1   total reduction in diameter (diameter
       value)
Zzzz1   total length to be cut
Xxxx2   depth of cut per pass (diameter value)
Zzzz2   length to tool retraction
Zzzz3   length of tool advance
F   selection of in feed 0 for feed in (G01) 1
    for rapid feed (G00)

Using this canned cycle, the component shown
in Figure 9.32 would require the following
block:

N0030G81;30;60;4;60;0;1*

It is important to note that for this cycle the
dimensions are *not* addressed, and a 0 (zero)
must be programmed if there is no value to be
entered. The numbers must be separated by a
semicolon (;). The two values Zzzz2 and Zzzz3
are required if the profile contains re-entrant
sections.

Repeated drilling with dwell (G82) This cycle
is similar in operation to G81, but in addition
has a dwell when the tool reaches its programmed position on the Z axis. The dwell
enables the tool to finish cutting. This is suitable
for drilling a 'blind' hole or spot facing, and on
lathes for turning to a shoulder.

The programming of a G82 cycle is similar to
G81, but in addition will require the duration of
the dwell to be programmed. The control system
may automatically apply a particular default
value, such as 4 seconds, but it is possible to
program other values.

Peck drilling (G83) This cycle is used for the
drilling of 'deep' holes. A deep hole is regarded
as one whose depth is more than five times its
diameter. The drill advances to a programmed Z
value and withdraws to the surface at a rapid
rate. The tool then advances at a rapid rate to the
previous programmed Z value, changes to cutting feed rate, advances the next 'peck', and
withdraws to the surface at rapid rate. It then
continues the sequence until the hole is drilled.

A program to drill the holes in the component
shown in Figure 9.31 but 30 mm deep would be:

N100G00X25.0Y20.0Z2.0M03*

position of centre of hole A. Spindle on
clockwise

N105G83X25.0Y20.0Z32.89Z8.0Z6.0F200*

confirms X and Y coordinates. Drills hole
with pecking action (see below)

N110X40.0Y20.0*

moves to hole B and drills hole with pecking
action

N120X55.0Y20.0*

moves to hole C and drills hole with pecking
action

N125G00X100.0Y−80.0Z10.0M00*

drill retracts to Z10.0 and moves to safe
position

In block N105 there are three Z values programmed: the first (Z32.89, includes point angle

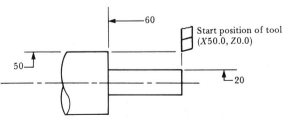

Figure 9.32 Component for turning cycle G81

185

allowance) is the total Z movement required; the second (Z8.0) is the depth of the first peck; the third (Z6.0) is the distance of subsequent pecks. The sum of the pecks does not have to equal the total depth, as the last peck will only be sufficient to advance to the total depth programmed. In this example, to advance to Z32.89 there would be the first peck of 8, plus $4 \times 6 = 24$ pecks, plus the last peck of 0.89.

Tapping (G84)   This cycle is suitable for tapping operations. The tool is advanced at a feed rate which is synchronized with the rotation of the spindle. When it reaches its programmed Z position the spindle is reversed and the tool is returned to its starting position at the same controlled feed rate. The tool is repositioned and the cycle recommences.

The block format for G84 is the same as G81, but in addition it is essential to program the correct spindle speed. The spindle has to advance a distance equal to the pitch of the thread to be cut per revolution of the spindle. Therefore

$$\frac{\text{feed (mm) per minute}}{\text{Revolutions per minute}} =$$
$$\text{pitch of thread to be cut (mm per rev)}$$

If the screw to be cut has a pitch of 1.5 mm, and a feed rate of 200 mm/min is satisfactory, the spindle speed to be programmed would be:

spindle speed (rev/min) =
$$\frac{\text{feed per minute}}{\text{pitch}} = \frac{200}{1.5} = 133$$

The cycle is mainly used when tapping, but may be used when screw cutting on a lathe. On some turning centres G33 is used for screw cutting as well as G84.

The Audit turning centre fitted with ANC control has both available, but recommends G84 in preference to G33 (acknowledgement is again made to J. C. Holt for the information). The cycle requires the following format:

$$\text{G84;A;B;C;D;E;F;G;H*}$$

where

G84   activates cycle

A	total thread depth (diameter value)
B	total thread length to be cut
C	depth of cut for first pass (diameter value) (see below)
D	lead of thread
E	taper expressed as a diameter value
F	pull-out length (not to exceed one pitch) (see below)
G	number of clean-up passes (not to exceed 9)
H	compound infeed (see below)

The depth of cut C for first pass has to be calculated to ensure that the number of passes during actual thread cutting does not exceed 48. C is determined from

$$C = A/\sqrt{N}$$

where A is the total thread depth and N is the number of passes (48 maximum).

If the pull-out length $F = 0$, the tool pulls out at rapid traverse.

The compound infeed H takes various values: for plunge infeed, $H = 0$; for Whitworth form, $H = 55$; and for metric form, $H = 60$.

It is important to note that the values must be separated by a semicolon (;) and are *not* addressed. Zero must be entered if there is no value, i.e. not a tapered thread. An example of programming using this cycle is given in Chapter 12 (in example program 3).

Boring (G85)   This cycle is suitable for boring a hole with a single-point tool mounted in a boring bar. With this cycle the spindle continues rotating, but the tool is fed in and out at the same controlled feed rate. The block format for G85 is the same as for G81.

Boring with tool disengagement (G86)   In this boring cycle, when the tool reaches the Z coordinate the rotation of the spindle is gradually slowed until the tool is at a desired orientation, and then the spindle is stopped. A small sideways movement on the X axis occurs to take the tool point away from contact with the work, and an automatic rapid withdrawal is then activated. This procedure prevents the tool tip rubbing against the work surface. The block format for G86 is the same as for G81.

Drilling with chip breaking (G87)   In this cycle the movements are similar to G83. However, pecks do not occur; the drill withdrawal is only a small distance, typically less than 2 mm. The stop and withdrawal occurs throughout the drilling operation. The block format for G87 is the same as for G83.

Drilling with chip breaking and dwell (G88)   In this cycle the only difference from G87 is that a dwell occurs when the tool reaches its programmed position.

Boring with dwell (G89)   With this cycle the only difference from G85 is that a dwell occurs at the bottom of the bore.

Canned cycle cancel (G80)   This code cancels G81 to G89. The canned cycles can also be cancelled by any of the mode of operation G codes G00, G01, G02, G03 or G06.

9.15 Subroutines

There are two types of subroutine: loops and macros. The essential difference between them is that a loop has to be programmed after the block which calls it up (activates the loop), whereas a macro has to be defined (programmed) before the block which calls it up (activates the macro). A macro may contain a loop, and a loop may contain a macro provided the macro has been defined before the loop is called.

Looping

Looping is a technique that instructs the control unit to read a set of blocks in the part program, and cause them to be executed as many times as indicated by the programmer. The dimensional information in the loop generally has to be in the incremental mode. The use of looping relieves the programmer of having to repeat the blocks in the loop in the program.

It is possible to loop through (repeat) the blocks in a loop any number of times. However, once the looping has been repeated the number of times specified in the program, and the program has gone beyond the last line of the sequence required, then another loop has to be created. It is possible to have a loop within a loop.

Looping can be used in part programs for most machines. The following form of looping statement is available on the Bridgeport CNC Series 1 machine with BOSS 5 or 6 control (acknowledgement is made to Bridgeport Ltd):

$$=Na/b$$

The equal sign = is essential; it instructs the control unit to read the blocks. Without the equal sign looping will not take place. Na is the sequence number of the block at the end of the loop; b is the number of times the loop is to be repeated.

The programming of the component shown in Figure 9.31 using looping is given below. A comparison of the difference in length between this and the program given previously for the G81 canned cycle will make clear the advantages of looping.

N100G00X25.0Y20.0Z2.0M03*
   position of centre of hole A. Spindle on clockwise
N105G91G81X0Y0Z14.89F200*
   confirms X and Y coordinates as incremental units and feeds downwards 14.89 mm
=N110/2*
   calls up loop to block N110 twice
N110X15.0Y0*
   moves to hole B and drills hole, then C and drills hole
N115G90G00X100.0Y−80.0Z10.0M00*
   drill retracts to Z10.0 and moves to safe position

The sequence number in the loop call line is the line number at the end of the loop. If ten holes at the same spacing were required, it would only be necessary to change the 2 in the loop call line to 9, i.e. = N110/9.

Another technique of calling up a loop-type subroutine is used on the Audit CNC lathe with ANC control. With this control the loop can be repeated up to 99 times. The subroutine can be

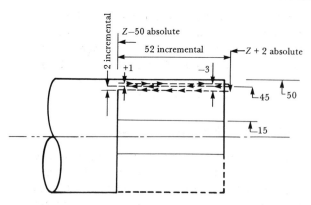

Figure 9.33 Looping operation on turning centre

written at the end of the program, using block numbers that are not used in the main program (acknowledgement is made to J. C. Holt for the essential details). The subroutine is called by the use of the address letter L followed by the block number at the start of the subroutine, followed by the number of times the subroutine is to be repeated. The end of the subroutine is programmed by LO (the letter L followed by 0). The following example program is for the component shown in Figure 9.33:

%*

Program start character.

N0010G00G71G90G95X100.0D00Z100.0M08
(S400T1)*

G95 selects feed rate in mm/rev. Tool positioned at X100 and Z100 coordinates. D00 cancels any tool offsets. The brackets (parentheses) are used to provide the operator with information on spindle speed and tool required; the control unit ignores anything between the brackets. The asterisk * indicates end of block.

N0020₄.45.0D01Z2.0D02F0.2*

Tool positioned at X45.0 and Z2.0 ready for a 2.5 mm depth of cut. D01 and D02 are calling up tool offsets on the X and Z axes respectively. Feed rate of 0.2 mm/rev programmed.

N0030L010007*

Calling up a looping subroutine to start at block number 0100, to be repeated 07 times.

N0040G00X15.0Z2.0*

Positioning operation; tool moves to X15 after the completion of the looping subroutine.

N0050G01Z–50.0F0:05*

Turning operation; tool moves to Z50. Feed rate changed to 0.05 mm/rev.

N0060G00X100.0Z100.0M30*

Tool positioned clear of work. M30 is end of program and return to start of program.

N0100G91G01Z–52.0*

Start of loop (N0100). G91 selects incremental programming. G01 linear interpolation (turning operation), length turned is 50 mm along the Z axis. The tool does not move on the X axis, and starts at Z2 and moves –52 mm on the Z axis (incremental).

N0110G00X1.0*

Tool moves 1 mm on the X axis away from the work.

N0120Z52.0*

Tool returns 52 mm on the Z axis to 2 mm clear of end of work.

N0130X–3.0*

Tool advances on the X axis from 1 mm clear of work to 2 mm into work.

N0140G90*

Absolute programming reselected.

N0150L0*

End of loop.

When the sequence from N0100 to N0150 has been carried out seven times, the program will continue at N0040 with a finish turning operation. In the loop the movements are incremental, and in seven steps of 2 mm depth of cut a total of $7 \times 2 = 14$ mm (28 mm on diameter) will be removed from the starting diameter position of X45.0. The diameter will be reduced from 45.0 to 17.0 in the loop. In block N0040 the tool will be positioned at X15, and a finish turning operation will take place in block N0050.

Macros

A macro is a series of consecutive lines for a sequence of operations that have to be repeated in different sections of the part program. It is effectively a short program within the main part program. A macro may be called up (used) any number of times. It is recommended that the macros required are written at the start of the part program. A number is assigned to the macro, which is then stored until called for; the macro is *not* executed at the time of definition (writing). Different control systems use different procedures for programming macros.

Macros are only generally required on milling and drilling machines and do not have as much application on lathes. This is because once a machining operation has occurred on a lathe, the material is removed, whereas on milling and drilling machines it is possible to use macros for a sequence of spot drilling of holes, and the same sequence can be used for drilling holes at the same position to a greater depth. They can also be used for milling operations where the depth of cut (Z axis) can be changed. For this purpose it is possible to program words with variable values in the macro and insert the required values when the macro is called.

The definition of a macro provided by the Bridgeport CNC series 1 machine with BOSS 5 or 6 control is of the form:

#1*

    #1 statement defines start of macro number 1

N1G00G71G91X20.0Y50.0*

    positions tool to (X20, Y50) from current position

N5G01Z*F100*

    The * between Z and F indicates a Z axis variable value

N10G00G90Z1.0*

    tool retracts to 1 mm above surface of work

$*

    $ terminates macro

Note: The use of the asterisk to designate both a variable and end of block could result in uncertainty and therefore the carriage return or line feed characters are more suitable for the end of block character.

The macro is *not* activated until the call statement of = #n is programmed, where n is the number of the macro. If the macro contains a variable, the value of the variable has to be entered every time the macro is called. In the above example, the first time the macro is called

the value of the Z coordinate has to be inserted, such as = #1Z*–3.0; the next time it is called in the program the value could change, for example = #1Z*–10.0. A macro can contain a number of variables; the variables would have to be entered in the correct order.

9.16 Other programming aids

The following programming aids – mirror imaging, scaling, rotation and preset absolute registers – can considerably reduce the work when manually part programming. These programming aids are also referred to as cutter transformations, as the cutter path is transformed from one position to another. It is essential to study the programming manual for the control system to determine the rules to be applied for the efficient implementation of all the features.

Mirror imaging
Mirror imaging is a technique mainly used for milling or drilling operations, in which the signs (+ or −) of the coordinates on either or both the X and Y axes are reversed. An example of mirror imaging is shown in Figure 9.34. By choosing a suitable position, the pattern of numbers can be considered to be repeated about a point of symmetry. It is necessary to program a subroutine for one pattern of numbers and to call up the subroutine four times, after specifying the code or instruction for activating the aid. It is essential when applying the mirror imaging facility that the centre of the spindle is at the axis of symmetry.

On some of the older control systems, mirror imaging is provided as an optional facility activated by switches, one for each axis. One problem with this method is that the complete program will be mirror imaged.

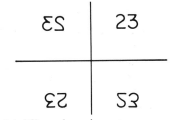

Figure 9.34 *Mirror imaging*

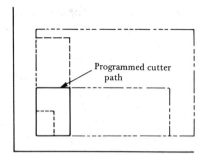

Figure 9.35 Scaling

Scaling

This technique can be used for milling, drilling and turning operations. The amount of movement of the cutter can be increased or decreased by a scale factor, on both axes or on an individual axis, as shown in Figure 9.35. The programmed size is considered to be full size, and components of the same shape but larger or smaller can be produced.

Rotation

This technique is mainly required for milling and drilling operations. It enables a cutter path to be rotated by an angle, and repeated if required at other angles, as shown in Figure 9.36. This makes programming of complex shapes relatively easy.

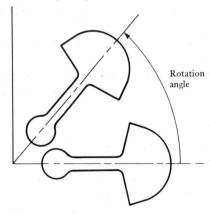

Figure 9.36 Rotation

Preset absolute registers

Although this can be used for turning operations, its main application is for milling and drilling. The preparatory function widely used

is G92. It enables the zero datum position of the X, Y or Z axes to be changed temporarily for some machining operations. One application is where there are a number of parts clamped to the machine table, all requiring identical machining operations. A program is written for one component which is used for the setting up of the tools and the establishing of the X, Y and Z zero datums.

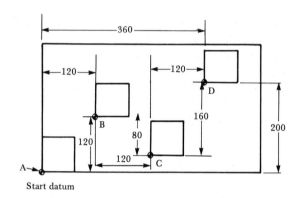

Figure 9.37 Preset absolute registers

Figure 9.37 represents a view of a machine table with four components clamped at different positions. A part program for the machining of the component has to be available, and for the purpose of explaining preset registers the program is called macro 1. To preset the registers so that the machine setting point is changed to the desired position, it is necessary to specify the X, Y and Z coordinates of the old origin with reference to the new origin. If it is not necessary to preset the registers for any axis, they may be omitted. The complete part program for the machining of all four components would be:

Macro 1

  first component would be machined
N95G00X120.0Y120.0Z2.0*

  centre of tool positioned at point B: datum at A
N100G92X−120.0Y−120.0*

  datum is transferred to point B: with reference to B, point A is −120 on the X axis and −120 on the Y axis

Macro 1

second component would be machined

N105G00X120.0Y−80.0Z2.0*

centre of tool positioned at point C: datum at B

N110G92X−120.0Y80.0*

datum is transferred to point C: with reference to C, point B is −120 on the X axis and 80 on the Y axis

Macro 1

third component would be machined

N115G00X120.0Y160.0Z2.0*

centre of tool positioned at point D: datum at C

N120G92X−120.0Y−160.0*

datum is transferred to point D

Macro 1

fourth component would be machined

N125G00X−360.0Y−200.0Z2.0*

centre of tool positioned at point A: datum at D

N130G92X360.0Y200.0*

datum is transferred back to point A

N135

this would be the work changing block

9.17 Writing the part program

When all the factors have been considered it is then possible to start entering details on the operation sheet, the tool detail sheet and the part program sheet. There are no standard forms for these sheets; most machine tool manufacturers will normally provide details of a suitable layout for the sheets for their machine and control system. Typical forms are used in Chapters 11 and 12.

It is important that careful consideration be given to the actual machining sequence. A balance has to be obtained between the prime consideration of safe working conditions, and the minimum time being taken up in:

Tool movements The tool path should be the shortest path practical, due consideration being given to providing safe and adequate working clearances when the tool or work is moving at rapid feed rates. The time taken to move 3 mm at 60 mm/min feed rate is only 3 s, but providing a 3 mm clearance by stopping movement at rapid

feed does give operators a chance to see that the tool will not collide with a rapidly rotating chuck or workpiece. Tools should not travel back over paths at slow feed rates. Apart from the time taken, it is detrimental to the tools if they are allowed to rub on the work.

Tool changing A particular tool should complete all the machining operations at one loading; there should be no need for the same tool to be loaded twice.

Changing speeds and feeds Where possible a constant cutting speed should be programmed.

9.18 Checking (proving) the program

Before the program is loaded into the control unit, it is recommended that it is checked. If the program is to be stored on punched tape, the printout produced while the tape is being punched can be checked visually to see if there are any obvious mistakes. If the program is to be stored on magnetic (floppy) disc or hard disc, or in the computer's memory, it may be necessary to produce a printout of the program for checking purposes. It is useful for the printout to accompany the tool detail sheet etc. to the machine. The obvious mistakes can include, for example:

(a) The capital letter O being entered instead of zero 0
(b) A decimal point omitted or a comma entered by error
(c) A negative sign omitted on a Z coordinate
(d) A G or M code omitted, or entered incorrectly
(e) Typing errors.

It will be unlikely that errors due to mathematical mistakes, or to misinterpretation of drawing details, will be eliminated by reading the printout. It is also possible that other errors, such as linear interpolation being entered instead of point-to-point operations, may not be noticed. In addition, the values of the arc centre offsets in circular interpolation movements may have been entered incorrectly. The only way in which these types of error can be detected is by 'proving' the program. There are three different methods of proving a program: graphics, dry run and stepping.

Graphics

There are two forms of graphic displays: on a plotter, and on the monitor screen of a computer.

Suitable computer software programs are required to try out the part program on a computer or a plotter, where the movement sequence contained in the program can be simulated on the screen or drawn on a two-axis plotter. On a plotter, different colour pens can be used to represent different tools and/or different depths. One problem with the use of a plotter is that for three-dimensional work the actual amount of movement in the third dimension cannot be checked.

There are computer software programs with which it is possible to obtain a three-dimensional view of the results of the tool movements on the monitor screen, and to print out the image on a dot matrix printer. An example of this method will be found in Chapter 13.

Any corrective action needed in the programming of the movement of the tools or work can then be taken before the part program is loaded into the machine, thus saving expensive machine time.

Dry run

This method can be an alternative to the graphics test, but does take up machine time. The program has to be loaded into the control unit. The dry run is mainly to determine if there are any incorrect movements in the operations resulting from the errors stated previously. The program is run through one block at a time with the feed rate override set at the minimum value; the machine setter observes the movements, and is ready to stop the machine if an incorrect movement appears to be taking place. This will ensure that the tools do not collide with any obstructions and that the tools follow the desired programmed path.

The program is used to control the machine either without a workpiece being loaded in the vice or work holding device, or without tools being mounted. Alternatively, a type of foam plastic can be used in place of the actual work material to save damage occurring to the tools due to any incorrect feed movements.

A dry run will enable the movements to be checked, but not the actual material removal or the size of the work produced.

Stepping

After a dry run, and especially if the program contains a complex machining operation, it is desirable that the program is run on the machine at reduced feed rates. A workpiece is machined, but the program is 'stepped' through with only one operation at a time being carried out. The feed rates are adjusted on the manual override control until satisfactory cutting conditions exist.

The graphics and dry run methods will only check the tool path; the first component must be checked to ensure that it meets the designed specification. When the first component is successfully produced the program is said to be *verified*, and can be stored for future use.

Questions

9.1 Discuss the various methods of creating part programs, giving a valuation of each method as to its advantages and limitations.

9.2 What are (a) centre line drawings (b) base line drawings (c) labelled drawings?

9.3 Which of the above drawings are most suitable for NC programming? Give reasons for your choice.

9.4 Explain the difference between the absolute and incremental methods of dimensioning drawings.

9.5 Detail the factors to be considered when deciding on the position of the datum.

9.6 Specify the factors related to the component that should be considered when writing part programs.

9.7 Explain how a tool library can help when writing a part program.

9.8 Specify the different cutting tool materials available, detailing their advantages and limitations.

9.9 Explain how the spindle speed and feed rates are determined for (a) milling (b) drilling and (c) turning.

9.10 What are the factors to be considered when determining spindle speeds and feed rates?

9.11 What is the machine setting point? Explain how its position can be determined.

9.12 With the aid of examples, explain how the coordinates for changing the work and changing the tools can be determined.

9.13 What is the difference in the movement that results when moving from one set of X, Y, Z coordinates to another set of X, Y, Z coordinates when in (a) positioning mode (b) linear interpolation mode?

9.14 What information has to be supplied in order to machine circular curves?

9.15 Explain how the feed rate is determined when milling circular curves on the outside of work.

9.16 Detail the difference in the dimensional information required when machining a 180 degree curve in: (a) single-quadrant circular interpolation (b) multiquadrant circular interpolation.

9.17 What information has to be supplied in order to machine parabolic curves?

9.18 What are polar coordinates? Give an example of how they could be programmed.

9.19 Explain how cutter diameter compensation can reduce the amount of programming.

9.20 Explain how cutter diameter compensation can be applied in a program.

9.21 What is tool nose radius compensation? Explain why it may be important.

9.22 What are canned cycles? Illustrate your answer with six different canned cycles.

9.23 Explain how the canned cycle G81 can reduce the amount of programming when drilling five holes.

9.24 Explain with examples the difference between (a) looping and (b) macros.

9.25 With the use of examples, describe the techniques of (a) mirror imaging (b) scaling (c) rotation.

9.26 Explain how the position of the datum point can be changed temporarily in the program.

9.27 Describe three different techniques of proving a part program.

Chapter Ten

Setting up NC machine tools

10.1 Setting-up procedure

The setting up of the workpiece and tools on numerically controlled machine tools follows the same general procedure as on manually controlled machine tools. The work has to be positioned and the tools have to be set. However, for economic reasons it is essential that non-machining time of CNC machines is kept to a minimum and the setting up of the tools and interchange of work must be carried out efficiently and quickly.

As with manually operated machines, it is necessary to ensure that the tools are sharp and held securely in their tool holders. The various methods of holding tools are explained in Chapter 3. The tools must be well supported to resist the cutting forces generated during material removal. When drills are to be resharpened they must be correctly sharpened on a drill grinding machine and not by hand. If turning tools are made of high-speed steel they should be ground on a tool grinder so that all the angles are true and the tool profile, especially the nose radius, is constant. It is essential to ensure that the tools selected are those specified by the programmer and are not ones chosen by the operator. Clamp tip tool holders with indexable tool tips enable the tips to be replaced quickly and accurately. It is important that any cemented carbide or ceramic tools are the correct grades.

There is also an additional requirement in the setting of the tools and work on numerically controlled machine tools. It is essential to establish that the coordinates of the work/tool relationship provided by the position transducers of the machine correspond with the dimensional coordinates of that point contained in the part program. For convenience when the work/tool relationship is at the datum point of the work the readout of the position on the control panel should be zero. This is because all the information for the movement of the tools and work provided in an absolute program will be to coordinates that refer to that point and checking of the part program is easier.

The establishing of the correct relationship between the tools and the work is necessary on at least three different occasions:

(a) During the initial setting-up stage of different components.
(b) When cutting tools are replaced after regrinding.
(c) After every switching on of power to the machine or control unit, because when power is switched off from the machine all stored part program information is usually lost. There are control systems which have battery backup to provide power to maintain memory systems. For these controls all program details and tool settings will be retained in the memory.

If the datum is not established correctly, tools and work will not be in the correct relationship as required by the program, and movement of the tools or work will not take place at the relevant position.

There are two basically different techniques of establishing that the work/tool relationship is correct (a) floating zero; (b) machine reference datum.

Floating zero

When the power is switched on for machines where the position transducers generate a signal of the *absolute* type, the position transducers will read the actual position of the carriages from the machine zero. This reading will be stored in the memory of the control unit, and will be displayed on the digital readout of the control panel. The reading has to be reset to the required value when the work/tool relationship is at the position which coincides with the selected program datum. Normally the value at the datum is zero. This provision of being able to reset to zero at any position is referred to as the 'floating zero' facility. With manually created part programs the input values of the different dimensional words in the part program will normally be with reference to this established zero.

When the power is switched on for machines that have position transducers that generate a signal of the pulse or incremental type, the digital readout of the position of the tables will display zero. This is because there will not have been any movement that will have energized the sensing elements of the position transducers. When the tables are moved during the setting-up stage, the readout will display the amount of movement during the setting up. The amount of movement will be relative to the position the tables were at when the machine was switched on. This reading has to be reset to the required value (conveniently zero) when the position of the tools is coincident with program zero datum, as for the absolute transducers. The fact that the digital readout is zero when the machine is switched on can be useful if it is necessary to switch off the machine for any reason, such as

overnight, and production of the same component is to be continued the next day. If the tables and tools are positioned so that they are at the zero datum position when the power is switched off, they will be correctly positioned when the machine is switched on, and time can be saved in resetting the tools.

Machine reference datum

On some machines, after switching on the power it is necessary to input a command that causes all the carriages to move along the guideways to the machine reference datum. This is a position whose X, Y and Z coordinate values (machine based) are known. Usually the carriages are moved to the machine setting point to establish the work datum. The amount of movement is input to provide the control with the data on where the program datum is in relation to the machine reference datum.

A similar technique can be used on machines that do not have the machine datum command. With this technique the carriages are driven manually to the extremity of the slides from where the machine movements originate. Position transducers of the absolute type would read zero at this location. Incremental transducers are usually automatically reset to zero at the machine reference datum. From this location it is possible to specify the location of the program datum as above.

Different makes of machine tools may require slightly different setting-up procedures from those described in the following paragraphs. However, it is repeated that the *essential* requirement for all machines is that the tools have to be set to the correct relationship with the work before the program can be run and the component machined, i.e. *the datums have to be established.*

The programmer must provide the following information on the program detail sheet:

(a) How and where on the machine the work should be set
(b) The method of holding the work
(c) Details of the tools
(d) If the speeds are manually controlled, infor-

195

mation on what the spindle speeds should be.

If the work will be repeated at some future date, it is useful to take a photograph of the set-up. Cameras that give 'instant' photographs can be used very effectively. The photograph can be stored with the program details.

10.2 Setting milling and drilling machines

These machines have at least three axes (X, Y and Z) on which the work/tool relationship has to be established. Establishing the X, Y datums involves setting the work relative to the centre of the spindle. Establishing the Z axis datum is relative to the end of the tools or the gauge plane. In the majority of cases it is more convenient to determine the X, Y relationship first.

Setting X and Y datums on milling and drilling machines

The various work holding methods are described in Chapter 3. The position of the work on the machine table has to be such that when the centre of the spindle is in line with the X, Y program datum position this coincides with the value displayed by the machine readout. The setting-up procedure is as follows:

(a) The work blanks must be checked to confirm that they are of the correct size and are prepared with the datum edges as detailed by the programmer. If the work blanks are oversize, problems can occur with the tool taking non-programmed cuts during rapid approach movements and work positioning movements. The programmer will allow for components of maximum size during positioning operations and ensure that the programmed position of the tool is clear of the work after a rapid feed movement and before power cutting feed is applied. As stated in Chapter 9, in some firms the operators of the numerically controlled machines are responsible for preparing the blanks.

(b) The work has to be mounted on the work

table so that the datum edges are correctly aligned with the table movement. This is because all the coordinates are at right angles, or at a known orientation, to the datum edges.

(c) The work must be positioned so that the zero datum of the work is in the planned position on the machine table as specified by the programmer. This position is referred to as the machine setting point (MSP).

(d) The work has to be securely clamped to a fixture or gripped in a vice as decided by the programmer. Jigs are not used on numerically controlled machine tools. (Note: jigs hold the work and guide the cutting tools, and are mainly used for drilling operations on manually controlled machines.)

(e) The type and position of the clamps used to hold the work must not be changed from that specified by the programmer, otherwise problems can be caused by the clamps being in the path of the tool during rapid positioning movements. The programmer must provide the machine setter with the information on the types of clamp and their position.

(f) The workpiece has to be supported to prevent it deflecting during clamping, or moving under the forces created during cutting. If the work is held in a vice the direction of cutting should be towards the jaws rather than parallel with the vice jaws (see Section 3.13).

Through holes, pockets or milling around part of the outside of the component will require the component being raised to clear the work table. It may be necesssary to drill some of the holes in the component held in a vice at one set-up, and use the holes for the holding of the component on a fixture or mounting plate for machining around the outside of the work at another set-up. If the component has to be machined around the outside and there are no suitable holes, or if no holes are required for the functioning of the component, it may be necessary to discuss with the design office the possibility of providing holes in the component for work holding purposes.

If the workpiece is a casting without convenient features for work holding, it may be necessary at the design stage to have lugs added to the casting.

Aligning the datum edges

It is essential for the work to be located, supported and clamped as specified by the part programmer. A dial test indicator can be used in the standard technique of checking the alignment of the datum edges of the work with the movement on the X and Y axes. The dial test indicator is clamped in a fixed position on the spindle head of the machine. The position of the machine table and spindle head are adjusted until the plunger of the indicator can be brought into contact with a reference edge which can be a straight edge held in a vice or placed against the locating pins in a grid plate or base plate, or with the work itself if there are two suitable edges that can be used.

To check the alignment of the work edges with the movement of the table, the plunger of the dial indicator is placed in contact with the reference edge and the table is moved under manual control. When the edge is correctly aligned the needle of the dial test indicator should have minimum deviation when the edge being tested is moved past the plunger. The movement of the datum edges on the X and Y axes is controlled manually during the setting up by utilizing special buttons on the control unit. These buttons cause the table to move continuously while the button is depressed. Additionally, there may be a facility provided which allows the spindle to be moved either continuously or in known increments, typically of 25.0, 10.0, 1.0 or 0.01 mm (or the imperial equivalents). The smallest movement (0.01 mm) is commonly known as *jog*.

Alternatively an electronic handwheel can be used to control the movement. An electronic handwheel is a small unit which may be fitted permanently on to the machine or mounted on a flexible cable linked to the control unit. It is possible to select the particular axis to be moved by the axis address keys on the machine control unit. The handwheel unit contains electronic circuitry or an electrical unit, such as a resistor over which a wiper blade moves. As the blade is moved over the resistor the change in resistance causes an imbalance in the control which results in movement occurring on the selected axis. When the handwheel is rotated the table will move either forward or backward depending on the direction of rotation. In order to provide for sensitivity of control it is possible to change the rate of movement when the handwheel is rotated.

Setting the datum point (machine setting point)

The work also has to be set on the machine table so that, when the X and Y zero program datums of the work are directly under the centre line of the spindle, the reading from the position transducers on both the X and Y axes read zero or the particular value specified by the programmer. This work table position is known as the machine setting point. The calculation of the position of the machine setting point was explained in Chapter 9. Most machines have scales fixed to the X and Y tables which can be used for positioning the work tables. On some machines the values of the coordinates of the centre point are stated on the machine.

The programmer will provide details of which part of the work has been chosen as the datum. When the work table is at the machine setting point (X and Y zero datum of the work), the readings of the position transducers can be set to zero using a switch or other facility on the control unit panel or set to the required value using the numeric keys on the control panel.

If grid plates, pallets or base plates are being used, the components must be accurately located and aligned on the pallets or base plates to the part programmer's specification. If the work is a complex shape and is to be held on a pallet, the position of the datum edges can be determined using straight edges placed against the locating pins. It is also important that the pallet or grid plate is located accurately on the machine table at the position specified by the part programmer.

There are a number of techniques for determining the location of the work on the work table such as using a setting mandrel, a wiggler, an optical device or a spindle mounted touch trigger probe.

Using a setting mandrel

One method of establishing the relationship of the work on the machine table to the centre of the work spindle utilizes a setting mandrel of a known diameter. The setting mandrel is mounted in a tool holder that is loaded in the spindle nose. With manual control of X, Y and Z axis movement, the mandrel is lowered so that the end of the mandrel is clear of the work and below the edge of the work. With manual control of movement the work table is slowly advanced on either the X or the Y axis towards the mandrel, first in steps of 0.1 mm and then in steps of 0.01 mm (jog), until (as shown in Figure 10.1) the datum edge of the work is in contact with the mandrel. Feeler gauges are used to determine the contact. The size of the mandrel and feeler gauge are known, and so the position of the edge relative to the centre of the spindle can easily be calculated. The edge of the work can be positioned under the centre of the spindle by moving the work table the required distance (half the diameter of the mandrel plus the feeler gauge). The above procedure has to be carried out for both X and Y datum edges of the work. The amount of movement of the work table can be determined with reference to the machine scales or digital readout.

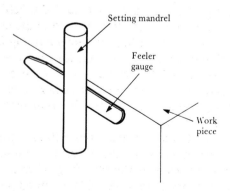

Figure 10.1   Setting X *and* Y *datums using setting mandrel and feeler gauge*

Using a wiggler

A tool known as a 'wiggler' or 'wobbler' can be used instead of the setting mandrel. A wiggler consists of a parallel shank which is held in a chuck mounted in the spindle nose of the machine. At the end of the shank is a ball joint mounting and by removing the knurled nut different styli can be fitted. A wiggler and attachments are shown in Figure 10.2. Stylus B

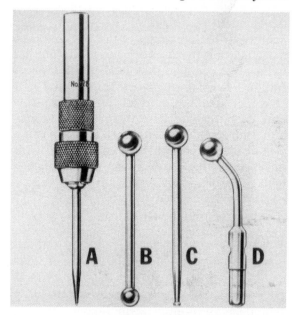

Figure 10.2   Wiggler and attachments (courtesy L. S. Starret)

has a spherical section of 6 mm diameter at each end; if one end is fitted into the balljoint, then when the spindle rotates the free end of the stylus will tend to fly outwards under centrifugal action. When the free end of the stylus is brought into contact with the edge of the work, it is prevented from flying outwards. The friction between the spherical end of the stylus and the work results in the end in contact with the work rolling across the edge of the work. As the work table is moved towards the spindle centre the eccentric movement of the spherical section of the wiggler reduces until finally it will tend to 'creep' up the side of the work. When the wiggler rotates without the spherical section creeping up the side of the work the centre of the spindle will be half the diameter of the spherical section of the stylus from the edge of the work. With the spindle stationary stylus A can be used for lining up a rough edge of work. Stylus C can be used instead of stylus B for positioning. Stylus D can be used for the holding of a small dial test indicator; for this application the spindle has to be stationary.

Using an optical centre finder

An alternative method of setting the work in position on the work table uses an optical centre finder. A typical centre finder is shown in Figure 10.3. The graticule in the centre finder

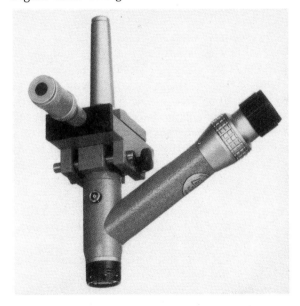

Figure 10.3   Optical centre finder (courtesy Trimos Sylvac Metrology)

has cross-lines, and the instrument is made so that the cross-lines are in the centre of the spindle. The edges of the work can be observed through an eyepiece and using manual control to move the work tables the edges are positioned directly under the cross-lines.

The work first has to be aligned and secured to the table; then the X and Y datums are established by positioning the machine table so that the datum corner of the work is in the position shown in Figure 10.4. To assist in setting up, the image of the workpiece is magnified usually about ten times. There are centre finders which

have small projection screens which make the image of the work easier to see. If the top edge of the work is not square but has a radius a special sighting piece can be used.

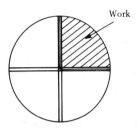

Figure 10.4   Image in optical centre finder when setting X *and* Y *datums*

10.3   Use of sensor probes on milling and drilling machines

Sensor probes, which includes 'touch trigger' and proximity probes can be used on most CNC machining or turning centres and are proving to be extremely beneficial in reducing time spent on tool setting, locating and checking work size.

As explained in Chapter 5, touch trigger probes do not measure but when the probe's stylus touches a surface a signal is generated. As a result of the signal any movement that resulted in the contact being made is stopped, and also the reading of the position transducers when the contact was made is registered within the control unit.

Although benefit can be obtained using touch trigger probes on manually operated machines, the greatest advantage to be gained in the use of these probes in setting up and operating CNC machine tools is with the control software that has been specially written for their use. The software is in the form of a number of canned cycles (G codes) created for different machine control systems.

Touch trigger probes are either spindle mounted or table mounted. Spindle mounted probes are used in the checking of work and move to contact the work; table mounted probes are used for checking tools and the tools move to the probe.

Spindle mounted probes

The probe assembly is mounted in a modified tool holder. The concentricity of the stylus of the probe to the machine spindle has to be calibrated with a datuming procedure when the probe is first mounted on the machine.

In use the holder with the probe assembly is loaded in the spindle nose either manually or by the use of automatic tool changers. An automatic tool changer can be seen loading a probe holder into the machine spindle in Figure 10.5; a

Figure 10.5   Automatic loading of spindle probe (courtesy Renishaw Metrology)

table mounted probe can also be seen in the lower right on the machine table. Spindle mounted probes can be used for a number of purposes, such as:

(a) Aligning, establishing and checking the X, Y and Z datums
(b) Checking that a workpiece has been loaded, and actual position of work
(c) Checking that work blanks are of the correct size

(d) Checking that a hole has been originated before a tapping or reaming operation
(e) Checking the position of a hole before a boring operation
(f) Checking (inspecting) the finished work.

The canned cycles consists of a sequence of movements. To make the initial contact the first movement programmed is at a rapid feed rate. This is followed by an automatic confirmation of contact checking operation with the work and probe first separating and then advancing slowly to confirm the position of the probe's contact.

The use of the CNC machine tool to inspect components which have just been cut on the same machine requires that the accuracy of the machine and its position transducers be checked periodically by measuring work of known size or special master test pieces. Alternatively, the accuracy of the results can be verified by measuring the same component on a coordinate measuring machine. It must be remembered that when probes are used there are no forces causing stresses and deflections in the constructional members etc., and the accuracy of the measurement of the movement of the carriages is mainly dependent on the accuracy of the transducers themselves.

Obviously the time spent in using a CNC machine tool for inspecting components must be limited but can be justified. Typically when the work is large and difficult to handle it may be beneficial to measure the first-off component on the machine using touch trigger probes. This will save the problem of relocating the component on the machine table if corrective machining is needed. The use of in-cycle gauging techniques where the checking of the size of only one important feature produced by each tool on every component will provide information for updating tool offsets. The increase in production time will be negligible compared with the increase in the number of accurate components produced. It is possible to use the optional block skip character to cancel the checking of some components whose results are proving consistently satisfactory.

Aligning and establishing work datums

Spindle mounted probes provide an accurate method of establishing the relationship of the datum edges of the work to the machine datum. A job contact probe being used to determine the position of vice jaws is shown in Figure 10.6.

Figure 10.6   Aligning datum edges with job contact probe (courtesy Renishaw Metrology)

Touch trigger probes that are interfaced to the control can be used more effectively if available. After the loading of the probe assembly the spindle is moved towards the vice until the centre of the sphere on the end of the probe's stylus is below the edge of the jaw. The table is moved towards the stylus and when the jaw contacts the stylus the probe's LED lights and the setter notes the read out of the position on the control panel. The alignment of the jaws can be checked by touching the jaws at one end and noting the reading of the transducers; the table is moved away from the probe and across so that the probe can be brought back into contact with the other end of the jaws; the reading being noted upon contact. If the jaws are aligned the readings should be the same.

With touch trigger probes using the canned cycles the in-built confirmation of contact sequence can be automatically activated and the reading of the position transducers is stored within the machine control unit.

For establishing the work datum the vice jaw will be half the diameter of the spherical section of the stylus from the the centre of the spindle. As with the test mandrel and wiggler both X and Y datums must be established.

Checking work datums

To reduce the need for precise positioning of successive pieces of the work on the machine table it is possible to determine the location of the work by contacting the edges of the work used as the datum and noting the reading of the transducers at a number of contacts. The difference between the actual values and the expected values can be entered as an offset compensation. If the datum edges are not correctly aligned it is possible for the control to allow compensation for the misalignment. A similar technique can be used when the work coordinates stored are relative to the machine datum. The probe can be used to check other features of the work (bores etc.) if they are used as the datum. A touch trigger probe being used to check the position of a bore in a casting is shown in Figure 10.7. The probe is positioned so that the stylus is in the bore and then brought into contact with the bore; the confirmation of contact sequence is activated and the reading of the position transducers registered. The table is moved until the probe contacts the other side of the bore and the reading is registered; the machine table is moved back half the difference in the readings of the position transducers to place the probe in the centre of the hole. The machine table is moved at right angles to the previous movement until the probe contacts the side of the hole and the position at this point is registered. The machine table moves diametrically across until the probe contacts the other side of the hole. The four readings are used by the control to calculate the location of the centre of the bore.

Checking work blank size

The set length from the gauge plane to the stylus tip is known and the probe is brought into contact with the work as shown in Figure 10.8

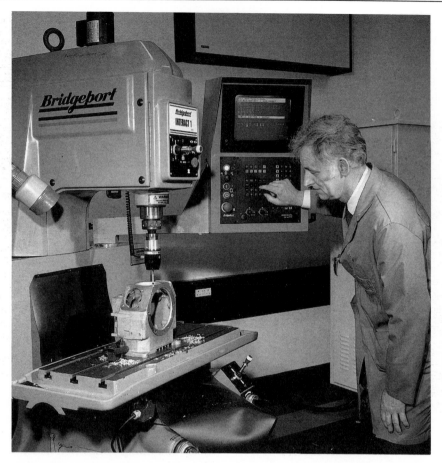

Figure 10.7   Locating casting with touch trigger probe (courtesy Renishaw Metrology and Bridgeport Machines)

and the size of the work can be determined. If the work is larger in size than programmed an additional machining operation is automatically included so that any excess material is removed; or if the work is smaller, machining operations can be cancelled.

The same procedure can be used to check the size of machined work as actually shown in Figure 10.8.

Checking presence and depth of hole

The work is positioned so that the centre of the hole is in line with the centre of the spindle. The probe is advanced towards the work and programmed to move the programmed depth of the hole. With a correctly drilled hole contact will be made when the probe reaches the bottom of

the hole and the reading of the transducers noted and registered within the control. If work without a hole, or a hole not correctly drilled, is positioned under the probe, contact will be made at the surface of the work or at the incorrect depth and a different reading of the transducers will be registered and the program will be stopped.

Checking size of work

To check the position of the edge of a workpiece and its distance from another edge, the machine table is moved so as to place the stylus to the side of one of the edges to be checked. The spindle with the probe assembly is advanced so that the point of the stylus is below the surface of the work and beside the edge as shown in Figure

Figure 10.8   Spindle mounted probe Z axis positioning (courtesy Renishaw Metrology)

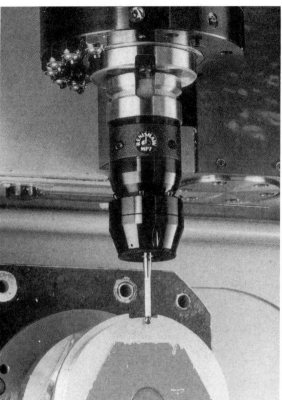

Figure 10.9   Spindle mounted probe X and Y axes positioning (courtesy Renishaw Metrology)

10.9. The confirmation of contact sequence will be activated and the position of the machine table on the particular axis is noted. The diameter of the contact sphere on the probe is known, and therefore the position of the edge of the work on the machine table can be calculated. The probe is raised and moved to the other edge to be checked, and the same procedure followed. The difference in the reading of the position of the machine's table at the two positions less the diameter of the stylus is the distance between the edges.

Checking size of hole
To check the diameter of a hole, the same sequence of movements in making four contacts in the bore is used as that for determining the position of the centre of the bore; except that the control uses the readings to calculate the

diameter of the bore not its centre position. Allowance is made for the diameter of the contact sphere on the probe's stylus. Figure 10.10 shows a probe being used to check the diameter of machined bore. The centre distance between holes can be checked by determining the position of the centre of the holes as detailed previously. A hole diameter can be increased but the position of an existing hole cannot be changed. Incorrect programmed hole positions can be changed by the control for future components.

10.4   Setting milling and drilling tools

If the machine has a tool storage facility from which tools can be automatically selected, the tools have to be placed in the correct numbered

Figure 10.10   Inspecting a component (courtesy Renishaw Metrology)

location as allocated by the programmer. If the tools have a built-in identification which can be read by the sensing mechanism, the loading of the tools in specific locations is not so important. See Chapter 3 for information on tool identification. If the machine has a tool turret the tools have to be mounted in the correct station. Tool setting on machining centres using multi-point tools is mainly establishing the relationship of the end of the tool relative to the Z datum as it is not possible to change the diameter of milling cutters and drills.

Setting the tools to the Z datum
The tools used will be of different lengths, and every tool has to be set so that its programmed movement takes place at the relevant part of the work. This requires that when the point or end of each tool is in contact with that part of the

work or is at the location selected by the part programmer to be the Z zero datum, the readout in the control unit is also zero for that particular tool.

There are different methods of ensuring that the correct programmed Z movement of each tool occurs at the required part of the work:

(a)   Determining the Z offset values
(b)   Presetting length of tools
(c)   Use of sensor probes.

Moving the spindle
During the setting up the tools have to be advanced on the Z axis; this movement of the tools in the spindle is manually controlled by utilizing the jog facility or electronic hand-wheel.

Determination of Z offset values
This method can only be used if the control system has the facility of storing offset values. As explained in Chapter 3, if the machine tool is of the type where the work spindle has limited axial movement it is necessary to adjust the position of the work table during setting up in order to accommodate work of different heights. The vertical position of the work table is usually manually adjusted during the setting up and the work table is clamped in position when the component is being machined and is not moved. As also explained in Chapter 3, on some machines the vertical movement of the work table can be programmed as well as the axial movement of the spindle. If this is available, the vertical movement of the work table is programmed as a W axis.

Usually the tools are changed, either manually or automatically, when the spindle is fully retracted; this position is frequently referred to as the 'home' position.

Where the Z offset value is determined on the machine, it is advisable to select the longest tool to be set first, to ensure that there will be sufficient room between the end of the tool and the surface of the work when the spindle is in its fully retracted position. When the longest tool is loaded into the spindle nose, its tool number has to be input to the control unit. After the tool is loaded, the spindle nose should be moved under manual control approximately 10 mm from its home position. The table with the work is then moved vertically until the end of the tool is at the location which the programmer has selected to be the Z zero datum.

The initial 10 mm movement is to ensure that there is room for the longest tool to move upwards, so that any Z positive dimensions can take place. The end of the tool must be clear of the work during positioning operations. If a tool other than the longest tool is used for setting the vertical position of the work table, it is possible that there will not be sufficient space between the work and the end of the longest tool.

The amount of movement of the spindle from the home position must be entered in the memory of the control unit. The amount can be stored either by entering its determined value using the keyboard of the control unit, or by the use of a special button on the control unit. Pressing the button will instruct the control unit to read the amount that the Z axis spindle has moved from the home (fully retracted) position and store that value in the memory. The amount of movement of the tool from the spindle's home position is stored in the control unit's memory as an 'offset' value which effectively establishes the Z zero datum for that tool.

It is common practice for the part programmer to select the finished top surface of the work to be the Z zero datum, because it is convenient for programming and it is relatively easy to position the end of the tools in contact with the surface. For work that has a rough or uneven surface, setting blocks of the required height may be used.

There are a number of techniques for judging when the point of the tool is in contact with the setting block or workpiece. The techniques, which depend upon the skill of the setter in judging the contact, are:

(a) Observing the end or point of the tool marking the surface (tool rotating).
(b) Using feeler gauges between the end or point of the tool and the surface (tool *not* rotating).

There is a small instrument which can be used for setting the tool position on the Z axis. The instrument comprises a dial indicator mounted inside a cylindrical body, as shown in Figure 10.11. The instrument is placed on the machine

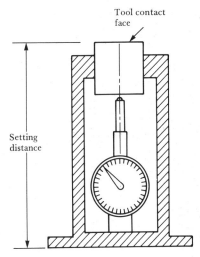

Figure 10.11   Tool setting instrument

table or workpiece datum face. The tool is brought into contact with the setting face of the instrument, and gradually moved until the dial indicator reads zero. The tool point is then a specified distance from the work table or datum face.

After the longest tool has been set, the other tools have to be loaded in succession into the spindle nose. The relevant tool number is entered into the control unit using the control unit's keyboard. The tools are moved under manual control from the home position until the end of the tool is in contact with the work or setting block. The amount of movement from the home position is the length offset of each tool. The offset is stored in the control's unit memory, and will automatically be applied when that tool is called up in the part program. It is essential that the vertical position of the machine table is not changed once it has been set for the longest tool.

Preset tools to specified lengths
Preset tools can be used on any machine, but the advantage of using preset tools may be small

when it is possible to determine the tool length offsets on the machine, or when only a limited number of tools have to be changed manually. The technique of presetting is particularly suited for:

(a) Machines where the complete spindle head moves rather than the spindle moving axially

(b) Machines with table mounted sensor probes

(c) Automatic tool changing.

Set length

Presetting the length of tools involves measuring the distance from the gauge plane to the tip of the tool (as shown in Figure 10.12) and adjusting the length to a specified value known as the set length. The gauge plane is at a particular diameter of the taper shank; this ensures that all tools fit into the same position in the spindle nose. For each size of taper shank the gauge plane is a constant diameter. The adjustment of the set length of a drill, and mill etc. is generally provided by either a screwed backstop in the holder which can be moved so that the tool can be set to the desired length, or a screwed adjustment on the body. The amount of adjustment of the set length varies with the tool holders made by different manufacturers, and

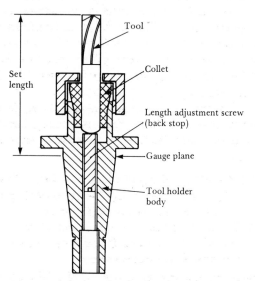

Figure 10.12   Set length adjustment

can be up to 30 mm for the backstop type and up to 3 mm for the screwed body type. Care has to be paid to the length of the actual cutting tools (drills, end mills etc.), as tools of the same diameter made by different manufacturers can be of different length.

Using preset tools it is necessary to establish the distance from the home position to the Z axis datum with only one tool; for convenience the longest tool is usually selected. The tool selected for establishing the Z axis datum is loaded into the spindle nose. The head is moved under manual control until the point of the tool is at the location selected by the programmer to be the Z zero datum. The transducer monitors the position of the tool spindle head along the Z axis. The reading of the control unit for the Z axis is then set to zero at this position, using the same procedure as explained for determining the Z offset values.

When programming the rapid approach of each preset tool to the work, the programmer has allowed for the difference between the set lengths of each tool and the set length of the tool used for establishing the Z datum. If the longest tool is specified by the programmer to be used for establishing the Z datum, a tool which is a different length will have to travel the difference in their lengths to reach the face of the work. If the longest tool is tool number 1 and it has a set length of 150 mm, and if tool number 2 has a set length of 125 mm, the programmer will have allowed for the difference in lengths (i.e. 25 mm) when programming the movement of tool number 2. If the control has been set to Z zero when tool number 1 is in contact with the work surface, as shown in Figure 10.13, tool number 2 will have to move to $Z-25.0$ before contacting the work surface. This is because when the spindle nose is in the home position, the gauge plane of all tools will be at the same location, but the ends of the tool will protrude different lengths from the gauge plane. It is the position of the gauge plane on the spindle nose which is controlled.

As stated previously, the surface of the work is the most commonly used location for the Z zero datum. However, some machines may require

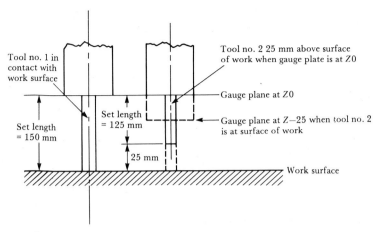

Figure 10.13   Gauge plane movements

the Z zero datum to be set a fixed distance above the surface of the work. Setting blocks are used to establish the Z zero datum. A suitable fixed distance for the gauge plane to be above the surface of the work for some machines is 200 mm. This is because most tools up to 20 mm in diameter have a set length less than 200 mm.

If the longest tool has a set length of 150 mm, and the Z datum is set 200 mm above the surface of the work, the longest tool will have to move to Z−50 before touching the work surface. The other tools will have to move an additional amount before they contact the surface of the work. This additional amount is the difference between their set lengths and the set length of the longest tool.

Where preset tools are used it is essential for the setter to ensure that the length of every cutting tool is the correct set length. Special tool presetting fixtures are used to set the lengths of the tools.

Tool presetting fixtures

The presetting of tools is a practice that has been in use for a number of years for such applications as setting tools on 'in-line' boring bars, and using tool setting templates for setting multiple tools engaged in roughing operations on capstan lathes and automatic machines.

Tool presetting fixtures for numerically controlled machines have measuring instruments that are used to establish the size and position of the cutting edge of the tools in relationship to tool holders or tool turrets. The lengths of the tools are adjusted to the stipulated set lengths. The main advantage of the use of preset tooling is that non-cutting time is saved, machine utilization is increased, and productivity is improved by reducing the set-up time. If there are a number of numerically controlled machine tools available it is desirable that all tool presetting should be carried out in a tool crib near the machines. It is not economic to provide identical presetting fixtures for each machine setter. However, if there are different machines, and the setter is provided with the presetting fixture, the tools could be set for the next workpiece when the machine is cutting the present work.

There are a large number of setting fixtures available, ranging in price (1990) from £8000 plus, but they can be classified mainly as either mechanical or optical. Each presetting device has its own peculiarities, but generally they all require:

(a) The initial setting of the datum faces of the fixture, and measuring the relationship of the tools to that datum position; or
(b) Establishing the size of one particular tool, and measuring the difference between that tool and all the other tools.

Presetting fixtures for milling and drilling consist of a dummy machine spindle nose in which the tool holders are mounted, and accurate measuring devices, using proximity contacts, for tool point position setting. For drilling and milling machines the use of the fixtures is mainly concentrated on setting the tools to the desired lengths. The simplest measuring instrument is a vernier height gauge, the contact with the end of the tools being with either the conventional blade or a dial test indicator. The fixture shown in Figure 10.14 can be used for setting the lengths of the tools and also for setting boring tools for machining work to desired diameters. The diameter of drills and solid milling cutters may be measured on the fixture, but cannot normally be changed except by regrinding. The clamped tip type of milling cutters using indexable inserts can be checked to ensure that all of the inserts protrude the correct amount. The information on the diameter of the tools is extremely important, and is required if tool diameter compensation is in the part program.

Figure 10.14   Tool presetter (courtesy Trimos Sylvac Metrology)

Figure 10.15   Tool presetter with optical projection head (courtesy W. Frost Engineers)

Figure 10.15 shows a presetting fixture with an optical projection head used for setting drills, milling cutters and single-point tools. There are different tool holding heads that can be mounted on the unit for use on a range of machines. The fixture is linked (interfaced) to a computer, which may be linked either to the machine's control unit or to the computer on which the part program is stored. If the tool setting unit is linked to the machine's control, the details of the set lengths can be stored automatically in the library which may be part of the control unit's memory. Therefore the movement required from the home position for individual tools to contact the work surface is known, and the Z coordinates in the part program only have to be from the Z datum. When the tool is selected during the program, allowance will automatically be made for the amount of movement of the gauge plane of individual tools from the home position to a clearance plane above the work contact position. When the tool setting unit is linked to the computer where the part program is stored, the part program is automatically modified to include a rapid approach movement for each tool from the home position to a clearance plane as required. The length of the rapid approach movement for each tool would be dependent on the measured length determined on the fixture.

A tool library, which contains all the details of the tools available, will be very helpful and time saving. If a tool library is used it is essential that the tools be set to the listed lengths.

Use of spindle mounted probes – *Z* axis datum

Spindle mounted probes provide an accurate method of establishing the relationship of the surface of the work to the tool's home position, i.e. fully retracted. The position of the gauge plane on the spindle at which the stylus of the probe is deflected (as shown in Figure 10.8) is stored in the control unit's memory as being the *Z* zero datum. The relationship of the set lengths of the probe and tools is known and stored in the memory of the control unit. When a tool is selected during the program, allowance will automatically be made for the amount of movement of the gauge plane of individual tools from the home position to a clearance plane above the work contact position.

10.5 Use of table mounted sensor probes for tool setting on machining centres

Canned cycles with preparatory codes (G codes) have been created for table mounted probes to activate the movement of the machine table and spindle. As stated previously, tools move to the probe for checking; the probe is clamped at a permanent position on the machine table where it does not cause obstruction. The position of the probe is stored in the control. A probe on a horizontal spindle machine is shown checking the diameter of a shell mill in Figure 10.16. To determine the diameter of cutters such as shell mills, end mills etc., they are programmed to rotate slowly in reverse while the tool is moved first on the *Z* axis and then movement occurs on the *X* and *Y* axes so that the periphery of the cutter can be brought into contact with one side

Figure 10.16   Table mounted probe, horizontal spindle (courtesy Cincinnati Milacron)

of the cube as shown in Figure 10.16. The size of the stylus (cube) is known together with the location of the centre line of the stylus relative to the centre of the spindle and so the diameter of the cutter can be determined. Although it is not possible to change the diameters of solid multi-point tools it is essential for their diameter to be known so that compensation can be applied for any cutters that are not to the exact size expected by the programmer.

A table mounted touch trigger probe for a vertical spindle machine is shown checking the length of an end mill in Figure 10.17. As can be seen more clearly in Figure 10.17 the stylus contact is of cube form. To determine the length of a multi-point cutter the table is moved so as to place the centre of the tool in line with the stylus. With the cutter rotating slowly in reverse

Figure 10.17 Table mounted probe, vertical spindle (courtesy Renishaw Metrology)

the tool spindle moves along the Z axis to contact the probe. The relationship of the position of the probe stylus to the machine axes is known; therefore the tool length can be established from the difference between the probe position and the machine spindle position. The length of the tool so determined can be automatically entered into the part program.

One big advantage to be gained by the use of table mounted sensors is that it is not necessary to determine the set lengths of the tools, as the length of each tool can be checked quickly and accurately.

Table mounted sensors as shown in Figure 10.18 can be used for detecting tool breakage. Two techniques are used. One is similar to the setting of the tool's length. The tool is brought into contact with the sensor and the actual length of movement compared with the previous length when the tool was set up. Any significant difference in the length of the movement can send a signal to stop the program for operator attention. The other technique is to bring the tool to a sensor which has proximity transducers or a jet of air to detect the presence of the tool.

Figure 10.18 Tool breakage sensor (courtesy Cincinnati Milacron)

10.6 Tool length compensation

Tool length compensation is the term used when amending the offsets for any tool by manually inputting data using buttons on the keyboard of the control panel. This is extremely useful if the size of the work produced is incorrect; the Z axis datum of any tool can be adjusted to remove any errors. The tool length offsets determined during the setting-up stage as explained in Section 10.4 can also be changed using the tool length compensation facility. It is also possible to enter tool length compensation by manual data input (MDI) using the control unit's keyboard.

10.7 Cutter diameter compensation

If cutter diameter compensation is being applied within the program, the setter may have to enter the details of the diameter of cutters being used.

The information on the diameter of the tool,

which the computer in the control system has to have to calculate the required compensation, has to be input into the control unit's memory before the operation in which the compensation is to be applied. The input of the tool diameter can be part of the program (at the beginning of the program) or can be input by the operator using manual data input (MDI) through the control unit's keyboard or as above by the use of sensor probes.

10.8 Setting turning centres or CNC lathes

These machines have a minimum of two axes (X and Z) on which the work/tool relationship has to be set. The problems associated with the setting up of numerically controlled turning centres are different from those with milling and drilling machines. On the three axis machine the work is clamped to the machine table and has exactly the same movement as the table. The size of the work after machining, and the position of the various features of the work, are thus controlled by the movement and position of the table. By contrast, except for sliding head machines, on lathes it is the tools that move on a minimum of two axes, and the diameter and length of the work is therefore dependent on the position of the cutting point of the tools.

Setting the work

The majority of work on turning centres originates from bars of stock size and the setting of the work is mainly establishing the distance from the face of the collet or jaw chuck to the end of the bar. The amount of protrusion of the work has to be set on the machine by the setter to values supplied by the programmer. If the work is parted off on the machine this will be the length of the component plus the width of the parting off tool and the minimum clearance between the parting off tool holder and the face of the collet or chuck. For some components a facing off allowance of 0.5 mm may also be allowed. A bar stop usually mounted in the tool post or turret is set to limit the forward movement of the bar from the chuck and so establish the length of the work.

If the work is a billet or a casting or forging it is usually located against a backstop in the work holder. It is essential to use the work holding equipment specified by the programmer. Details of work holding methods are described in Chapter 3.

Turret mounted probes can be used to check the position of the end face of the work as shown in Figure 10.19. The probe can be seen at the eight o'clock position on the right hand turret.

Work datums on turning centres

On the majority of machines the centre line of the work is taken to be the X zero datum, and the end of the work furthest from the chuck or work holder to be the Z zero datum.

With these machines the control system has to be set so that when the tools are in contact with the finished end face of the work, that position is interpreted as being Z zero. Also the control system has to be set to zero when the point of the tools is on the centre line (X zero datum).

On other machines the Z and X coordinates may be stated with reference to the end face of the spindle nose and the centre of the spindle axis respectively.

On some machines there is a home position for the tools, the coordinates of which are supplied by the manufacturer of the machine, and all movements of the tools have to be calculated by the programmer from this position.

10.9 Setting of lathe tools

The setter has to ensure that the tools are positioned in the correct turret stations specified by the programmer. The overhang of the tool from the tool holder must be a minimum or that specified by the programmer.

There are four techniques of setting turning tools so that they are in the correct relationship to the work:

(a) Setting on the machine
(b) Use of a presetting fixture
(c) Qualified tooling
(d) Use of sensor probes.

Figure 10.19   Turret mounted probe on a twin turret mill-turning centre (courtesy Renishaw Metrology)

Setting on the machine

In this case the position of each tool has to be established in relation to a reference tool which is the first tool set. To establish the Z zero datum of the reference tool, the tool is mounted in the tool holder and moved by manual control until its cutting edge makes contact with the finished end face of the work, or a setting piece known as a reference gauge. This position together with the tool number is then stored in the control's memory.

The X zero datum is established by the following steps:

(a) Using manual control, the reference tool is moved to touch a known diameter of the work, or the reference gauge.
(b) The tool's identification number is entered in the control unit's memory using buttons or keys on the control panel.
(c) The diameter with which the tool is in contact is measured.

(d) Without changing the position of the tool on the X axis, the measured diameter of the work is also input to the control unit using buttons or keys on the control panel.

All the other tools to be used are loaded in sequence and, using the same procedure as for the reference tool, are brought into contact with the same end face and diameter that was used for the reference tool. The control unit automatically stores their position relative to the reference tool position as an offset value (see Section 10.10).

10.10   Tool offsets

Tools can be of a different shape and thickness, and the length of the tools protruding from the tool post can vary. It is therefore obvious that, as shown in Figure 10.20, the centre line of the tool holder will be at a different position on the machine when the cutting point of the different

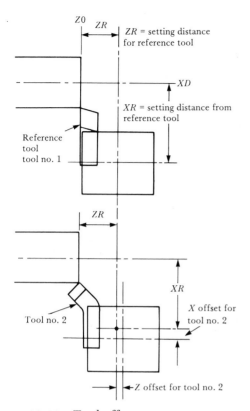

ZR = setting distance for reference tool

XD

XR = setting distance from reference tool

Reference tool
tool no. 1

ZR

XR

X offset for tool no. 2

Tool no. 2

Z offset for tool no. 2

Figure 10.20   Tool offsets

tools is in contact with the same part of the work. When the reference tool is mounted, the position of the centre of the tool holder is taken as the check position. The difference in the positions of the centre line of the tool holder for different tools is referred to as the tool offset, and it will vary for different tools. When a number of tools are being used, it is necessary during the setting-up stage to determine the position of the cutting point of each tool in relation to the zero datums of the work in order that the tool offsets are known and stored in the memory of the control unit. If it is found that the size of the work (diameter or length) actually being produced is not satisfactory, it is possible to edit the value of the offset for the tool so that work of the correct size can be produced.

When the tool offset value is amended to compensate for tool wear, this is frequently known as tool wear compensation. However, if the value of the offset has to be changed, it is

extremely likely that the tool will already have deteriorated so much that the tool will have to be replaced.

Presetting of tools

Turning tools can be preset individually, or the complete turret can be taken off the machine and mounted on the presetting fixture especially made for the machine. It is reported that it is possible to preset turning tools so that two tools set to the same standard will produce work within 0.005 mm.

When the tools are set individually, the tool holder has to have a precise and repeatable means of locating on the presetting fixture and on the machine tool. Each tool has to be first preset for height so that when they are mounted on the machine the top of the cutting tip of the tool is on the centreline of the spindle. A unit similar to that shown in Figure 10.21 can be used for setting the height of the tools. The fixture shown in Figure 10.22 has a dummy tool post, and the reading of the dial indicators is set using a standard setting piece of known size. The tool and holder are mounted on the fixture and the position of the point of the tool is determined from the reference point using the dial indicators. The amount of overhang of the tool, which influences the diameter of the work being turned, can be adjusted and set using tool adjustment screws. It is essential that the tools are set exactly as prescribed by the programmer. It is not practicable to have a wide setting range for the front edge of the tool; this means that there may have to be more adjustments to the tool offsets for length than for diameters.

The presetting fixture can be an optical device such as that shown in Figure 10.15.

Qualified tooling

All qualified tools are made so that the cutting point of the tool is a fixed distance from the side and end of the shank of the tool. In Figure 10.23, tool B is back, front and end qualified; tool F is qualified from the front and end; and tool Q is qualified from the back and end.

When the tools are mounted in the tool

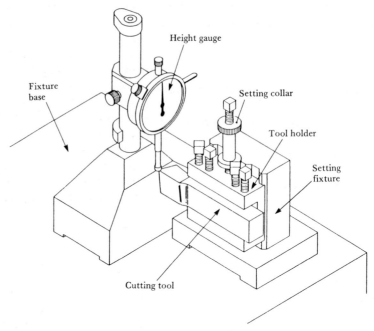

Figure 10.21   Tool height presetter (courtesy Dicksons Engineering)

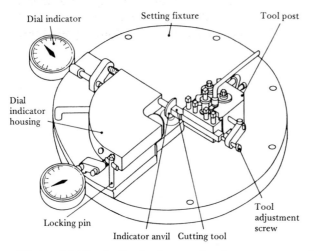

Figure 10.22   Lathe tool presetter (courtesy Dicksons Engineering)

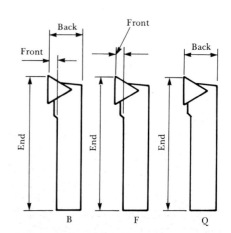

Figure 10.23   Qualified tooling

holder, they have to be located from the same side and end of the shank that they are qualified from. Tool B can be located from either the back or front with the end. When the tools are set in the tool holder the cutting point is in a specified relationship in the tool holder geometry. One advantage of this method is that it provides the facility to change a tool during the operating cycle without the necessity of resetting.

Semi-qualified tools have setting studs screwed into the ends and side of the shank, so that the tool tip can be repositioned correctly after regrinding.

Qualified tools are more expensive than standard tools, and when the control has tool offsetting facilities the advantage of the ease of resetting is marginal.

Use of sensor probes

The use of touch trigger probes similar to those used on setting tools on milling and drilling machines has reduced the amount of time spent on adjusting the position of the tools. For maximum efficiency when the tools are mounted on the turret or tool post the distance from a reference point on the tool mount such as the centre of the turret to the tip of the tool on both the Z and X axes should be standardized for all tools to be used. These distances should be stored in a tool library.

The probe assembly is positioned at a fixed point on the machine during the setting up of the tools as shown in Figure 10.24. The contact point of the probe is located on the machine at known coordinates. The tools are brought up in sequence to touch the probe and as each tool contacts the probe the coordinates at which contact is made can be noted or stored within

the machine's control unit. When all the tools have been checked the probe assembly is removed and its mounting position covered by the cover that can be seen suspended on a chain on the left of the Figure 10.24.

For turning centres in an FMS cell the tool details will be stored in the control unit and the cycle for tool checking will use the distances as a first value for the movement of the tool to the probe. The probe is mounted permanently on the machine as can be seen in Figure 2.18. The difference between the actual reading of the transducers and the programmed value is stored as a tool offset and applied automatically by the control.

10.11   Tool nose radius compensation

If tool nose radius compensation features in the part program, it is essential during the setting-up stage to check that the tools used have the nose radius specified by the programmer. A nose radius can only be changed by grinding and can only be effectively measured by using an optical projector with a set of radius graticules.

Figure 10.24   Touch trigger probe for tool setting (courtesy Renishaw Metrology)

Questions

10.1   Explain how clamp tip tool holders with indexable inserts can reduce the time of resetting tools.

10.2   When is it necessary to establish the correct relationship between the tools and the work?

10.3   Why is it essential to establish the correct datums?

10.4 What is the floating zero facility? Explain how it can be used.

10.5 Discuss the setting-up procedure required on NC machine tools.

10.6 What is the Z offset value? Explain how it can be determined for each tool to be used.

10.7 Explain why it is more convenient to use the longest tool as a reference for establishing the Z offsets on a vertical spindle machining centre.

10.8 What is the set length of a tool? Why is it important?

10.9 How is the set length of a tool determined and set?

10.10 What is a sensor or touch trigger probe? Explain the difference in the use of (a) spindle mounted and (b) table mounted probes.

10.11 Describe three different methods of setting the work to the machine setting point on machining centres.

10.12 Describe how the datum positions are established on turning centres.

10.13 What are qualified tools?

10.14 What are the advantages and limitations of using qualified tools?

10.15 Discuss the advantages and limitations of presetting tools or establishing tool offsets on the machine.

10.16 What is an electronic handwheel, and how is it used?

Chapter Eleven

Part programming for milling and drilling work

11.1 Example program 1

This example has only point-to-point and linear interpolation operations. The component shown in Figure 11.1 is to be produced on a 2CL milling machine (two axes continuous path and one axis linear interpolation control). The machine has a work table movement of 500 and 300 mm in the X and Y axes respectively. The Z axis movement is obtained by the spindle moving axially 150 mm.

The following standard preparatory and miscellaneous functions (G and M codes) are available. They should be the minimum provided in the control system of most machine tools.

Preparatory functions
G00　point-to-point
G01　linear interpolation

G02　circular interpolation clockwise
G03　circular interpolation counter-clockwise
G04　dwell
G17　circular interpolation X and Y plane
G18　circular interpolation X and Z plane
G19　circular interpolation Y and Z plane
G40　cutter diameter compensation off
G41　cutter diameter compensation left
G42　cutter diameter compensation right
G70　inch units
G71　metric units
G80　cancels canned cycles
G81　drilling cycle
G82　drilling cycle with dwell (spot facing)
G83　deep hole drilling
G84　tapping cycle
G90　absolute programming
G91　incremental programming
G92　preset absolute registers
G94　feed rate, millimetres per minute
G95　feed rate, millimetres per revolution
G96　cutting speed, metres per minute
G97　spindle speed, revolutions per minute

Miscellaneous functions
M00　program stop
M01　optional program stop
M02　end program
M03　spindle on clockwise
M04　spindle on counter-clockwise
M05　spindle off
M06　tool change
M07　mist coolant on

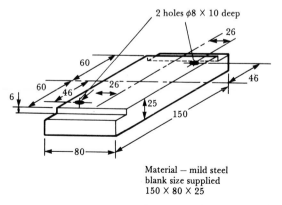

Material — mild steel
blank size supplied
150 × 80 × 25

Figure 11.1　Component for example program 1

M08 flood coolant on
M09 coolant off
M25 spindle home
M30 end of tape

Before beginning to write the part program it is necessary to decide on such factors as the following, which are not in order of precedence but are all interrelated:

(a) Position of the zero datums
(b) Sequence of operations
(c) Work holding method, and the machine setting point
(d) Speeds and feeds for the tools to be used.

11.2 Position of the zero datums

For the convenience of establishing the coordinates on the X, Y and Z axes, it is beneficial for the component to be redrawn and redimensioned to satisfy the requirements of base line dimensioning. Figure 11.2 meets these requirements. For this example the X and Y zero datums can be conveniently established at the bottom left-hand corner of the workpiece. The Z zero datum will be the top surface of the work.

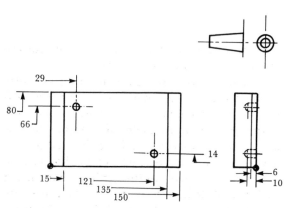

Figure 11.2 Coordinate dimensioning

11.3 Work holding method and machine setting point

It is intended that the component will be held in a vice with the centre of the component over the centre of the work table. Figure 11.3 shows the location of the component on the work table. It will be seen that the programmed zero datums ($X0$, $Y0$) at the lower left-hand corner of the component will be at *machine* coordinates of $X175$ and $Y110$. This is the machine setting point.

Coordinates for work changing
The first component will be placed in the vice when the machine is under manual control during the setting up of the machine. When the machine is under program control, new work blanks will replace the finished workpieces at the end of all the machining operations. To provide sufficient space to carry out the work change, the machine table should be moved away from the tool in the spindle and towards the front of the machine or to a position nearest and convenient to the operator. This gives the operator a clear area to work in so that the components can be changed without risk to the operator's hands from the tools. The swarf would have to be cleared away before the vice jaws are opened. Obviously for safe working practice the main tool spindle would be programmed to stop rotating in the last block before work changing.

The machine setting point is $X175$ and $Y110$, and therefore the maximum possible movement of the machine table towards the right and front of the machine is -175 mm on the X axis, and $300 - 110 = 190$ mm on the Y axis. To allow some latitude in the positioning of the vice on the machine table and to provide clearance to prevent the override microswitches being activated, the program coordinates selected for a safe position for changing the work are $X-150.0$ and $Y170.0$. The position of the centre of the spindle relative to the machine table at work changing is shown in Figure 11.3.

11.4 Sequence of operations

The sequence of operations has been designed for a number of components to be machined:

(a) Load 10 mm diameter spot drill.
(b) Drill two conical holes (spot drill).

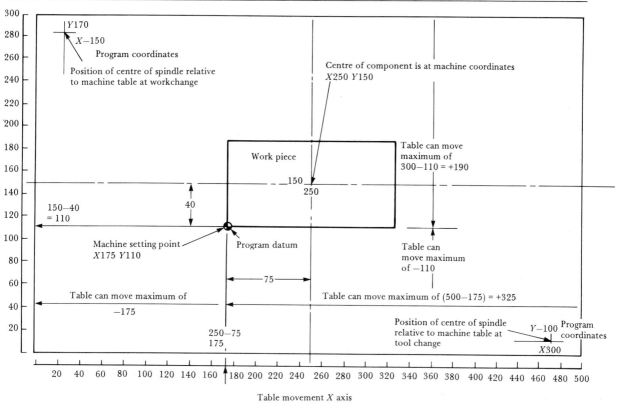

Figure 11.3 Machine setting point

(c) Load 8 mm diameter drill.
(d) Drill two holes.
(e) Load 20 mm diameter end mill.
(f) Mill two end steps.
(g) Remove work from vice, and replace with fresh blank holding on long sides, locating against stop in stepped jaw of the vice.

Spot drilling

This operation produces a small conical hole that will provide a starting location for the actual drill which will be used to originate the hole. A spot drill is shown in Figure 11.4a; it has a 90 degree point angle, and can be used for a range of hole diameters. A spot drill has a very short flute length and an increased web thickness to provide extra strength to prevent bending under thrust load. Also, because the spot drill is so rigid, there is less tendency for it to flex and wander when it contacts the surface of the work before starting to cut.

A 10 mm spot drill can be used for all holes between 4 and 8 mm in diameter. The spot drill has been programmed to produce a conical hole with a maximum diameter of 10 mm, which is 2 mm larger than the hole to be drilled. As shown in Figure 11.5, the 8 mm drill locates on its outside diameter in the conical hole created by the spot drill, and not on its point, because the drill used for the 8 mm hole has a point angle of 120 degrees. This prevents the drill wandering and reduces the tendency for it to chatter or vibrate before it starts to cut along the two edges of its angled point. In addition, because the maximum diameter of the cone produced by the spot drill is larger than the 8 mm drill, a small countersink is left on the edge of the hole and no burr is formed at the surface of the work.

219

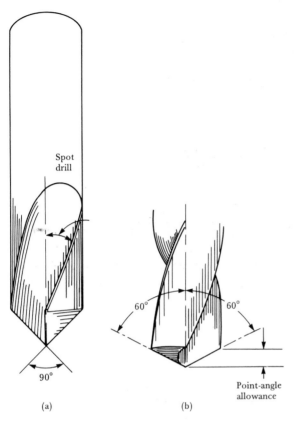

Figure 11.4(a) Spot drill; (b) point angle allowance

Because the end of the spot drill has a 90 degree angle, the point angle allowance for a 10 mm maximum diameter countersink is:

$$PAA = 5 \times \tan 45° = 5$$

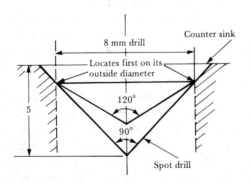

Figure 11.5 Guiding of drill

Note that tan 45° = 1, and therefore the PAA for the spot drill for any hole is the radius of the hole to be drilled plus 1 mm. A spot drill is only used for producing conical holes, and not for normal drilling operations.

Drilling operation

The Z datum is the surface of the work, and the Z offset for a drill is established when the point of the drill contacts the work surface. The drawing requires the holes to have a parallel portion 10 mm deep; to this must be added a point angle allowance (PAA). As shown in Figure 11.4b, the PAA for the 8 mm drill which has a point angle of 120 degrees is:

$$PAA = 4 \tan 30° = 2.3$$

The depth of the hole to be programmed is therefore $10 + 2.3 = 12.3$.

Milling the steps

Because the steps are only 6 mm deep, their milling can be completed in one operation. If the width of the steps had been important then it would have been advisable to take a roughing cut, leaving 1 mm on the width for a finishing cut.

11.5 Speeds and feeds for the tools to be used

The tools are as follows:

Tool no. 1: spot drill, 10 mm diameter with two flutes

Tool no. 2: 8 mm diameter high-speed steel twist drill with two flutes

Tool no. 3: 20 mm diameter end mill with five teeth

A cutting speed of the order of 30 m/min is suitable for mild steel when using high-speed steel drills and milling cutters. For material removal a feed rate of 0.1 mm per tooth is possible.

To calculate spindle speed:

$$\text{cutting speed (m/min)} = \text{circumference (m)} \times \text{spindle speed (rev/min)}$$

$$= \frac{3.1416 \times \text{diameter (mm)} \times \text{rev/min}}{1000}$$

Hence

$$\text{spindle speed (rev/min)} = \frac{\text{cutting speed (m/min)} \times 1000}{3.1416 \times \text{diameter (mm)}}$$

Therefore for a cutting speed of 30 m/min,

$$\text{spindle speed (rev/min)} = \frac{30 \times 1000}{3.1416 \times \text{diameter (mm)}} = \frac{9549.2}{\text{diameter (mm)}}$$

The spindle speeds for the three tools are therefore as follows:

Tool no. 1: 9549.2/10 = 954.92 rev/min
Tool no. 2: 9549.2/8 = 1193.65 rev/min
Tool no. 3: 9549.2/20 = 477.46 rev/min

For practical reasons, it is more sensible that these spindle speeds be modified to the following:

Tool no.	Calculated spindle speed (rev/min)	Selected spindle speed (rev/min)	Actual cutting speed (m/min)
1	954.92	950	29.845
2	1193.65	1200	30.159
3	477.46	500	31.416

The actual cutting speeds are acceptable and well within the capability of the tools to provide satisfactory tool life.

The feed rate is given by

$$\text{feed rate (mm/min)} = \text{feed per tooth} \times \text{no. of teeth} \times \text{spindle speed}$$

Hence the feed rates for the three tools are as follows:

Tool no. 1: $0.1 \times 2 \times 950 = 190$ mm/min
Tool no. 2: $0.1 \times 2 \times 1200 = 240$ mm/min
Tool no. 3: $0.1 \times 5 \times 500 = 250$ mm/min

Note: When writing the part program with X, Y and Z coordinates programmed in the same block, it must be remembered that, when the machine is in the positioning mode (G00), the movement of the tables will occur differently from when the machine is in the linear interpolation mode (G01) (see Chapter 9).

Coordinates for tool changing
The machine has manual tool changing facilities, and responds to a tool change operation (M06) by the spindle stopping and retracting fully to its home position. M06 also turns the coolant off if it is flowing.

The work table should be positioned so that the tools can be removed and replaced in the spindle nose without interference from the work or vice. To provide this facility the machine table has to be moved away from the operator towards the left and the rear of the machine. This will also enable the operator to be able to reach for the tool without having to lean over the work.

To determine the programmed coordinates of a safe position for tool changing it is necessary to consider the position of the work on the work table. With reference to Figure 11.3, the machine setting point is at machine coordinates of X175 and Y110. There is a maximum possible movement in the direction required of $500 - 175 = 325$ mm on the X axis and 110 mm on the Y axis.

To provide clearance to prevent the override microswitches being activated, it is recommended that the maximum possible movement should not be programmed. The coordinates chosen for the tool change are X300 and Y−100. These will position the table so that there will be clearances of 25 and 10 mm respectively on the X and Y axes to the microswitches. For a tool change the machine table will move so that the position of the centre of the spindle relative to the machine table will be as shown in Figure 11.3.

On machines which have automatic tool changing facilities it may not be necessary to calculate the coordinates for tool changing. The control system may automatically position the tool spindle at the correct position ready for the tool transfer mechanism when a tool change (miscellaneous code M06) is programmed. Where it is necessary to calculate the coordinates it is essential that the spindle head is moved so that there is adequate clearance for the transfer mechanism to operate without interference.

11.6 Construction of part program 1

The part program is usually written on special work sheets. However, here each block is printed with its explanation following.

%*
  %   program start and tape rewind stop character
  *   end of block symbol

N1G00G71G90X300.0Y−100.0S950T1M06*
In this block the spot drill is loaded and the spindle speed set.
  N1   sequence (block) number 1
  G00   places the machine in the positioning mode (maximum feed rate)
  G71   specifies input to be in metric units
  G90   defines absolute programming
Note: G00, G71 and G90 are modal.
X300.0Y−100.0 are the X and Y coordinates of a safe position for tool changing.
  S950   states required spindle speed
  T1   tool number 1 to be used (spot drill)
  M06   selects tool changing facility
Note: it is not necessary to program a Z coordinate, as the Z axis will automatically move to a fully retracted position when M06 is programmed for tool changing.
  *   end of block symbol

N5X29.0Y66.0Z2.0M03*
Positioning operation.
  N5   sequence (block) number 5
X29.0Y66.0Z2.0 are the coordinates of position of centre of hole A; movement will occur on the X and Y axes first and then on the Z axis to Z2.0. The tool will advance until its point is 2 mm above the surface of the work after moving on the X and Y axes, all at maximum feed rate. It is advisable for there to be a small clearance (the approach) between the end of the tool and the surface of the work to provide a distance for the tool cutting conditions to be established before the tool starts cutting. It can be dangerous for the tool to advance at a rapid feed rate on to the surface of the work.
  M03   spindle rotation clockwise
  *   end of block symbol

N10M08*
  M08   flood coolant on
Normally only one M code can be entered in one block, and the coolant must be on before drilling starts.

N15G01Z−5.0F190*
Spot drilling operation
  N15   sequence (block) number 15
  G01   places the machine in linear interpolation mode, requires a feed rate to be programmed

Z−5.0 tool point advances so that its point is at a depth of 5.0 mm below the surface of the work

F190 feed rate of 190 mm/min is modal and remains active for all operations until changed

* end of block symbol

N20G00X121.0Y14.0Z2.0*

Positioning operation.

N20 sequence (block) number 20

G00 places the machine in positioning mode

X121.0Y14.0Z2.0 are the coordinates of hole B. Because the tool has penetrated the work, it will retract to 2 mm above surface of work at maximum feed rate *before* moving on the X and Y axes. The 2 mm will provide a clearance so that the tool does not touch the surface of the work when moving at the fast feed rate on the X and Y axes. The table will move to (X121.0, Y14.0) *after* moving on the Z axis

* end of block symbol

N25G01Z−5.0*

Spot drilling operation: this operation is the same as N15.

G01 cancels G00 and it is necessary to enter a Z coordinate. The feed rate previously entered in N15 is acceptable and does not need to be changed.

* end of block symbol

N30G00G71G90X300.0Y−100.0S1200T2M06*

Tool changing operation

When M06 is programmed both the spindle is stopped and the coolant turned off automatically by the control system.

The 8 mm drill is loaded and the spindle speed changed to 1200 rev/min.

It is advisable to program G00G71G90 at all tool changes as it may be necessary to advance to the block after changing a broken tool.

Except for the spindle speed and tool number, the remainder of the block is the same as N1.

* end of block symbol

See Figure 11.6 for the sequence of movements in blocks N1 to N30.

N35X29.0Y66.0Z2.0M03*

Positioning operation.

The point of the 8 mm drill is positioned 2 mm above the centre of hole A.

M03 spindle start

* end of block symbol

N40M08*

Flood coolant on.

N45G01Z−12.3F240*

Drilling operation.

In this block the 8 mm drill is fed so that the parallel portion of the hole extends to a depth of 10 mm below the surface of the work; the additional 2.3 mm is the point angle allowance.

The feed rate is changed to 240 mm/min before cutting starts.

* end of block symbol

N50G00X121.0Y14.0Z2.0*

Positioning operation.

The 8 mm drill first retracts to 2 mm above the surface of the work before the table moves until the drill is positioned above the centre of hole B.

* end of block symbol

N55G01Z−12.3*

Drilling operation.

This is a machining operation where hole B is drilled to the required depth (Z−10.0), as explained in N45. The feed programmed in block N45 is active and acceptable.

* end of block symbol

N60G00G71G90X300.0Y−100.0S500T3M06*

Tool changing operation.

The 20 mm end mill (T3) is loaded and the spindle speed changed to 500 rev/min. The spindle is stopped and coolant turned off automatically by the control system. See block number N30 for reasons for G00G71G90.

* end of block symbol

The sequence of movements in blocks N35 to N60 is essentially the same as in blocks N1 to N30, shown in Figure 11.6.

N65X5.0Y−12.0Z2.0M03*

Positioning operation.

In this operation the centre of the 20 mm end mill is positioned so that the left-hand step can be cut. The tool is positioned so that there

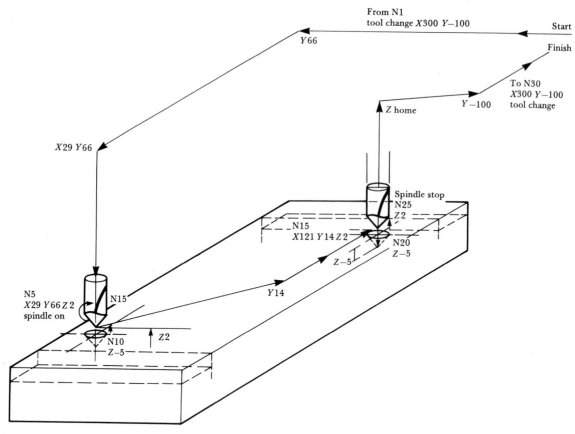

Figure 11.6 *Sequence of movements for drilling, blocks N1 to N30. Also blocks N35 to N60, except Z-12.3 instead of Z-5.0*

is a clearance of 2 mm between the outside diameter of the milling cutter and the edge of the work. This is for safe working practice and will prevent any possibility of the tool feeding into the work at rapid feed rate. If the size or finish of the step was important it would have been advisable to take a roughing and a finishing cut. A suitable X coordinate for a roughing cut would be X4.0, leaving a 1 mm finishing cut. The need for a finishing cut is dependent on the effect of the cutting action on the work material during material removal and not on lack of accuracy of positioning control system.

M03   spindle start
*   end of block symbol

N70M08*
Flood coolant on.
N75G01Z−6.0F250*
Machining operation.
In this block the end mill is fed down to the depth required (assuming that a finishing cut is not necessary) and the feed rate for the end mill is changed to 250 mm/min. Although the tool is clear of the work, for safe working practice it is advisable that the tool is fed down at a cutting feed rate.
*   end of block symbol
N80Y81.0*
Milling operation.
The left-hand step is cut by the end mill being fed at 250 mm/min along the Y axis until its

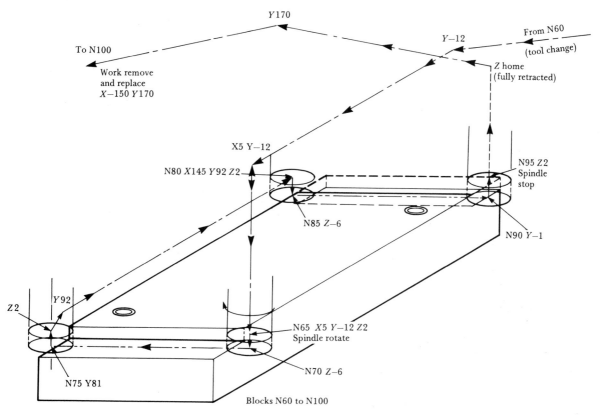

Figure 11.7 Sequence of movements for milling, blocks N60 to N100

centre is 1 mm clear of the edge of the work; this ensures that the full diameter of the cutter has cleared the edge of the work. The feed rate programmed in N75 is active and acceptable.

*   end of block symbol

N85G00X145.0Y92.0Z2.0*

Positioning operation.

The end mill will first retract to 2 mm above the surface of the work, and then move so that its centre is 10 mm past the edge of the right-hand step and 12 mm away from the edge of the work, all at maximum feed rate.

*   end of block symbol

N90G01Z−6.0*

Machining operation.

The end mill is fed down to 6 mm below the surface of the work.

*   end of block symbol

N95Y−1.0*

Milling operation.

The right-hand step is cut by the end mill being fed at 250 mm/min along the Y axis until its centre is 1 mm clear of the edge of the work. The feed rate programmed in N75 is active and acceptable.

*   end of block symbol

N100X−150.0Y170.0M02*

Work changing operation.

This is the end of the program as indicated by M02; the Z axis will move so that the spindle is fully retracted before the table moves to a safe position for changing the work.

The miscellaneous function M02 will cause the tape to be rewound to the start or cause the control system to move automatically to the commencement of the program. M02 will also

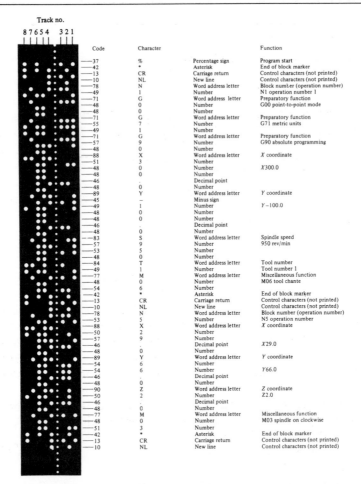

Track no.	Code	Character		Function
	37	%	Percentage sign	Program start
	42	*	Asterisk	End of block marker
	13	CR	Carriage return	Control characters (not printed)
	10	NL	New line	Control characters (not printed)
	78	N	Word address letter	Block number (operation number)
	49	1	Number	N1 operation number 1
	71	G	Word address letter	Preparatory function
	48	0	Number	G00 point-to-point mode
	48	0	Number	
	71	G	Word address letter	Preparatory function
	55	7	Number	G71 metric units
	49	1	Number	
	71	G	Word address letter	Preparatory function
	57	9	Number	G90 absolute programming
	48	0	Number	
	88	X	Word address letter	X coordinate
	51	3	Number	
	48	0	Number	X300.0
	48	0	Number	
	46	.	Decimal point	
	48	0	Number	
	89	Y	Word address letter	Y coordinate
	45	–	Minus sign	
	49	1	Number	Y –100.0
	48	0	Number	
	48	0	Number	
	46	.	Decimal point	
	48	0	Number	
	83	S	Word address letter	Spindle speed
	57	9	Number	950 rev/min
	53	5	Number	
	48	0	Number	
	84	T	Word address letter	Tool number
	49	1	Number	Tool number 1
	77	M	Word address letter	Miscellaneous function
	48	0	Number	M06 tool chante
	54	6	Number	
	42	*	Asterisk	End of block marker
	13	CR	Carriage return	Control characters (not printed)
	10	NL	New line	Control characters (not printed)
	78	N	Word address letter	Block number (operation number)
	53	5	Number	N5 operation number
	88	X	Word address letter	X coordinate
	50	2	Number	
	57	9	Number	
	46	.	Decimal point	X29.0
	48	0	Number	
	89	Y	Word address letter	Y coordinate
	54	6	Number	
	54	6	Number	Y66.0
	46	.	Decimal point	
	48	0	Number	
	90	Z	Word address letter	Z coordinate
	50	2	Number	Z2.0
	46	.	Decimal point	
	48	0	Number	
	77	M	Word address letter	Miscellaneous function
	48	0	Number	M03 spindle on clockwise
	51	3	Number	
	42	*	Asterisk	End of block marker
	13	CR	Carriage return	Control characters (not printed)
	10	NL	New line	Control characters (not printed)

Figure 11.8 Punched tape

result in the spindle being stopped and the coolant turned off.

*   end of block symbol

See Figure 11.7 for the sequence of movements in blocks N65 to N100.

There is no particular reason why the holes should be drilled first and the milling second; the milling could quite effectively have been carried out first. It is obvious that the hole positions should be spot drilled before being drilled. If the component is large the position at which milling commences should be considered; to save time the starting point should be nearest to the tool change position.

A copy of a section of the tape for the first two blocks is shown in Figure 11.8. An explanation of the decoding of tape is given in Chapter 7.

11.7 Example program 2

The component is shown in Figure 11.9, and it is assumed that the same machine is used as for example program 1. The program is suitable for the Bridgeport Series 1 CNC Machine with BOSS 5 or 6 control (it may be necessary to remove the M03 and M08 codes, which may not be implemented on this machine). Details of the operation sequence, tools to be used, spindle

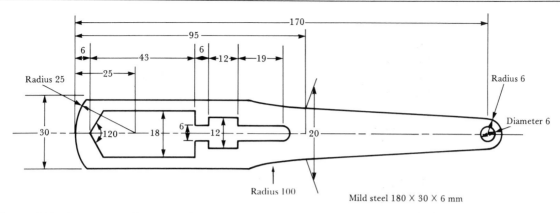

Figure 11.9 *Component for example program 2*

speeds etc. will be found on the operation sheet in Figure 11.10.

The component has been chosen to illustrate the calculation of arc centre offsets, as well as to provide an example of programming circular interpolation movements, subroutines (macros), and additional preparatory and miscellaneous functions. The drawing requires sharp corners in the internal shape; this is a practical impossibility by end milling. Therefore small holes are drilled that slightly overcut into the corners, to ensure that there is no radius left in the corners after milling. This type of alteration in design would have to be approved by the designer.

In order to complete one component on one set-up, the first operation requires the blank to be gripped in stepped vice jaws and located against an end stop. The drilling and milling of the internal shape is completed. For the milling of the external form, the work is removed from the vice and located on a holding block, using the 18 mm wide rectangular section of the internal shape for location. The component is clamped on to the holding block with a top plate and also through the 6 mm diameter hole at the end of the component. The holding block is then gripped in the lower half of the vice jaws, locating against the end stop. The holding block is high enough for the component to be raised clear of the vice jaws to enable the end mill to pass around the work without any interference.

Calculation of position of corner clearing holes

It is necessary to drill holes of 3 mm at every internal corner such that the hole overlaps into the corner by 0.02 mm, providing the clearance required in the corner. For programming purposes the coordinates of the centres of the holes have to be determined.

The calculations of the coordinates of the corner (labelled Pt1) in Figure 11.11 are:

X axis: $x_1 + 6 = (9/\tan 60°) + 6 = 11.196$
Y axis   6.0

Using Figure 11.12, the coordinates of the hole centre (labelled Pt2) are:

X axis: $x_2 + 11.196 = (1.48 \cos 60°) + 11.196 = 11.936$
Y axis: $y_2 + 6$ $= (1.48 \sin 60°) + 6$ $= 7.282$

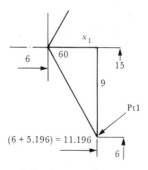

Figure 11.11 *Calculation of corner (Pt1) coordinates*

Operation schedule

Part: Example no. 2   Part no.: Figure 11.9   Work material: mild steel $180 \times 30 \times 6$ mm

Machine: Bridgeport   Control: Boss 5

Details of work holding: stepped vice jaws and holding block

Machine setting points: $X250$ $Y150$

Tool materials: HSS   Cutting speed roughing: 30–80   Finishing: 80

Feed rate roughing:*   Finishing:

* 3 mm drill 0.04 mm/tooth   End mill 6 mm 0.05 mm/tooth (4 teeth)
   6 mm drill 0.08 mm/tooth   20 mm 0.05 mm/tooth (6 teeth)

Op. no.	Details of operation	Feed	Spindle speed	Details of tools
1	Change tool	—	—	Tool no. 1 6 mm spot drill
2	Spot drill 11 holes	240	3000	
3	Change tool	—	—	Tool no. 2 3 mm drill
4	Drill 11 holes	240	3000	
5	Change tool	—	—	Tool no. 3 6 mm drill
6	Drill 2 holes	240	1500	
7	Change tool	—	—	Tool no. 4 6 mm end mill
8	Mill internal form	150 V 300 H	1500	150 = vertical down feed 300 = horizontal feed
9	Tool change and load work auto block	—	—	
10	Mill external form	150	150	
11	Remove work and load new piece	—	—	

Figure 11.10 Operation schedule

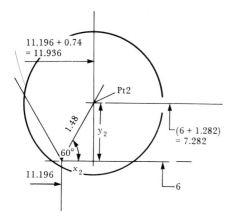

Figure 11.12 *Calculation of centre (Pt2) of 3 mm diameter drill*

The coordinates of the centres of the corner clearing holes at the various internal corners that are required for programming purposes are given in Table 11.1. These coordinates are shown in Figure 11.14, except for point 12.

The rectangular components of the centre of the holes at the 90 degree corners shown in Figure 11.13 are:

X axis: $x_3 = 1.48 \cos 45° = 1.047$
Y axis: $y_3 = 1.48 \sin 45° = 1.047$

Table 11.1 *Coordinates for corner clearing holes*

Point	X coordinate	Y coordinate	Point	X coordinate	Y coordinate
2	11.936	7.282	8	56.047	19.953
3	7.48	15.0	9	65.953	19.953
4	11.936	22.718	10	65.953	10.047
5	47.953	22.953	11	86.0	15.0
6	47.953	7.047	12	170.0	15.0
7	56.047	10.047			

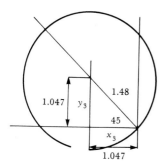

Figure 11.13 *Calculation of centre of 3 mm diameter drill at 90 degree corners*

Calculation of position of centre of end mill

A 6 mm drill will be used to originate a hole at points 11 and 12. To mill out the internal shape a 6 mm diameter end mill is used. Figure 11.15 is used to determine the coordinates of the centre of the 6 mm end mill at the 120 degree corners (points 13 and 16).

The remainder of the path of the centre of the 6 mm end during the milling of the internal shape is calculated in a similar way as above, and the points are given in Table 11.2. These coordinates and the tool path are shown in Figure 11.16.

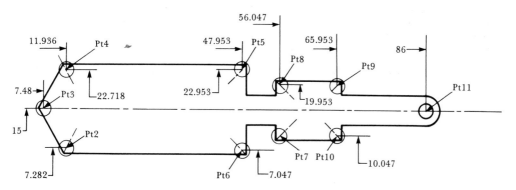

Figure 11.14 *Coordinates of centre of 3 mm diameter holes*

Table 11.2 *Coordinates for end milling*

Point	X coordinate	Y coordinate	Point	X coordinate	Y coordinate
13	9.464	15.0	21	58.0	15.0
14	46.0	15.0	22	58.0	18.0
15	46.0	9.0	23	64.0	18.0
16	12.928	9.0	24	64.0	12.0
17	9.464	15.0	25	58.0	12.0
18	12.928	21.0	26	58.0	15.0
19	46.0	21.0	11	86.0	15.0
20	46.0	15.0			

Calculation of points of tangency

It is necessary to determine the points of tangency on the outside shape. These are the points where the 100 mm radius, and the 6 mm radius at the end of the component, blend with the tapered section. These points (29, 30, 32, 33) are shown in Figures 11.17a–c. The drawing coordinates of these points have to be determined for programming purposes.

First it is necessary to determine the angle at

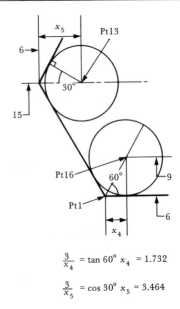

$$\frac{3}{x_4} = \tan 60° \quad x_4 = 1.732$$

$$\frac{3}{x_5} = \cos 30° \quad x_5 = 3.464$$

Figure 11.15 Calculation of centre (Pt13 and Pt16) of 6 mm diameter end mill

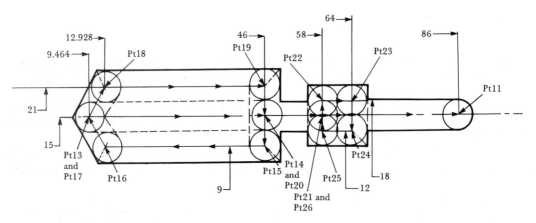

Figure 11.16   Coordinates of path of centre of 6 mm diameter end mill

which the sides slope. Using Figure 11.17a, the angle can be calculated: 4/75 = tan A, and therefore A = 3.053 degrees.

The coordinates of the points of tangency can be calculated using Figure 11.17b. The rectangular components of the point of tangency where the width is 20 mm are:

X axis: $x_6 = 10 \sin 3.053° = 0.533$
Y axis: $y_6 = 10 \cos 3.053° = 9.986$

The rectangular components of the point of tangency shown in Figure 11.17c where the width is 12 mm are:

X axis: $x_7 = 6 \sin 3.053° = 0.32$
Y axis: $y_7 = 6 \cos 3.053° = 5.992$

The drawing coordinates of the points of tangency are therefore as in Table 11.3. These coordinates are shown in Figure 11.20.

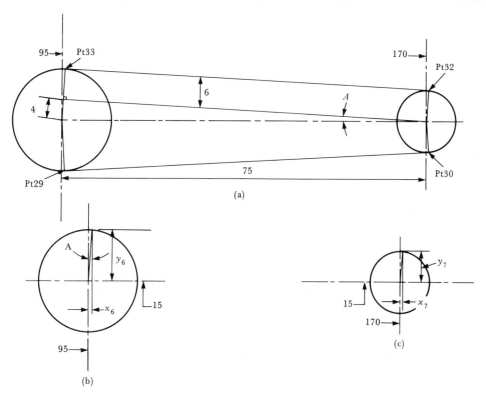

Figure 11.17 Points of tangency

Table 11.3 *Coordinates for points of tangency*

Point	X coordinate			Y coordinate		
28	100.859 − 31.696 =	69.163			0	
29	95.0 + 0.533 =	95.533	15.0 − 9.986 =	5.014	
30	170.0 + 0.32 =	170.32	15.0 − 5.992 =	9.008	
31	170.0 + 6.0 =	176.0	15.0 + 0 =	15.0	
32	170.0 + 0.32 =	170.32	15.0 + 5.992 =	20.992	
33	95.0 + 0.533 =	95.533	15.0 + 9.986 =	24.986	
34	100.859 − 31.696 =	69.163		30.0	
35	25.0 − 20.0 =	5.0		30.0	
36		0	15.0 + 0 =	15.0	
37	25.0 − 20.0 =	5.0		0	

Calculation of arc centre offsets

As can be seen in Figure 11.18, a radial line from the centre of the blending radius is inclined at angle to the normal of 3.053 degrees. The rectangular coordinates of the centre of the blending radius are calculated as follows:

X axis: $x_8 = 110 \sin 3.053° = 5.859$

Y axis: $y_8 = 110 \cos 3.053° = 109.844$

The drawing coordinates of the centre of the blending radius points (38 and 39) are shown in Table 11.4.

It is necessary to calculate the coordinates of the points at which the blending radius meets

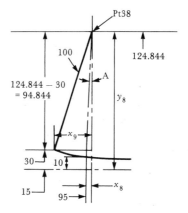

Figure 11.18 Coordinates of art centre offsets

231

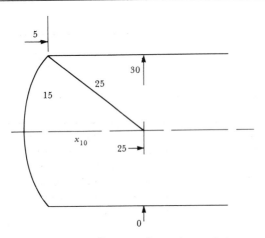

Figure 11.19 *Coordinates of meeting points*

Table 11.4 *Coordinates of centre of blending radius*

Point	X coordinate	Y coordinate
38	95.0 + 5.859 = 100.859	15.0 + 109.844 = 124.844
39	95.0 + 5.859 = 100.859	109.844 − 15.0 = 94.844

the edge of the component. Using Figure 11.18 it can be seen that:

X axis: $x_9 = \sqrt{(100^2 - 94.844^2)} = 31.696$

The drawing coordinates of these points (28 to 34) are shown in Table 11.3.

The rectangular coordinates of the points at which the 25 mm radius at one end of the component meets the edge are shown in Figure 11.19. It can be seen that:

X axis: $x_{10} = \sqrt{(25^2 - 15^2)} = 20.0$

The drawing coordinates of these points (35 to 37) are shown in Table 11.3.

The arc centre offsets are defined as the distances from the start of the arc to the centre of the arc. It is very important to remember that this definition of arc offsets applies to single-quadrant interpolation, and it must be applied accurately. The offsets on the X and Y axes have the addresses I and J respectively. To calculate the offsets it is convenient to subtract the drawing coordinates of the start of the arc from the

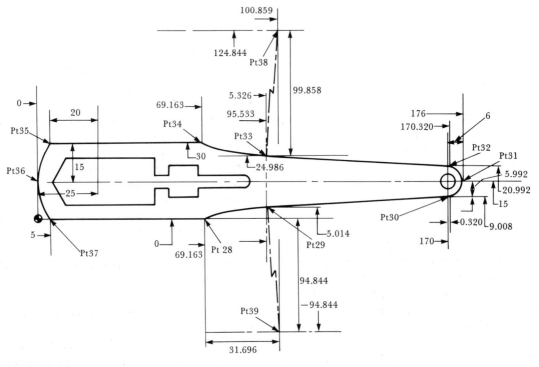

Figure 11.20 *Base line (coordinate) dimensioning*

Table 11.5 *Arc centre offsets*

From point	To point	X coordinates of		Arc offset	Y coordinates of		Arc offset
		start of arc	*centre of arc*	*I value*	*start of arc*	*centre of arc*	*J value*
28	29	69.163 −	100.859 =	31.696	0 −	94.844	= 94.844
30	31	170.320 −	170.0 =	0.32	9.008 −	15.0	= 5.992
31	32	176.0 −	170.0 =	6.0	15.0 −	15.0	= 0
33	34	95.533 −	100.859 =	5.326	24.986 −	124.844	= 99.858
35	36	5.0 −	25.0 =	20.0	30.0 −	15.0	= 15.0
36	27	0 −	25.0 =	25.0	15.0 −	15.0	= 0

drawing coordinates of the centre of the arc. Note that on most control systems the I and J values are unsigned, i.e. have no negative signs. Thus when machining the arcs the offsets are as in Table 11.5.

Figure 11.20 shows all the coordinates for the milling of the outside shape.

11.8 Construction of part program 2

%
Program start character.
N1G00G71G90X200.0Y−100.0S3000T1M06

Tool changing operation. Tool no. 1 spot drill.
N5X11.936Y7.282Z2.0M03

Positioning operation at Pt2. Spindle on; tool 2 mm above work surface.
N10M08

Coolant on; normally only one M code is inserted in one block.
#1

Start of definition of macro 1.
N15G81X11.936Y7.282Z*F240

Start of G81 canned cycle (drilling cycle). Confirmation of X and Y values of Pt2. The Z* is used to indicate that the amount of Z movement has to be defined (entered) when the macro is called.
N20X7.48Y15.0

Rapid positioning move to Pt3, then drill feeds down on Z axis.

N25X11.936Y22.718

Rapid positioning move to Pt4, then drill feeds down on Z axis.
N30X47.953Y22.953

Rapid positioning move to Pt5, then drill feeds down on Z axis.
N35X47.953Y7.047

Rapid positioning move to Pt6, then drill feeds down on Z axis.
N40X56.047Y10.047

Rapid positioning move to Pt7, then drill feeds down on Z axis.
N45X56.047Y19.953

Rapid positioning move to Pt8, then drill feeds down on Z axis.
N50X65.953Y19.953

Rapid positioning move to Pt9, then drill feeds down on Z axis.
N55X65.953Y10.047

Rapid positioning move to Pt10, then drill feeds down on Z axis.
N60X86.0Y15.0

Rapid positioning move to Pt11, then drill feeds down on Z axis.
N65X170.0Y15.0

Rapid positioning move to Pt12, then drill feeds down on Z axis.
$

End of macro may be referred to as *termac*.
=#1Z*6.0

The equal sign (=) calls (activates) macro 1. The Z*6.0 defines the amount of Z movement (6.0 incremental).

N70G00G71G90X200.0Y−100.0S3000T2M06

Tool changing operation G00 cancels G81. Tool no. 2 is 3 mm drill. Coolant and spindle turned off by M06.

N75X11.936Y7.282Z2.0M03

Positioning operation to Pt2. M03 spindle on.

N80M08

Coolant on.

= #1Z*10.0

The equal sign (=) calls (activates) macro 1. The Z*10.0 defines the Z movement (10.0 incremental).

N85G00G71G90X200.0Y−100.0S1500T3M06

Tool changing operation. Tool no. 3 is 6 mm drill.

N90X86.0Y15.0Z2.0M03

Positioning operation to Pt11 to 2 mm above surface.

N95M08

Coolant on.

N100G81X86.0Y15.0Z11.0F240

Start of drilling cycle G81, confirmation of coordinates of Pt11. Drill breaks through. Z11.0 is incremental.

N105X170.0Y15.0

Drilling of hole at Pt12.

N110G00G71G90X200.0Y−100.0S1500T4M06

Tool changing operation G00 cancels G81. Tool no. 4 is 6 mm end mill.

N115X9.464Y15.0Z2.0M03

Positioning operation to Pt13, to 2 mm above work surface.

N120M08

Coolant on.

#2

Start of definition of macro 2.

N125G01Z*F150

G01 linear interpolation, feed down of 6 mm end mill to Z value to be specified when macro is called.

N130X46.0F300.0

Linear interpolation to Pt14.

N135Y9.0

Linear interpolation to Pt15.

N140X12.928

Linear interpolation to Pt16.

N145X9.464Y15.0

Linear interpolation to Pt17.

N150X12.928Y21.0

Linear interpolation to Pt18.

N155X46.0

Linear interpolation to Pt19.

N160Y15.0

Linear interpolation to Pt20.

N165X58.0

Linear interpolation to Pt21.

N170Y18.0

Linear interpolation to Pt22.

N175X64.0

Linear interpolation to Pt23.

N180Y12.0

Linear interpolation to Pt24.

N185X58.0

Linear interpolation to Pt25.

N190Y15.0

Linear interpolation to Pt26.

N195X86.0

Linear interpolation to Pt11.

$

End of macro.

=#2Z*−3.0

Macro 2 activated; Z*−3.0 specifies Z movement to Z−3.0 absolute.

N200G00X9.464Y15.0Z2.0

Positioning operation to Pt13, 2 mm above work surface.

=#2Z*−6.0

Macro 2 activated; Z movement to Z−6.0 absolute.

N205G00X200.0Y−100.0Z10.0

Positioning operation prior to tool change.

T5//20.0

Specifies cutter diameter of 20 mm.

N210G00G71G90X200.0Y−100.0S500T5M06

Tool changing operation. Tool no. 5 is 20 mm diameter end mill. The component is removed from the vice and secured to a holding block which is clamped in the vice jaws.

N215X0Y0M03

Positions tool at X and Y datums (Pt0), tool fully retracted.

N220G92X0Y−5.52M08

Preset absolute registers; sets zero datum 5.52 mm on Y axis, to accommodate change in position of work due to securing on holding block.

N225G00X−3.0Y−15.0Z10.0

Positions centre of tool clear of work, preparatory to applying cutter diameter compensation.

N230G01Z2.0F150

Tool feeds down to Z2.0; work is held above vice jaws on holding block.

N235G42X−3.0Y−15.0

Establishes cutter diameter compensation to the right of work in the direction of cutting.

N240X−3.0Y0F150

Tool moves so that the edge of the cutter is on Y0.

N245X69.163Y0

Linear interpolation to Pt28.

N250G02X95.533Y5.014I31.696J94.844

Clockwise circular interpolation to Pt29.

N255G01X170.32Y9.008

Linear interpolation to Pt30.

N260G03X176.0Y15.0I0.320J5.992

Counter-clockwise circular interpolation to Pt31.

N265G03X170.32Y20.992I6.0J0

Counter-clockwise circular interpolation to Pt32.

N270G01X95.533Y24.986

Linear interpolation to Pt33.

N275G02X69.163Y30.0I5.326J99.858

Clockwise circular interpolation to Pt34.

N280G01X5.0Y30.0

Linear interpolation to Pt35.

N285G03X0Y15.0I20.0J15.0

Counter-clockwise circular interpolation to Pt36.

N290G03X5.0Y0I25.0J0

Counter-clockwise circular interpolation to Pt37.

N295G01X8.0Y0

Linear interpolation in preparation for cancelling of cutter diameter compensation.

N300G40X8.0Y−15.0

Cancels cutter diameter compensation.

N305G00G90X0Y0Z10.0

Positioning operation to re-establish datum zero.

N310G92X0Y5.52

Resets datum zero to original location.

N315G00X−100.0Y100.0M02

Work changing position; coolant and spindle turned off by M02 which specifies end of program.

E

This is the tape rewind character required on the Bridgeport machine.

Assignments

Figures 11.21–11.27 are of components to be produced on a machine which has the following specification:

(a) Tape format:

N03G02X+/−042Y+/−042Z+/−042
I042J 042K 042F04S4T03M02*

(b) Spindle speed steplessly variable and automatically controlled between 30 and 3000 rev/min

(c) Manual tool change.

For each component, carry out the following:

1 Draw the component dimensioned in the form suitable for the requirements of numerical control. Give reasons for the choice of datum. (All drawing dimensions are in millimetres). metres.)

2 Create a sequence of operations to produce the component, selecting tools and calculating feed rates and spindle speeds. Use the values for feed and cutting speed given in Chapter 9.

3 Produce a part program for the component (the heading of a typical part programming sheet is shown in Figure 11.28).

4 Provide details for the setting of the machine.

5 Draw the tool path using graph paper.

6 Select the required preparatory and miscellaneous functions from the following:

Preparatory functions
G00   point-to-point
G01   linear interpolation
G02   circular interpolation clockwise

G03   circular interpolation counter-clockwise
G04   dwell
G17   circular interpolation X and Y plane
G18   circular interpolation X and Z plane
G19   circular interpolation Y and Z plane
G30   cancel mirror image
G31   reverses programmed direction of X axis
G32   reverses programmed direction of Y axis
G40   cutter diameter compensation off
G41   cutter diameter compensation left
G42   cutter diameter compensation right
G70   inch units
G71   metric units
G80   cancels canned cycles
G81   drilling cycle
G82   drilling cycle with dwell (spot facing)
G83   deep hole drilling
G84   tapping cycle
G85   boring cycle
G86   boring cycle with tool disengagement
G87   deep hole drilling cycle with chip breaking
G88   as G87 plus dwell

G89   boring cycle with dwell
G90   absolute programming
G91   incremental programming
G92   preset absolute registers
G94   feed rate, millimetres per minute
G95   feed rate, millimetres per revolution
G96   cutting speed, metres per minute
G97   spindle speed, revolutions per minute

Miscellaneous functions
M00   Program stop
M01   optional program stop
M02   end program
M03   spindle on clockwise
M04   spindle on counter-clockwise
M05   spindle off
M06   tool change
M07   mist coolant on
M08   flood coolant on
M09   coolant off
M25   spindle home
M30   end of tape

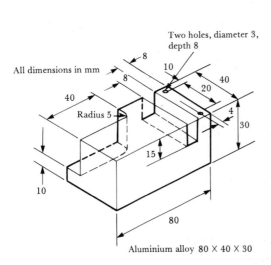

Figure 11.21

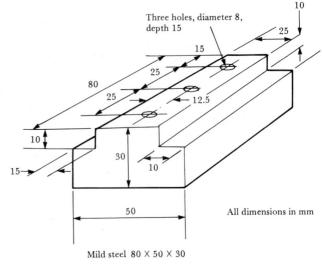

Figure 11.22

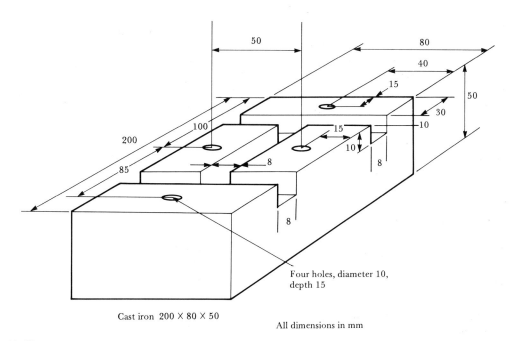

Four holes, diameter 10,
depth 15

Cast iron 200 × 80 × 50

All dimensions in mm

Figure 11.23

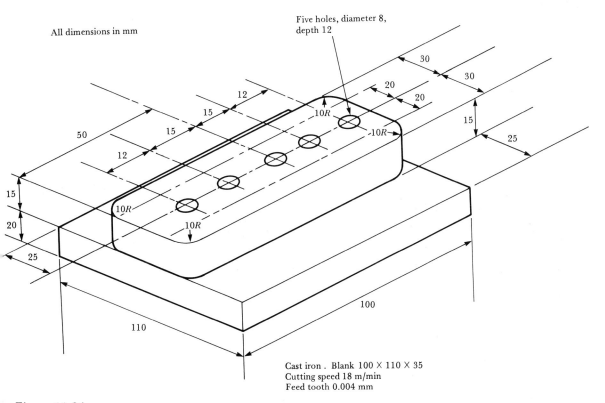

All dimensions in mm

Five holes, diameter 8,
depth 12

Cast iron . Blank 100 × 110 × 35
Cutting speed 18 m/min
Feed tooth 0.004 mm

Figure 11.24

All dimensions in mm

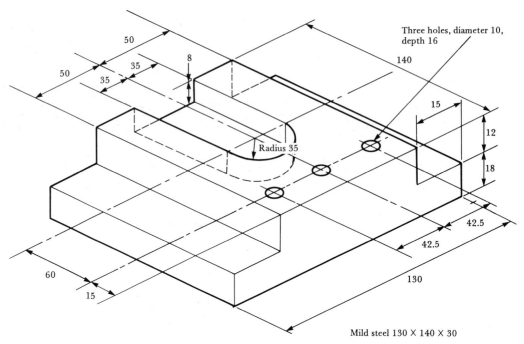

Figure 11.25

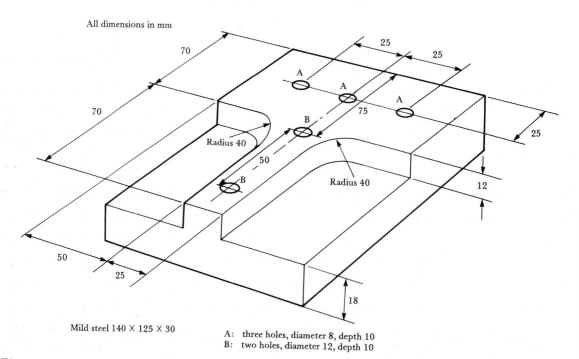

Figure 11.26

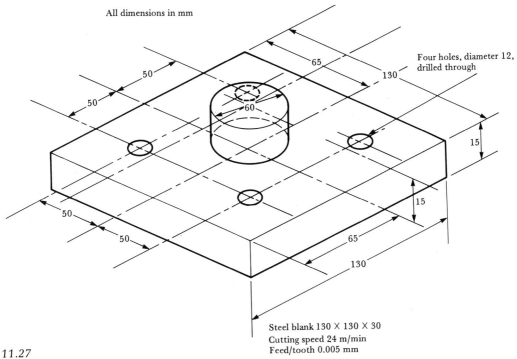

All dimensions in mm
50
50
65
130
Four holes, diameter 12, drilled through
60
15
15
50
50
65
130
Steel blank 130 X 130 X 30
Cutting speed 24 m/min
Feed/tooth 0.005 mm

Figure 11.27

Part programming sheet

Sheet no. ________

Part ________ Part no. ________ Programmer ________

Machine ________ Control ________ Date ________

Operation	Block no.	Preparatory function	Coordinate	Coordinate	Coordinate	Arc centre offset Axis	Arc centre offset Axis	Feed rate	Spindle speed	Tool no.	Misc. function	EOB
	N	G	X	Y	Z	I/J	J/K	F	S	T	M	

Figure 11.28 Typical part programming sheet

Part programming for lathe work

Chapter Twelve

Part programming for lathe work

12.1 Example program 3

In this example of a part program for a lathe, it is assumed that the component shown in Figure 12.1 is to be produced on a lathe which has positioning and continuous path control (linear and circular interpolation) on both the X and Z axes of movement. It is also assumed that the following standard preparatory and miscellaneous functions (G and M codes) listed on the drawing are provided by the control system (these are the minimum that are normally available on most turning machines):

Preparatory functions

G00 point-to-point
G01 linear interpolation
G02 circular interpolation clockwise
G03 circular interpolation counter-clockwise
G04 dwell
G18 circular interpolation X and Z plane
G40 tool tip radius compensation off
G41 tool tip radius compensation left
G42 tool tip radius compensation right
G70 inch units
G71 metric units
G80 cancels canned cycles
G81 parallel turning cycle
G82 parallel turning cycle with dwell
G84 screw cutting cycle
G90 absolute programming
G91 incremental programming
G92 preset absolute registers
G94 feed rate millimetres per minute
G95 feed rate, millimetres, per revolution
G96 cutting speed, metres per minute
G97 spindle speed, revolutions per minute

Miscellaneous functions

M00 program stop
M01 optional program stop
M02 end program
M03 spindle on clockwise
M04 spindle on counter-clockwise
M05 spindle off
M06 tool change
M07 mist coolant on
M08 flood coolant on
M09 coolant off
M30 end of tape

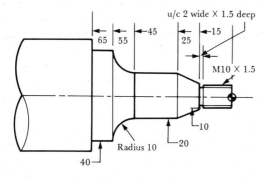

Figure 12.1 Component for example program 3

The program is based on the Audit CNC machine, and can also be used on the Boxford CNC lathes fitted with ANC control. With some minor alterations which are indicated in the explanation section, such as changing the details of the screw cutting operation, the program is suitable for a number of lathes. It is important to note that the cutting tools on the Audit CNC machine operate at the rear of the machine.

The following details must be considered before beginning to write the part program:

(a) Position of the X and Z zero datums
(b) Size of work blanks
(c) The sequence of operations
(d) The tools to be used
(e) Speeds and feeds.

These details are not in order of importance or in any particular sequence, but have to be considered in relation to each other. For example, the position of the zero datums can be affected by the size of blanks and to some extent by the sequence of operations.

12.2 Position of the zero datums

For this example there are only two axes (Z and X) of movement to be considered. With reference to Chapter 4 it can be seen that the Z axis, for a lathe, is collinear (in line) with the main spindle axis and aligned along the bed of the lathe. Movement of the tool along this axis affects the length of the different features of the work. It is convenient to locate the Z zero datum at the end of the workpiece furthest from the chuck (work holder), because this face is readily accessible. This also enables the Z coordinates to be the lengths of the various features of the work, and in absolute programming the units would be entered as negative (minus) values.

The X axis is at right angles to the Z axis and parallel to the cross-slide. Movement of the tool on this axis will result in a change in the diameter of the work being machined. The X zero datum is conveniently located at the centre line of the component because, when working in absolute dimensions, the value of the X coordinates will be the actual radius of the work.

However, a number of controls require the X value to be a diameter value and not a radius; all X coordinates in the entire program will be diameter values.

12.3 Size of work blanks

For this component the work blanks are long enough to be held securely in a collet chuck, and there is no parting-off operation to be carried out. For consecutive components it is essential that the material protruding from the work holder is the same length. To ensure that the correct length protrudes from the face of the chuck, a blank tool held in the tool turret is used as a stop bar.

12.4 Sequence of operations

It must be emphasized that it is possible for any component to be produced in a number of different ways, each method having certain advantages when compared with the other methods. The main criterion in deciding on the method to be adopted, apart from safety of the operator and machine, is that the part must be produced at an economic cost on the appropriate machine tool.

At the start of the setting-up process it will have been necessary for the work material to have been placed in the work holder, and the first component would have been machined during the setting-up. At the end of the machining sequence for the second and additional components, the tool would be withdrawn to a safe position clear of the machining area, so that the work can be removed from the collet chuck in safety. Immediately after each completed workpiece is removed from the work holder, the new work blank would be loaded. Thus the changing of the workpieces is normally the last operation.

Before the start of the machining operations it will be necessary to ensure that the correct length of the work protrudes from the collet. As stated previously, this will be accomplished by using a blank tool as a work stop. The first two operations are to load the blank tool and then to move the blank tool into position with the spindle stopped. For the second operation, the collet would have to be released so that the end of the bar can be brought into contact with the stop. The next operation is to change the tools. If the machine has an automatic tool changing facility then the desired tool number or tool station would have to be entered in the program. To save repeated changing of the spindle speed when turning the different diameters, spindle speeds have been selected which although not ideal are practical. The sequence of operations to be programmed is as follows:

1 Load stop for work, tool no. 1.
2 Adjust work to correct length.
3 Load tool no. 2, facing tool.
4 Face off end of work at Z zero datum.
5 Load tool no. 3, rough turning tool.
6 Rough turn outside diameters.
7 Load tool no. 4, undercutting tool.
8 Form undercut.
9 Load tool no. 5, finishing tool.
10 Finish turn outside diameters.
11 Load tool no. 6, screw cutting tool.
12 Screw cut M10 thread, 1.5 pitch.
13 Remove machined component and load new workpiece.

These numbered operations and other details of tools, speeds and feeds to be used are shown in the operation schedule in Figure 12.2.

12.5 Construction of part program 3

Each block is printed with its explanation following:

%*

Program start character. The asterisk * is the end of block character.

N0001G00G71G90G94X100.0D00Z150.0M00 (T1)*

Tool change operation at point-to-point G00 control.

N0001 sequence number. G00 point-to-point operation. G71 metric units. G90 absolute programming.

G94 requires feed rate in mm/min; by programming G94 it will be possible to move the tool slide under program control when the spindle is not rotating. If G95 (feed mm/rev) had been programmed the spindle would have to be rotating for the tool slide to move. No feed rate need be entered as the next block is a positioning operation, and a feed rate does not have to be entered for a G00 operation.

D00 cancels all tool offsets. The values of the tool offsets are established during the setting up stage, and may be modified as and when required. The tool offsets are stored in the memory or tool register in the control, and would be applied when called up (i.e. entered in the program). It is convenient to use odd numbers for radial offsets (X axis), and even numbers for axial offsets (Z axis).

X100.0Z150.0 are coordinates of a safe tool changing position. Note that the value of the X coordinate is a diameter value.

M00 causes the program to stop to allow the operator to change tools. No speed need be programmed as the spindle has to be stationary for work length adjustment. If automatic tool change facilities were available, M06 would be entered instead of M00. Message displayed in brackets is information for operator intervention: (T1) calls up tool no. 1 which is a blank tool acting as a work stop. After changing the tools the operator causes the program to continue by pressing the restart button on the control panel.

N0005X−2.0D01Z0D02M00*

Positioning operation.

D01 and D02 are tool offsets for tool no. 1 on the X and Z axes respectively.

M00 stops the program automatically to allow the operator to release the collet and move the end of the work forward to contact the stop. After adjusting the length of the work, and tightening the collet, the operator would cause the program to continue by pressing the restart button on the control panel.

N0010G00G71G90G95X100.0D00Z150.0M00(S800T2)*

Tool change operation.

At every tool change it is advisable to enter

Operation schedule

Part: Example no. 3   Part no.: Figure 12.1   Work material: alloy steel 500 × φ50

Machine: Audit CNC   Control: ANC

Details of work holding: collet chuck

Machine setting points: *X* axis = centre line   *Z* axis = end face*

Tool material: cemented carbide   Cutting speed roughing: 50–100   Finishing: 100

Feed rate roughing:0.2   Finishing: 0.1

* Work to protrude 75 mm from chuck face

Op. no.	Details of operation	Feed	Spindle speed	Details of tools
1	Change tool	—	—	Tool no. 1 blank tool
2	Adjust work to stop	—	—	
3	Change tool	—	—	Tool no. 2 facing tool
4	Face off	0.2	800	
5	Change tool	—	—	Tool no. 3 external turning tool
6	Rough turn o/d	0.2	800	
7	Change tool	—	—	Tool no. 4 grooving tool
8	Form groove	0.05	2000	
9	Change tool	—	—	Tool no. 5 finish turning tool
10	Finish turn o/d	0.1	1000	
11	Change tool	—	—	Tool no. 6 screw cutting tool
12	Screw cut thread	1.5	1000	
13	Change work	—	—	

Figure 12.2 Operation schedule

G00, G71, G90 or G91, and either G94 or G95 as applicable. Additionally, at every tool change, all tool offsets which have been called up should be cancelled with D00 to ensure that the correct values are used when they are entered in the program. G95 selects feed rate in mm/rev, required for linear or circular interpolation operations.

X100.0Z150.0 are coordinates of safe tool changing position.

Information for operator intervention is in brackets (S800 is spindle speed in rev/min; T2 is tool no. 2). After changing the tools, the operator selects the spindle speed, starts the spindle, and causes the program to continue by pressing the restart button on the control panel. The spindle speed of 800 rev/min used for the facing operation results in the cutting speeds at the diameters of 5 and 50 mm ranging from 12 to 125 m/min.

N0015X52.0D03Z2.0D04F0.2M08*

Positioning operation.

X52.0Z2.0 coordinates of position of tool.

D03 and D04 are the tool offsets to be applied to tool no. 2 on the X and Z axes respectively. Once called up, the tool offsets will apply for all operations while that tool remains in use or is cancelled by D00.

F0.2 feed rate of 0.2 mm/rev. The feed rate would be stored in memory and would apply until changed for all linear or circular interpolation operations.

M08 coolant on. Normally only one M code can be entered in one line, and it is necessary for the coolant to be on before the machining starts. Spindle speed and spindle on and off are controlled by operator. If the spindle speed can be controlled through the program then M03 (spindle start) would have to be entered in this block and an additional block containing M08 entered in the program.

N0020G01Z0*

Facing operation start.

Tool moved to Z0; the feed previously entered (0.2) is acceptable.

N0025X−1.0M09*

Facing operation finish.

Tool moved across the face of the work to X−1.0 at the feed rate of 0.2 mm/rev previously entered. The movement to X−1.0 will ensure that the face of the workpiece is flat, with no pip left at the centre.

M09 coolant turned off at the end of the operation.

N0030G00G71G90G95X100.0D00Z150.0M00 (S800T3)*

Tool change operation.

Spindle stopped automatically by control when M00 is entered (tool no. 3 at a spindle speed of 800 rev/min). Spindle turned on by operator after changing tools. The spindle speed of 800 rev/min used for turning all the diameters ranges from 27 to 125 m/min.

N0035X41.0D05Z2.0D06M08*

Positioning operation.

Tool positioned 2 mm clear of end of work ready to rough turn a diameter of 41 mm, leaving 1 mm for finishing. Feed of 0.2 mm/rev programmed in N15 acceptable.

M08 coolant on.

N0040G01Z−64.0*

Roughing operation: linear interpolation.

Tool feeds along to leave 1 mm on the face of the step for finishing. The feed entered in N15 is still active and suitable.

N0045G00X42.0*

Positioning operation.

Tool retracts 0.5 mm from work surface.

N0050Z2.0*

Positioning operation.

Tool returns to 2 mm clear of end of work.

N0055X30.0*

Positioning operation.

Tool is 2 mm clear of end of work and advances to diameter of 30 mm.

N0060G01Z−51.0*

Roughing operation: linear interpolation movement only along Z axis.

N0065X40.0Z−54.0*

Roughing operation.

Machining of taper by a two-axis linear interpolation movement. Nose radius compensation is not required for a roughing operation.

N0070G00Z2.0*

Positioning operation.

Tool returns to 2 mm clear of the end of work.

N0075X21.0*
 Positioning operation.
 Tool advances to a diameter of 21 mm.
N0080G01Z−45.0*
 Roughing operation: linear interpolation movement only along Z axis.
N0085G02X40.0Z−54.5I9.5K0*
 Roughing operation.
 Turning of radius. The arc centre offsets I and K are unsigned incremental values from the start of the arc to the centre of the arc. Nose radius compensation is not required for a roughing operation.
N0090G01X42.0*
 Roughing operation to clear tool: linear interpolation movement only along X axis.
N0095G00Z2.0*
 Positioning operation.
 Tool returns to 2 mm clear of end of work.
N0100X11.0*
 Positioning operation.
 Tool advances to a diameter of 11 mm.
N0105G01Z−14.0*
 Roughing operation: linear interpolation movement only along Z axis.
N0110X21.0Z−24.0M09*
 Roughing operation.
 Machining of taper by a two-axes linear interpolation movement.
N0115G00G71G90G95X100.0D00Z150.0M00 (S2000T4)*
 Tool change operation.
 Tool no. 4 at a spindle speed of 2000 rev/min; this results in a cutting speed of 47 m/min at 7.5 mm diameter.
N0120X14.0D07Z−15.0D08M08*
 Positioning operation.
 Tool advances to a diameter of 14 mm and to Z−15 mm.
N0125G01X7.5F0.05*
 Groove forming operation.
 Feed reduced to 0.05 mm/rev.
N0130G00X15.0*
 Positioning operation.
 Tool retracts to 15 mm diameter.
N0135G00G71G90G95X100.0D00Z150.0M00 (S1000T5)*
 Tool change operation.

Tool no. 5 at a spindle speed of 1000 rpm. This results in the cutting speed ranging from 15 to 81 m/min. The technique of applying nose radius compensation is detailed in Section 12.6.
N0140X10.0D09Z2.0D10F0.1M08*
 Positioning operation.
 Tool advances to a diameter of 10 mm and to Z2 mm. For finishing operations, feed rate is reduced to 0.1 mm/rev.
N0145G01Z−15.0*
 Finishing operation: linear interpolation.
N0150X20.0Z−25.0*
 Finishing operation: linear interpolation machining of taper.
N0155Z−45.0*
 Finishing operation: linear interpolation movement only along Z axis.
N0160G02X40.0Z−55.0I10.0K0*
 Finishing operation: turning of radius.
N0165G01Z−65.0*
 Finishing operation: linear interpolation movement only along Z axis.
N0170X52.0M09*
 Finishing operation to clear tool: linear interpolation movement only along X axis.
N0175G00G71G90G95X100.0D00Z150.0M00 (S1000T6)*
 Tool changing operation.
 Tool no. 6 for screw cutting at 1000 rev/min; this results in a cutting speed of 16 m/min. The limiting factor is the movement of the tool along the bed of the machine. At 1.5 mm/rev (pitch) the rate of tool movement is 1.5 m/min.
N0180X14.0D11Z2.0D12M08*
 Positioning operation.
 Tool advances to a diameter of 14 mm and to Z2 mm.
N0185G84;1.624;16.0;0.51;1.5;0;0;3;0*
 Screw cutting operation using G84 which is a canned cycle for screw cutting provided on the Audit by the ANC control (see Chapter 9 for full details).
 1.624 is the double depth of thread diameter value.
 16 is the incremental length of movement from the start position (Z2.0) to the undercut.
 0.51 is the depth of cut for the first pass,

calculated from $C = A/\sqrt{N}$, where C is depth of cut for the first pass, A is thread depth (diameter value) (=1.624), and N is number of passes (=10). The number of passes was selected as 10 as a practical value, consideration being given to having a reasonable first depth of cut and not spending too much time. 1.5 is the lead (pitch) of the thread.

The next two zeros are the amount of taper and pull-out length respectively. Since a groove has been formed, a pull-out length is not required. The tool will pull out at rapid traverse and return to start position.

3 is the number of clean-up passes; 3 is selected as a reasonable number.

The final zero indicates plunge feed; if it had been decided to have a compound infeed a figure of 60 would have been programmed. The amount of spindle speed is controlled by the operator, but the synchronization of rotational movement of the work and axial movement of the tool required for screw cutting is obtained through the transducers mounted on the spindle and tool carriage.

N190G00X100.0Z100.0M30*

The tool moves to a safe position to allow the operator to change the work. M30 is the end of program. The machine stops after all the commands contained in this block, and the program returns to the start of the program ready for the next component.

Figure 12.3 is a copy of the screen dump of the graphical simulation of the tool movements.

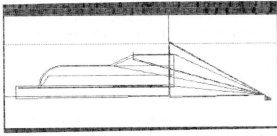

Figure 12.3 Screen dump of tool path

12.6 Tool tip radius compensation

Tool tip radius compensation (TTRC) is only necessary for finishing operations where ac-

curacy of size is of prime importance. A fuller explanation of the reasons for TTRC is given in Chapter 9.

The following section on the technique of applying tool tip radius compensation is based on information in the manual supplied with the Audit CNC machine with ANC control, and only applies to that machine and control. However, changes to a program will have to be made for other machines and controls; the programming manual issued with the particular control must be studied to obtain the correct programming procedure.

When applying TTRC on the Audit CNC machine, the tool path controlled is the path of the centre of the circle at the tool tip. The X and Z coordinates to be entered in the program are the same as when TTRC is not applied. The control automatically adjusts the position of the tool tip by the amounts $X = 2R$ on the X axis $Z = R$ on the Z axis, where R is the tool tip radius. This repositioning would result in a change of absolute dimensions in X and a positional error in the Z axis. To counter the adjustment made by the control to the position of the tool, the tool offsets established when setting up the tool must be edited by equal and opposite values, referred to as TTRC preset values. Alternatively, an additional pair of tool offsets can be added to the tool register, which are called up for those operations where TTRC is to be applied. The actual value of the TTRC preset will be influenced by whether the tool is to the right or left of the work when looking in the direction of the feed movement, and also by whether the operation is outside turning, facing or boring.

When the tool is to the left of the workpiece when looking in the direction of feeding, the preparatory function (G code) G41 must be entered in the relevant block. Code G42 must be entered when the tool is to the right of the workpiece when looking in the direction of feeding. As stated previously, the cutting tools on the Audit CNC machine operate at the rear of the machine. Consequently during normal turning on the outside diameter, the tool is to the right of the work when looking in the direction of feeding. To cancel the TTRC code G40 has to be entered; tool offsets should also be cancelled at the same time. In the same block as when activating TTRC (entering G41 or G42) it is essential to program a Z axis movement which leaves the tool clear of the work. It is recommended that a linear interpolation movement be used with a fast feed rate and not a point-to-point operation.

For the example, the tool offset in the tool register for the X axis has to be edited by the TTRC preset value of −2R, and for the Z axis the tool offset has to be edited by the TTRC preset value of −R. These tool offsets are given the numbers D19 and D20 respectively in the tool register and program.

In addition to these tool offsets, it is necessary to enter the size of the radius at the tool tip as a tool offset in the tool register; the number selected for this example is D18. This offset has to be entered in a block before it has to be used, together with code G42. The changes in the program from block N0135 are as follows:

N0135G00G71G90G94X100.0D00Z150.0M00 (S1000T5)*

    Tool change operation.
    Tool no. 5 at a spindle speed of 1000 rev/min.

N0140X10.0D19Z5.0D20M08*

    Positioning operation.
    Tool advances, and the required tool offsets (D19 and D20) are applied.

N0141G01F1000*

    This block is required to specify linear interpolation and to enter the fast feed rate.

N0142G42D18Z2.0*

    Activating TTRC, and calling up the offset (D18) where the tip radius is stored.

N0145X10.0Z−15.0F100*

    Linear interpolation operation at a feed rate of 100 mm/min: this is equivalent to 0.1 mm/rev.

The remainder of the blocks up to N0160 inclusive would remain the same, and then block N0165 has to be changed and N171 and N172 would be added.

N0165G01Z−64.0*

    Finishing operation: linear interpolation movement only along Z axis.

N0166G02X42.0Z−65.0I1.0K0*

    Circular interpolation movement so that the tool can move smoothly through the change of direction required at the sharp internal corner. A blending radius of 1 mm is created in the corner.

N0170X52.0*

    Finishing operation to clear tool: linear interpolation movement only along X axis.

N0171Z−64.0M09*

    In this block the tool is moved out of contact with the work as a precaution to ensure that there will be no conflict with the work when TTRC is cancelled.

N0172G40D00*

    TTRC and tool offsets cancelled.

The remainder of the program stays the same. It can be seen that there are a few differences to be made to a program when TTRC is applied. One important consideration is that sharp internal corners requiring an abrupt change of direction cannot be produced and that special care has to be taken in specifying workpiece contour definition.

Assignments

Figures 12.4–12.6 are of components to be produced on a machine which has the following specification:

(a) Tape format:

    N03G02X+/−042Z+/−042I042K 042F04S4T03M02*

(b) Spindle speed steplessly variable and automatically controlled between 30 and 3000 rev/min

(c) Manual tool change.

For each component, carry out the following:

1 Draw the component dimensioned in the form suitable for the requirements of numerical control. Give reasons for the choice of datum.

2 Create a sequence of operations to produce the component, selecting tools and calculating feed rates and spindle speeds. Use the

values for feed and cutting speed given in Chapter 9.

3 Produce a part program for the component.
4 Provide details for the setting of the machine.
5 Draw the tool path using graph paper.
6 Select the required preparatory and miscellaneous functions from the following:

Preparatory functions

G00 point-to-point
G01 linear interpolation
G02 circular interpolation clockwise
G03 circular interpolation counter-clockwise
G04 dwell
G33 thread cutting constant lead
G40 tool tip radius compensation off
G41 tool tip radius compensation left
G42 tool tip radius compensation right
G70 inch units
G71 metric units
G80 cancels canned cycles
G81 parallel turning cycle
G82 parallel turning cycle with dwell
G84 screw cutting cycle
G90 absolute programming
G91 incremental programming
G92 preset absolute registers
G94 feed rate, millimetres, per minute
G95 feed rate, millimetres per revolution
G96 cutting speed, metres per minute
G97 spindle speed, revolutions per minute

Miscellaneous functions

M00 program stop
M01 optional program stop
M02 end program
M03 spindle on clockwise
M04 spindle on counter-clockwise
M05 spindle off
M06 tool change
M07 mist coolant on
M08 flood coolant on
M09 coolant off
M30 end of tape

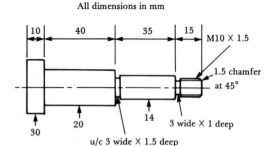

All dimensions in mm

Figure 12.4

Mild steel bar 30 diameter

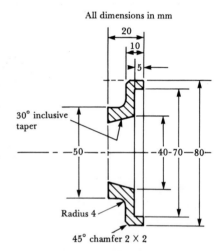

All dimensions in mm

Figure 12.5

Aluminium alloy billet
25 thick × 80

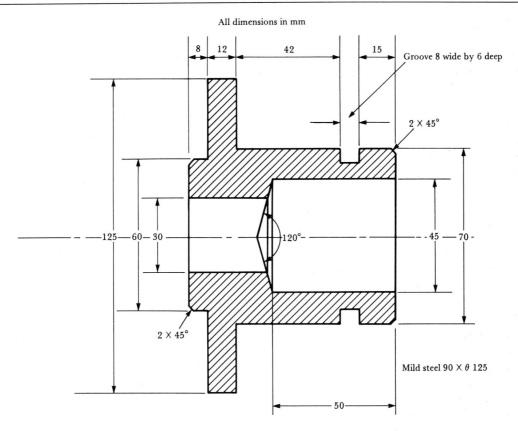

All dimensions in mm

Groove 8 wide by 6 deep

2 × 45°

2 × 45°

120°

Mild steel 90 × θ 125

Figure 12.6

Chapter Thirteen

Computer-aided part programming

13.1 Computer-aided programming

As explained in Chapter 7, the majority of the computers used in the control unit are specifically designed (dedicated) to utilize the information contained in the part program for the operation of the machine tool.

Using manual data input (MDI), if previously created word addressed part programs had not been stored and are fairly short it is possible to enter (type in) the programs at the keyboard of the control unit of most computer numerically controlled machines. This technique (MDI) of entering part programs in word address format can be extremely useful but its use is limited and not very practical for other than short programs.

One of the facilities provided with CNC machines is that it is possible to edit part programs stored in the memory of the computer using the keyboard of the control unit. One advantage of this facility is that part programs already created by any method can be optimized as machining conditions permit. The edited version of the program can then be saved on punched tapes or magnetic discs or transferred to the memory of a separate computer. This technique of entering part programs at the control, as above, only uses the computer in the control unit and not a separate computer which is used for computer-aided programming. For computer-aided programming it is more convenient to use computers which are desktop units and not part of the control unit. The name

'desktop computer' indicates a general purpose computer which can be used for many different applications. For example, when manually creating a part program in word address format, it may be necessary to calculate the coordinates of a large number of change points along a tool path which will define the shape of the component. For components requiring relatively few coordinates, an electronic calculator can effectively be used for determining the values of the point. However, for complex parts requiring the calculation of a large number of coordinates, desktop computers with special software considerably reduce the work. Provided software is available, desktop computers can be used in a number of other ways in computer-aided programming:

(a) Using a computer as a text editor, and entering the part program directly in word address format, and either producing a punched tape or down loading the program direct to the machine.

(b) Entering a part program into a desktop computer in word addressed format, and with the use of special computer software the work and tool movements resulting from the input details are shown graphically on the screen after every block, or after the complete part program is entered. When the part program is seen to produce the correct movements a punched tape is produced or the program is down loaded direct to the machine.

(c) Using part programming languages such as APT.

(d) Using graphical techniques to produce the shape of the component on the screen, the part program is produced within the computer with computer-aided machining (CAM) programs. This method can be referred to as Graphical Numerical Control (GNC).

(e) If the CNC machine tool has been fitted with a special control unit with suitable software; computer-aided programming can take place directly at the machine tool using conversational programming techniques. This method is also known as 'interactive programming' or 'dialogue programming'.

It is important to remember that a computer will attempt to follow input instructions exactly, and has no power or ability to evaluate the accuracy or practical logicality of the instructions. When expert system software has been developed for computer-aided machining or computers have been developed with the ability to make value judgements it may then be possible to depend on the computers to check whether input information is practical.

Expert systems are 'knowledge-based' programs in which information contributed by experts is available for use and provides guidance in particular applications. The knowledge is normally held in the form of rules such as 'If (condition) then (requirement)' and is manipulated by a reasoning mechanism which relates these general rules to particular cases.

13.2 Using the computer as a text editor

This is the simplest way to use a computer for part programming, and does not require any programs specific to the numerical control of machines. However, to use the computer for text editing it has to have either word processing software installed or a text editor program that is also used for writing batch files etc. This technique of using the computer is possible because most of the word processors and text editors use the same ASCII (American Standard Code for Information Interchange) codes to store the letters, numbers and other characters in the text, as those used on control systems for numerically controlled machines.

It is important to note that the part programmer first creates the program using appropriate programming sheets. The complete information (part program) for the manufacture of the component is typed in, and treated as text. The letters and numbers are shown on the visual display unit (screen) of the computer as the program is entered. Any errors observed can be eliminated, or the information can be edited before the program is stored. When the program is seen to be correct it can be stored on magnetic discs or sent direct to the machine, or transmitted to a tape punch; the computer only being used as a link to the method used for storing the program.

The special tape punch units which can be interfaced to the computer are much smaller than a teletypewriter but operate at a faster speed. However, the tape punch units do not usually have a paper printout facility, and it may be necessary to print out the part program on a separate printer connected to the computer.

There are a number of advantages of using a computer as a text editor, such as:

(a) The part program can be stored in the computer's memory or on disc.

(b) It is far more convenient to edit the part program before the tape is punched.

(c) A fresh punched tape can be produced each time the component has to be set up.

(d) The problems of identifying and storing the tapes are overcome, and new tapes are clean and not torn or damaged.

(e) It is possible to store an unfinished program and return to it later.

Special care has to be taken when interfacing the desktop computer to the control unit at the machine or to the tape punch, as there are a number of ways of making the electrical connections. One widely used system of electrical connections is the Electronic Industries Association (EIA) communication standard RS232C. This is serial interface, details of which can be obtained from standard reference books on data transmission. There can be up to eight or more wires in the linking cable, but the minimum connections required are:

(a) A wire for transmitting data from the computer to the machine (TXD)
(b) A wire for sending data from the machine to the computer (RXD)
(c) A wire to act as a common return (signal ground).

To send direct to the machine it is necessary for the control units of the machine tool to have an RS232C interface fitted as part of the electrical circuitry. The speed of transmission of the data has to be set for both transmitting from the computer and receiving by the control unit. The speed of transfer is measured in bauds, and the transfer rate can be set to values ranging from 110 to 9600 bauds. It is also necessary to ensure that the format of the byte being sent i.e. the number of data bits, odd even or null parity and number of stop bits is what the computer in the machine tool control unit is set up to receive.

13.3 Using the computer as a graphical simulator

The essential difference between using the computer as a text editor and using a computer as a graphical simulator is special computer software has to be available. This software will accept part programs written in word addressed format, and then show by graphical representation on the screen what the resulting movements of the tool or work would be when the instructions are input. Obviously it is possible for the part program to be edited.

The main advantage of this technique is that the software program enables the part program to be proved by using the machining instructions (X and Y movements etc.) to simulate the movement of the work or tool on the monitor screen or on a plotter. There are software programs which will show what the movement of the work or tool would have been after every block of information or after the complete part program has been entered. When the programmed movements are correct, the part program can be transmitted as explained in Section 13.2.

It is extremely important that the part programmer has a very clear idea of how and where the work will be held, because although the movements may be correct it is possible for the tool to clash with the work holder during rapid non-cutting movements.

13.4 Part programming languages

Programming languages were first developed in the late 1950s and early 1960s. They require special software, some of which is very expensive. This software contains routines for translating or handling the input information on the component and compiling (arranging) it into a sequential list of cutter movements to produce the required component. To run the part programming languages the computer has to have a large memory capacity. When the languages were first introduced the only computers available were physically very large and very expensive, and were installed in large firms or institutions. These provided a bureau-type service where part programs written in a particular format were produced on punched tapes or magnetic tapes for input to the machine tools.

The programming languages were generated so that part programs could be produced which would have been difficult or time-consuming to create manually. The part programmer has to be trained to use the language, as the information must be entered in an exact and specific form. All part programming language requires information which consists of such details as a description of the shape of the component, the path to be followed by the tool or work, and operating information.

There are two types of part programming language:

(a) Special machine or computer oriented languages, where the information is entered using special symbols. These require smaller computers but are limited in application.

(b) Universal i.e. machine or computer independent languages, where the information is input in English-like statements.

There are a number of universal part programming languages (computer programs) which are all fundamentally of a similar type of format. The names of the languages are acronyms of their full title, such as:

APT   Automatically Programmed Tools
CINAP   Cincinnati Numerical Automatic Programming
PICNIC PERA Instruction Code for Numerical Control (PERA is the Production Engineering Research Association of Great Britain)

APT (Automatically Programmed Tools)

This was one of the first of the part programming languages, and is generally accepted as being the most powerful. APT is reputed to have taken over one hundred man years (one hundred men one year) to develop and write. It is probably the most extensively used as it is able to cover a very wide range of applications. An APT part program requires the information in three sections:

1 *Operating information*
   (a) Identify the component (name or number for reference purposes).
   (b) Specify the machine tool and control to be used. (This is necessary as each control system requires a different post processor.)
   (c) Specify size and shape of the tools to be used, or select the tools from a tool library.
   (d) Specify the acceptable tolerance zone. (This influences the accuracy to which the coordinates have to be calculated.)
   (e) Detail the machine tool operating functions required, i.e. speed, feed and ancillary requirements such as coolant (flood or mist) etc. (required for specifying G or M codes).

2 *Geometric information*
   (a) Label all points, lines, curves and surfaces on the part.
   (b) Specify the 'profile data' or give the 'geometric statements' of the component. This is defining all points, lines, curves and surfaces, i.e. ellipses, parabolas etc., in programming language. These are specified in Cartesian coordinates and are quoted in reference to a set-up point or another convenient point.

Figure 13.1 *APT definition of a line joining two circles*

The general format of a definition of a line or surface is:

label (or symbol) = geometric surface/data and modifiers

Once the line or surface has been defined, it can be referred to by its label or symbol. An example of the definition of a line joining two circles (shown in Figure 13.1) is:

L3 = LINE/RIGHT,TANTO,C5,RIGHT, TANTO,C6

Translated, this means that L3 is a LINE on the RIGHT and tangent to (TANTO) circle 5 (C5) and on the RIGHT and tangent to (TANTO) circle 6. The reference to 'right' is used as if looking from the first circle labelled towards the second circle labelled. All the separate items – lines, points etc. – would have to be defined.

The APT language is very extensive. For example, it is possible to define the position of a point in 15 different ways, the position of a line in 18 different ways, and the position of a circle in 28 different ways.

3 *Motion information*
   This is the details of the machining sequence, i.e. the path to be followed by the tool. The path is usually given by referring to the label of the point etc. which is part of the geometry of the component. Although on vertical mill-

253

ing and drilling machines it is normally the work that moves on the *X* and *Y* axes, the movement described in the motion information is as if the tool itself is moving.

To define the position of a tool, use is made of three surfaces (see Figure 13.2):

(a) Drive surface: the tool moves along the drive surface, which corresponds to the contour being produced.
(b) Part surface: the part surface corresponds to the surface on which the end of the tool moves.
(c) Check surface: the check surface terminates the motion of the tool moving along a portion of the contour being produced.

The motion information also includes all the formulae for numerical or trigonometrical calculations. This reduces the amount of calculations which the programmer has to do.

A short extract from an APT program with an explanation of the terms is as follows:

CUTTER/20	use 20 mm diameter cutter
TOLER/0.1	tolerance zone 0.1 mm
FEDRAT/200	feed rate = 200 mm/min

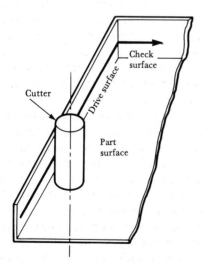

Figure 13.2 APT definition of drive, part and check surfaces

HEAD/1	use head 1
SPINDL/900	turn on spindle at 900 rev/min
COOLNT/FLOOD	turn on coolant, flood setting
PT1=POINT/100,125	definition of reference point PT1 as the point with the coordinates (X100, Y125)
SETPT=POINT/25,25	definition of a point called SETPT with the coordinates (X25, Y25)
FROM/SETPT	start the tool from the point called SETPT

The information provided in the three sections (operating, geometric and motion) was transferred on to punched cards, with one line of the program per card. The cards were then fed through a reader which input the information into a mainframe computer. This procedure was followed when large computers were used in order to reduce computer time taken and hence keep the costs as low as possible. The card punch unit was not part of the computer system, and the punching of the cards did not take up any computer time. The information provided in the three sections is converted by the programs (the compilers) into a form which is referred to as the cutter location data (CLDATA) or cutter location file (CLFILE). The cutter location data is not in a form suitable for inputting to the control unit of the machine as it is in a generalized format. This data has to be converted into the word addressed or other format acceptable to the particular machine tool. The conversion is accomplished with the use of another computer program which is called a *post processor*. There has to be a different post processor for different types of machine tools or different control systems.

Post processor

The post processor converts the generalized format into the required word address or other format by ensuring that the syntax is correct, namely that:

(a) The correct letters are used for the addresses.

(b) The numbers in the part program are arranged in the accepted format; leading or trailing zero suppression etc.

(c) The acceptable G and M codes are inserted at the correct places in the program.

(d) The coordinates for tool and work change are entered at the desired places.

(e) The movements required for the workpiece are checked to ensure that there is no collision between tools and the movements required are within the capacity of the machine.

The part program was usually supplied on a magnetic tape or a punched tape. However, since the development of computers which use the silicon chip, the details are stored either in the computer's memory and transmitted over the telephone lines or on magnetic discs.

It is obvious that to write efficient and economic programs the part programmer should have machining experience, and in addition should be fully acquainted with all the terms that can be used in the part programming language. These are defined in a manual which is issued with every part programming language. Some of the manuals are very extensive. The main advantage of the use of part programming languages over programming in word addressed format is that the calculations for all the coordinates are carried out in the computer's memory and the word addressed part program does not have to be typed, which is where a lot of errors are created.

The development of conversational programming and particularly the computer-aided machining techniques with graphic display facilities has tended to reduce the use of APT and other part programming languages.

13.5 Conversational Programming

Conversational programming is the term used to describe a technique for creating part programs available on some CNC machines which have the required control systems. The part program is created by the programmer inputting details via the keyboard of the control unit of the CNC machine tool that are the answers to a series of questions or prompts presented on the screen – hence the name 'conversational'. Alternative names for this technique are *interactive programming* or *dialogue programming*, because of the interaction or two-way involvement between the programmer and the machine.

As indicated previously, for conversational programming facilities the control units of CNC machines must contain suitable software and have control panels specifically designed for creating part programs. Certain of the keys on the keyboard and/or control panel usually have symbols and are used for inputting special commands with one press of the key. Other systems have touch sensitive screens where the commands are displayed on the screen. The programmer inputs the command by touching that part of the screen where the command is displayed. The software transforms the details input into a part program in the required format for the CNC machine. A post processor is not required to convert the details into a part program, because conversational programming is carried out at the control unit of a particular machine. The part program created is in the format required for the control, and will be stored and printed out in that format. A part program for a new component to be machined can usually be created while actual cutting is taking place on a different component. On most control systems it is possible to load a part program in the required format (word address etc.) created by other methods directly into the control. Usually the part program created by the conversational method will only be suitable for use on the same make of machine with the same control unit, but most controls have the facility to edit programs. The controls for conversational programming normally have the provision for storing a number of successful (verified) programs within its memory. It is also possible to store verified programs on magnetic discs etc. This provides the facility of having a verified part programs readily available if additional components have to be machined at a future date. On the majority of systems when the control is switched on a menu normally appears on the screen offering various options, a typical menu could be:

1 Load an existing program from memory or magnetic disc etc.
2 Create a new program
3 Edit an existing program.

This menu is the minimum that is normally provided, and the programmer selects the option by keying in the number of the option at the keyboard. The actual input details required for conversational programming techniques are simplified, and in the majority of cases the programmer is able to input details directly from the drawing instead of from a programming sheet. Normally a certain amount of initialization information has to be input before beginning to input machining details; the different parameters involved appear on the screen as prompts such as:

(a) Program name or number?
(b) The units of measurement (inches or millimetres)?
(c) Absolute or incremental programming?
(d) Size and type of cutting tool?
(e) Speeds and feeds?
(f) Coolant selection mist or flood?
(g) Tool change point?
(h) Clearance plane (a safe position to which the tools can return after a machining operation)?

The details that have to be entered for the machining operations are reduced and simplified by the selection of machining facilities provided by the control, such as:

(a) Roughing operations
(b) Area clearances
(c) Canned cycles
(d) Finishing operations.

The particular facility is selected by inputting commands via the keyboard of the control unit. Different systems use different methods of pro-

viding the details for creation of the part program as there is no standard method of entering the details.

One of the advantages to be gained from the introduction of control systems with conversational programming facilities is that they provide greater job satisfaction for skilled machinists. This is because these systems enable the machinist to be involved in decision making concerning the programming and operation of the machine, rather than merely following set-up instructions and loading the part program into the control unit when the program is provided by other means. Conversational programming is also of value where a firm has only one or two numerically controlled machines and the expense of a specialist part programmer may not be justified. Alternatively, the use of conversational programming techniques can be of particular benefit where the operation of a machine requires special skills as it can augment the expertise and practical ability of the setter/operator; especially where the machine tool is in a category such as a grinding machine or spark-erosion (EDM) machine. For these machines control systems have been designed to meet the particular requirements of the machine and the material removal technique.

Conversational programming of grinding machines

The Osai-Allen Bradley control developed with Jones and Shipman for cylindrical grinding machines requires the setter to work through an initialization or set-up procedure. Acknowledgement is given to Jones and Shipman PLC who have supplied certain details and the manual on which the following is based. Figure 13.3 shows the screen display at the change of set up. If the setter inputs Y (yes) to either prompts the next screen display such as that shown in Figure 13.4 establishes the workpiece datums. The procedure displayed on the screen establishes the relationship of the sides of the grinding wheel to the tailstock and workhead on the Z axis. There are other similar set-up procedures which establish the workpiece datums and the relationship of the wheel and

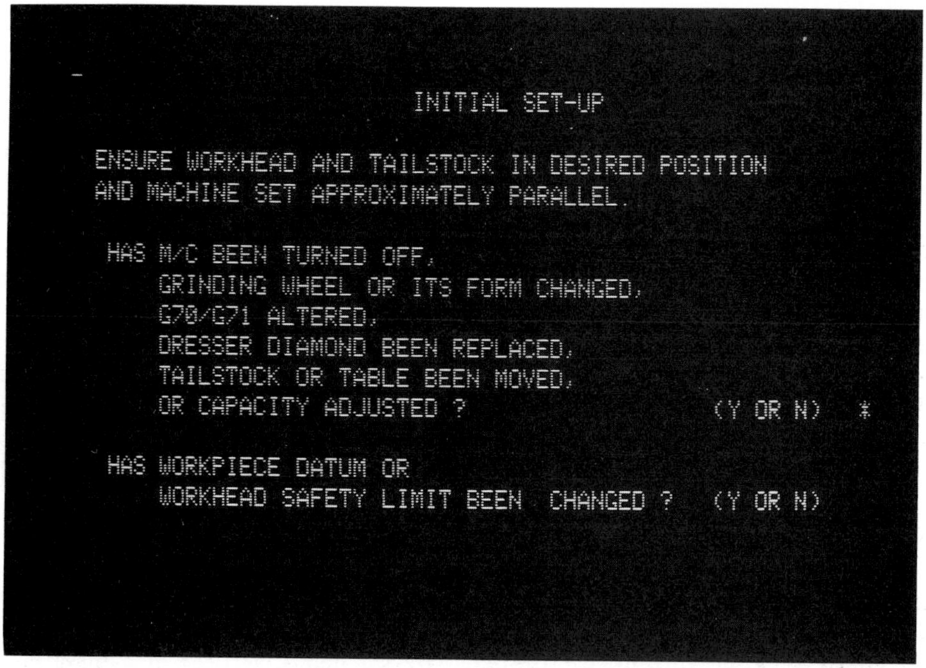

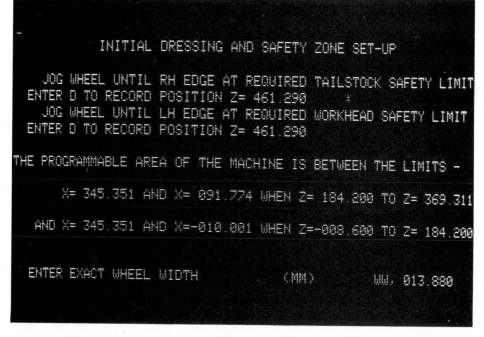

INITIAL SET-UP

ENSURE WORKHEAD AND TAILSTOCK IN DESIRED POSITION
AND MACHINE SET APPROXIMATELY PARALLEL.

HAS M/C BEEN TURNED OFF,
    GRINDING WHEEL OR ITS FORM CHANGED,
    G70/G71 ALTERED,
    DRESSER DIAMOND BEEN REPLACED,
    TAILSTOCK OR TABLE BEEN MOVED,
    OR CAPACITY ADJUSTED ?   (Y OR N)   *

HAS WORKPIECE DATUM OR
    WORKHEAD SAFETY LIMIT BEEN CHANGED ?   (Y OR N)

Figure 13.3   Screen display for initial set-up (courtesy Jones and Shipman PLC)

INITIAL DRESSING AND SAFETY ZONE SET-UP

JOG WHEEL UNTIL RH EDGE AT REQUIRED TAILSTOCK SAFETY LIMIT
ENTER D TO RECORD POSITION Z= 461.290   *
JOG WHEEL UNTIL LH EDGE AT REQUIRED WORKHEAD SAFETY LIMIT
ENTER D TO RECORD POSITION Z= 461.290

THE PROGRAMMABLE AREA OF THE MACHINE IS BETWEEN THE LIMITS -

    X= 345.351 AND X= 091.774 WHEN Z= 184.200 TO Z= 369.311

AND X= 345.351 AND X=-010.001 WHEN Z=-008.600 TO Z= 184.200

ENTER EXACT WHEEL WIDTH   (MM)   WW, 013.880

Figure 13.4   Initial dressing and safety zone set-up (courtesy Jones and Shipman PLC)

the dressing diamonds. Obviously it is extremely important that the setter enters the correct values of wheel width, etc., as the computer cannot correct this type of error.

The initialization (set-up) procedure has to be completed every time the grinding wheel is changed or the diamonds are changed. The setter uses manual jog controls to move the wheel head to the required positions so that the work/machine datums can be established. When the wheel is at the location points the setter completes the prompts displayed on the screen by pressing a key on the control panel. This is referred to as digitizing. The initialization procedure reduces the possibility of the wheel colliding with the work or any part of the machine.

The control uses standard G and M codes such as G70 and G71 for inch and metric dimensions respectively and M02 end of program or M04 start work spindle. Other options are applied to the practice of grinding, especially wheel forming and dressing. The screen display for dressing a surface grinding wheel is shown in

Figure 13.5. Similar displays are available for cylindrical grinding. The dressing cycle can be selected at any time by the setter if their expertise considers that it is required, the display offers various default values for acceptance by the setter. Jones and Shipman machines have a number of canned cycles which have been designed for grinding; such as G61 Contour generation grinding and G89 Combined shoulder and plunge grinding. These require the setter to complete prompts for the clearance plane, minimum and maximum diameter of work, table traverse points, radius of work and radius of wheel, feed rate, number of sparkout passes etc. Various default values are provided which the setter can accept or change.

Conversational control systems

There are control systems with conversational programming facilities that have been designed by the manufacturer of a machine tool to be fitted only to their machines. There are other control systems that have been designed and produced by a specialist manufacturer of con-

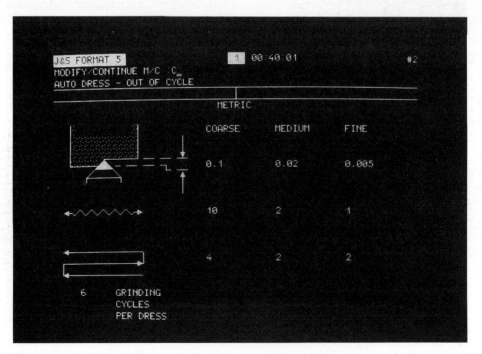

Figure 13.5   Screen display for dressing a surface grinding wheel (courtesy Jones and Shipman PLC)

trol systems that can be fitted to one category of machine tools, e.g. there are control systems for machining centres and control systems for turning centres.

When creating a new program on some systems using conversational programming for machining centres after the initialization details are input, the next item displayed on the screen is a wire frame box; the programmer specifies the dimensions of the blank workpiece. The box changes in size to conform to the details input and represents the blank workpiece. As each block of the machining details is entered a graphical representation of the movement of the point or centre of the tool is shown superimposed on the shape of the billet displayed on the screen of the control unit. The movement of the tool is shown in a different colour to the billet. As the machining details are entered the shape of the finished component gradually evolves on the screen. Figure 13.6 shows a control panel

that is used for conversational programming on a machining centre. On other systems after the initialization details have been input the shape of the billet has to be defined by the programmer inputting the required data. The shape of the finished component is gradually built up on the screen, showing all the holes, slots etc. superimposed on the shape of the blank. The machining commands are entered and a graphical representation of the tool or work movement is shown. If any error is observed the faulty block can be edited before proceeding with the input of further blocks.

On some controls, if information is entered which is incorrect a prompt question appears on the screen asking for confirmation.

When the final shape displayed on the screen appears to be correct it is necessary to run the program by stepping through each block on the machine to ensure that adequate and safe clearances have been provided. The first component produced must be measured to ensure that it conforms to the design specification.

For turning centres, normally the profile of only one side of the component is shown. Figure 13.7 shows a control panel of a turning centre with the programmer in the process of entering the information. With the increase in memory capacity now available in computers it is be-

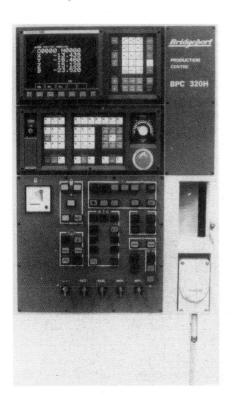

Figure 13.6   Control panel of CNC centre (courtesy Bridgeport Machines)

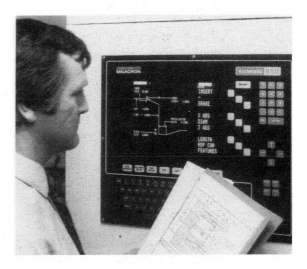

Figure 13.7   Control panel for conversational programming (courtesy Cincinnati Milacron)

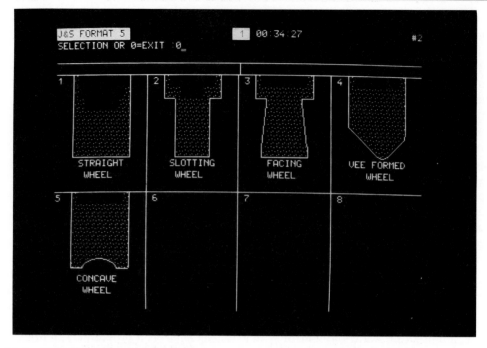

Figure 13.8   Grinding wheel tool library   (courtesy Jones and Shipman PLC)

coming practice for machining 'technology' to be presented on the screen. The technology, which helps the programmers when creating the part programs, might include:

(a) A library of tool shapes and sizes
(b) Suitable cutting speeds for a variety of tool and work materials
(c) Suitable feed rates for power available at cutting tool or for desired surface textures.

Figure 13.8 shows the shapes of the grinding wheel for use on a CNC surface grinding machine.

Another advantage of conversational programming is that the programmer does not have to remember special codes or words. If a canned cycle option is selected all the canned cycles available on the machine are listed from which a choice is made. To mill the pocket shown in Figure 13.9 on a machine with a Fanuc controller, the tool would be programmed to be positioned at the start point before selecting the cycle. The rectangular pocket milling facility would be selected by keying in a special com-

mand. The details in Table 13.1 appear on the screen, and the programmer has to use the keyboard to enter the coordinates etc. to the right of the equal sign (=) on the screen display as shown in the table. The explanations do not appear on the screen. (Information supplied by Fanuc UK Ltd.)

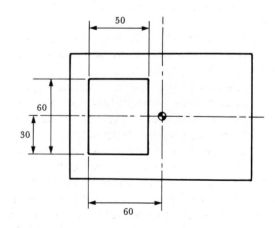

Figure 13.9   Rectangular pocket (courtesy Fanuc)

Table 13.1 *Conversational programming for component in Figure 13.9*

On screen			Entered by programmer	Explanation
U : X LENGTH	=		50.0	Incremental length of side
V : Y LENGTH	=		60.0	Incremental width of side
X : X CO-ORD	=		−60.0	Coordinates of corner of pocket in
Y : Y CO-ORD	=		−30.0	absolute units
R : R POINT	=		3.0	Clearance plane, Z axis
Z : Z POINT	=		−10.0	Depth of pocket, Z axis
W : CUTT'D DEPTH	=		5.0	Depth of cut
E : FEED RATE	=		50.0	Vertical feed rate in mm/min
T : TOOL ID NO	=		2	Tool number
Q : TOOL NAME	=		END MILL	Tool name
H : H OFFSET NO	=		2	Diameter offset number
D : D OFFSET NO	=		12	Length offset number
M : COOLANT	=	COOLANT M08		Coolant type and flowing
S : SPINDLE SPEED	=		800	Spindle speed in rev/min
F : ROUGH'G FEED	=		120	Milling feed rate in mm/min
J : FINIS'G FEED	=		100	Finishing feed rate in mm/min
C : FINISHING	=		2.0	Thickness of finishing cut
K : CUT DEPTH %	=		50.0	Percentage depth of cut

Heidenhain dialogue programming

A widely used interactive programming system is supplied by Heidenhain Ltd. This system is an example of a control developed by an independent manufacturer for use on other manufacturers' machine tools. On some of the Heidenhain controls the programs may be entered either in the ISO format (G and M codes) or the Heidenhain plain-language interactive dialogue.

The control panels of the Heidenhain control systems have a number of keys marked (labelled) with symbols dedicated for inputting specific commands in the dialogue technique, and also numeric keys for inputting tool numbers and coordinate values. Knowledge of the purpose of each key is gained with practice, but the labels printed on each key provide a memory aid. Figure 13.10 shows a typical control panel and monitor. The control panel is shown in the lower right of the figure. The monitor is displaying a plan view of a component in the process of programming. The unit in the lower left of the figure is a portable unit via which programs created on a desktop computer can be transferred to the CNC machine control. The control panels of other models of the Heidenhain control system have the same keys but they may be in a different arrangement. Touch probes and electronic handwheels are used for work and tool setting on CNC machines with Heidenhain controls as on CNC machines with other control systems.

Programs for Heidenhain control systems follow the sequence of:

1 The billet or blank size is entered
2 The tools to be used are defined
3 The tool to be used is loaded into the working station
4 The tool is moved at rapid rate to the start of the first machining sequence
5 The machining sequence is programmed
6 The tool is retracted to the tool change position.

The sequence 3 to 6 is repeated until all the machining operations are completed. As with other methods of creating part programs it is advisable that the sequence of operations with

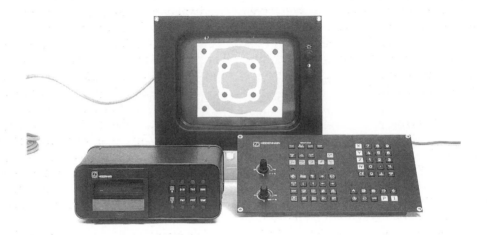

Figure 13.10 Control panel of Heidenhain TNC 355 control system (courtesy Heidenhain (GB) Ltd)

speeds and feeds to be used and full details on all the tools to be used are planned before commencing to enter a new program.

Acknowledgement is given to Heidenhain (GB) Ltd for the use of the manual on which the following information is based.

The entry of a new program starts with the key labelled 'PGM/NR' being pressed and a prompt 'PROGRAM NUMBER =' is displayed on the screen. The program number is entered using the numeric keys on the control panel of the machine. The number is displayed on the screen as it is entered, if no errors have been made in entering the program number it is accepted by pressing the key labelled 'ENT'. This results is another prompt 'MM = ENT/INCH = NO ENT' that requires the dimensional units to be specified. If the units are to be millimetres the key labelled 'ENT' is pressed which results in the first block being shown on the screen. Block numbers do not have an address and are automatically entered by the control; the first block is designated 0 'zero' and would appear on the screen as:

0 BEGIN PGM the program number MM

The block numbers are automatically created in sequence by the control. If graphic simulation of the machining operations is to be utilized it is necessary to enter the billet size using the key labelled 'BLK/FORM' and completing the prompts. Two views can be shown: a plan view (XY plane) and a side view (YZ).

All the tools to be used are entered in tool definition sequences which are programmed by the key labelled 'TOOL/DEF' being pressed, and a prompt requires the tool number to be entered using the numeric keys. When the tool number has been accepted a prompt 'TOOL LENGTH L?' has to be completed. The length required is the tool length compensation or offset value followed by 'TOOL RADIUS R?'. Upon entry a block is displayed such as:

1 TOOL DEF 1 L + 20.0
R + 5.0

A tool is not selected until the key labelled 'TOOL/CALL' is pressed and prompts for tool number, axis and spindle speed have been answered. Tool radius compensation is activated through the use of keys labelled 'RL' or 'RR' (compensation right or compensation left respectively) during the entry of each machining movement.

The entry of a program is continued by pressing the appropriate keys and completing the prompts as they appear on the screen. Machining operations are programmed either as single operations of straight lines and circular arcs or canned cycles. Linear interpolation is activated by pressing the key labelled 'L' and completing the prompts for:

'COORDINATES?'
'TOOL RADIUS COMP.; RL/RR/NO COMP.?'
'FEED RATE?'
'AUXILIARY FUNCTION M?'

Absolute coordinates are entered by first entering the address of the axis on which movement is required using the keys that are labelled X Y or Z etc. followed by the numeric value. Incremental and polar coordinates are entered by selecting the axis and then utilizing the keys labelled I and P respectively before the numeric value is entered. Linear interpolation movement can take place simultaneously on three axes. As stated previously, tool radius compensation is designated by RL or RR for tool right or left of the work respectively. The right and left relation is looking in the direction the tool is travelling.

Heidenhain controls using dialogue programming do not use preparatory functions (G

codes). However, M codes which are referred to as auxiliary functions are used. The M code is activated by entering the number at the M prompt. For the most widely used functions the codes used are the same as the standard codes in the ISO format i.e. M00 is used to activate program stop; and M06 tool change etc.

For circular interpolation movements there are four keys which can be used. The direction of movement is designated by entering + or − at the prompt 'DR'. − (minus) for clockwise and + (plus) for counterclockwise. Similar prompts to those shown above for linear interpolation appear for all the individual machining operations. Circular interpolation movement can take place on any two axes.

There are a number of canned cycles available which require different numbers of program blocks. When the key labelled 'CYCL/DEF' is pressed all the canned cycles available are displayed in sequence and the required cycle is selected by pressing the 'ENT' key. When the canned cycle is selected prompts appear on the screen using workshop terms that require the programmer to enter the required details. The canned cycle for peck drilling is canned cycle number 1 and requires the following six program blocks; the block numbers are assumed to start at block 50:

```
50 CYCL DEF 1.0 PECKING
51 CYCL DEF 1.1 SET-UP      −2.000   (clearance plane)
52 CYCL DEF 1.2 DEPTH      −15.000   (total hole depth)
53 CYCL DEF 1.3 PECKG       −7.000   (pecking depth; infeed per cut)
54 CYCL DEF 1.4 DWELL        4.000   (dwell time seconds)
55 CYCL DEF 1.5               F100   (feed rate in mm/min)
```

The coordinates are entered by the programmer and stored in the memory of the controller, and are used to create the part program. The text in brackets such as (clearance plane) is given here for information and does not appear on the

screen. A canned cycle for machining a rectangular pocket is cycle number 4 would display the following seven prompts; the block numbers are assumed to start at block 60:

```
60 CYCL DEF 4.0 POCKET MILLING
61 CYCL DEF 4.1 SET-UP          Set-up clearance
62 CYCL DEF 4.2 DEPTH           Milling depth?
63 CYCL DEF 4.3 PECKING         Pecking depth?
              F                 Feed rate for vertical movement (pecking)
```

64 CYCL DEF 4.4 X   First side length?

65 CYCL DEF 4.5 Y   Second side length?

66 CYCL DEF 4.6 F   DR   F = Feed rate for horizontal movement

DR = Cutter path rotation

DR − (minus) for clockwise or DR + (plus) for counterclockwise

Definition of the above terms is shown in Figure 13.11a and b which shows the start position. The tool would have to be programmed to be at the start position in a separate block before calling the cycle. Figure 13.12 shows the path followed by a tool in milling a pocket; the control system calculates the amount of step-over depending on the size of the cutter and the power available. The size (diameter) of the cutter determines the radius in the corners as there is no circular movement in the corners.

Heidenhain controls have the facility for programmers to develop machining cycles particu-

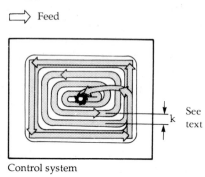

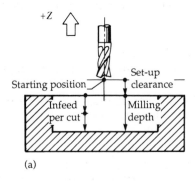

Figure 13.12   *Tool path pocket milling (courtesy Heidenhain (GB) Ltd)*

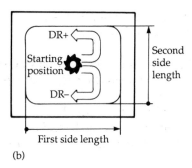

(a)

(b)

Figure 13.11   Start position for pocket milling: (a) side view ZX plane; (b) plan view XY plane (courtesy Heidenhain (GB) Ltd)

Figure 13.13   Control panel of TNC 415 control with graphical simulation of component (courtesy Heidenhain (GB) Ltd)

larly suitable for a firm's product and store them in the control memory. The selection of the machining facility required is either through the use of special buttons on the keyboard, or by making a selection from a menu displayed on the screen.

The last block in the program is entered by pressing the key labelled 'END'. A prompt requires the program number and the dimensional units to be confirmed. When all the details are entered the shape of the component can be displayed as shown in Figure 13.13 where the control panel and screen can be seen with the graphic display of a machined component. The control system 415 shown can be used away from the machine and when the program has been verified it is stored within the memory

of the control in the Heidenhain plain-language format, and either downloaded to a portable unit for transferring to the CNC machine, or via an RS232C link, or copied to a magnetic disc as a backup. A portable unit for transferring the program is shown in Figure 13.10. The control panel shown in Figure 13.13 has the special keys for Heidenhain plain-language dialogue input and also has a QWERTY type keyboard which can be used for other purposes.

The first 28 lines of a part program for a Heidenhain TNC 150 control in the Heidenhain plain-language format are shown below. The program is for the milling of the two end steps of the component previously manually programmed in Example program 1 in Chapter 11

0 BEGIN PGM 1000 MM	Start of program
	Blocks 1 to 3 define the tools
1 TOOL DEF 1 L+50,0 R+5,0	Tool 1 is a 10 mm dia spot drill with 90 degree tool point angle
2 TOOL DEF 2 L+45,0 R+4,0	Tool No 2 is a 8 mm drill
3 TOOL DEF 3 L+0,000 R+10,0	Tool No 3 is a 20 mm dia end mill
4 TOOL CALL 0 Z S+40,000	Ensures that any tool details stored in tool offset register are removed
5 L Z+50,000 R0 F9999 M	Blocks 5 and 6 tool change point linear interpolation (L) at rapid feed rate
6 L X+200,000 Y−50,000 RO F M05	(F9999) M05 is coolant on
7 STOP M00	M00 program stop for tool change
8 TOOL CALL 3 Z S+500,0	Load tool number 3 (20 mm dia end mill)
9 L X−11,000 Y−50,000 R F M03	Blocks 9–11 are positioning (linear interpolation (L))
10 L X−11,000 Y−1,000 R F M	operations; M03 spindle on
11 L Z+2,000 R0 F M08	
12 L Z−6,000 R0 F250 M	Tool advancing at feed rate to depth of step
13 L X+5,000 Y−1,000 R F M	Tool moving to start of left hand step
14 L X+5,000 Y+91,000 R F M	Machining step
15 L Z+2,000 R0 F9999 M	Tool retracting above surface of work at rapid feed rate
16 L Z+125,000 Y+91,000 R F M	Blocks 16–17 positioning operations for right hand step
17 L X+125,000 Y+92,000 R F M	
18 L Z−6,000 R0 F250 M	Tool advancing at feed rate to depth of step
19 CC X+135,000 Y+92,000	Blocks 19 and 20 are for positioning the end mill at the start of the right-hand step. CC is circle centre coordinates
20 C X+145,000 Y+92,000 DR− R F M	Circular interpolation (C)
21 L X+145,000 Y+9,000 R F M	Machining step
22 L X+150,000 Y+9,000 R F M	Blocks 22–24 are positioning operations to clear tool at rapid feed rate
23 L X+150,000 Y−1,000 R F9999 M	Positioning movements at rapid feed rate (F9999)

```
24 L Z+50,000 R0 FM
25 TOOL CALL 0 Z S+40,000          Details of tool No. 3 removed from memory
26 L Z+50,000 R0 F9999 M           Blocks 26 and 27 tool change position
27 L X+200,000 Y−50,000 R0 F M05
28 STOP M00                        Program stop for tool change
```

As stated previously, conversational programming can be of great help in providing skilled machinists with greater job satisfaction. However, conversational programming techniques do require the machinists to be trained in its use. When the firm has computer-aided drawing facilities and the components are fairly complex, considerable time may have to be spent in programming and it may be more economic to create the part program using other techniques, such as computer-aided machining away from the machine.

13.6 Computer-aided machining (CAM)

This technique is also referred to as 'Graphical Numerical Control' (GNC), or 'Geometric Part Programming' by the firms which supply computer systems and software for this purpose. The software can be quite expensive and in order to use certain of the CAM programs the computers are required to have a mathematics coprocessor and hard disc.

There are usually three stages in the production of the part program using the CAM method:

(a) Geometric definition of the shape of the component
(b) Machining instructions
(c) Conversion of the above details into a part program in the required format with a post-processor.

To produce efficient part programs by CAM requires knowledge of machining practice and techniques and familiarity with the various input commands. Unfortunately the commands have not been standardized, and each company uses its own devised commands to input details. There are different techniques for inputting the commands which are used in the first two stages. Additionally, with one type of system the commands are selected from a menu shown on the monitor screen by the entry of a number or letters either via the keyboard, or alternatively a mouse is used to move the cursor to the option required. The inputting device on the left of the keyboard shown in Figure 13.14 is a 'mouse'. A dual-screen computer system is shown where the coordinates, input commands, prompts etc. are listed on the right-hand screen and the shape of the component is displayed on the left-hand graphic monitor. The detail shown on the graphic screen on the left of the photograph is for a component to be produced on a turning centre. A single-screen system can also be used, where up to three sets of commands can be listed on a narrow band at the bottom of the screen and the shape of the component is displayed on the remainder of the screen.

Another type of computer system uses a specialist inputting device known as a 'digitizing tablet' on which the various commands

Figure 13.14   Computer-aided machining (courtesy Pathtrace Engineering)

required are laid out in pictorial form in a series of squares. The command required is selected by placing the cross wires of a coil on a freely movable inputting device known as a 'puck', over the square containing the command and pressing a button on the puck. Another method of selecting the command from the squares uses an instrument similar in shape to a pen. The point of the pen which is pressure sensitive and retracts when pressed, is placed on the command (square) required and depressed; a button on the pen is pressed to confirm the selection. Both the puck and pen are interfaced via a lead with the computer's central processing unit.

Expertise in using the inputting devices comes with practice, and most systems provide help facilities. Part programs can be created with CAM for practically all categories of CNC machine tools. The graphical representation of the geometry of a die to be produced on a wire spark erosion machine is shown at the bottom of Figure 13.15. The die produced is shown in the middle of the figure and the two plugs resulting from the movement of the wire are shown at the top of the figure. The die would require a machine with four axes of movement: X and Y axes for control of the bottom of the profile; and U and V for control of the top of the profile.

A graphical representation of the path to be followed by the punches to produce a component on a turret punch machine is shown in Figure 13.16. The figure shows four components, however only one drawing of a component would be created and this would then be copied or mirrored.

Stage 1   Geometric definition of the shape of the component

The graphical representation displayed on the screen is a visualization of the component to be machined and represents the finished work piece. The shape displayed is a series of lines on the X Y and Z axes, and is referred to as being bounded geometry because it shows the limits or boundaries of machining. The technique of generating the graphical shape depends on the complexity of the component. Two techniques

of creating the shape are enumerated geometric definition and digitizing.

Enumerated geometric definition

The shape is generated by specifying each geometric feature (such as lines and circular arcs) which make up the shape of the component; and inputting the necessary values of the coordinates of the change points or intersection of the geometric features. The coordinates can be with reference to a datum point (absolute) or relative. This method is suitable for components

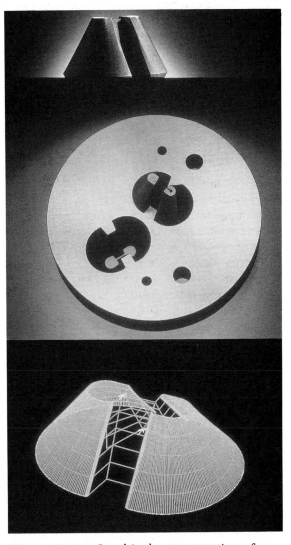

Figure 13.15   Graphical representation of spark erosion (courtesy Pathtrace Engineering)

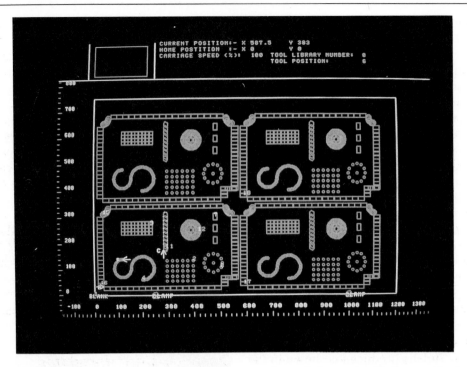

Figure 13.16   Graphical representation of turret punch work (courtesy Pathtrace Engineering)

for which drawings exist and whose shape consists of straight lines, circular arcs and curves which have coordinates which can be calculated from mathematical formulae. The geometric shape can be created:

(a) By the part programmer actually defining the shape of the component; or

(b) Using a drawing of the component prepared by computer-aided draughting (CAD) techniques.

Creation of drawings by the part programmer

There are different techniques used for creating drawings. It must be remembered that the part programmer is not designing a component but is inputting values of the coordinates or dimensions from an existing drawing. It must be noted that the specific shape the part programmer is creating is going to be produced on a particular machine tool such as a turning centre (lathe) where the component rotates and the machine has fundamentally two axes of movement (Z and

X), or on a machining centre (milling/drilling machine) where fundamentally there are three axes of movement (X, Y and Z). The different software for the graphical construction has been written for the creation of part programs to be used on the different types of machines, and is designed so that only the appropriate information need be entered. The geometric shape of the majority of components can be constructed of points, straight lines and circles which concurs with the machining modes of positioning (point-to-point), linear and circular interpolation. The drawing of the shape of the component can be started at any convenient point, but it may save time and avoid some problems later if the drawing is commenced at a point which will be a suitable zero datum. The dimensions of the various geometric features or contour elements of the component can be entered as a series of coordinates in either absolute values from the selected start position or relative (incremental) values. As each set of coordinates are entered the shape of the component is built up on the screen.

On a number of CAM computer systems various aids such as rotation and mirror imaging are available which reduce the amount of input information required.

As each element of the drawing is specified the computer usually labels (identifies) the intersections or direction change points with a number or a letter automatically. In some systems these labels may be called 'nodes'. The labels can be referred to when detailing the machining operations.

With most systems it is possible to obtain a copy of the screen display of the drawing on a dot matrix printer or plotter for record purposes. The graphical shape created on the screen is only suitable for the developmnent of part programs and is used for limiting tool movements; the copy is not intended to be used as dimensioned working drawing.

Each shape or feature of the component must be identified with a different name. If the component is to be produced on a machining centre and is composed of separate shapes or features that would require fundamentally different types of machining operations (drilling holes, milling slots and pockets etc.) each shape or feature on the drawing would need to be selected individually and identified by a different filename as shown in Figure 13.17. However, the same filename would be used to identify a feature that is to be machined with tools of different size such as spot drilling, drilling and tapping; or for roughing and finishing operations of the same feature.

External and internal profiles on turned work would need to be defined separately and identified by different filenames. The same filename would be referred to for a grooving operation or screw threading operations on a section of the profile. The particular section would be identified by either its 'node numbers' or its coordinates.

If the work blank is of complex form such as a casting or forging it is necessary to define its geometric shape and size, and identify it with a filename which is related to the work piece e.g. filename/b. If the work blank is a basic shape such as bar stock for turning operations it is generally possible to specify its size before detailing the machining operations.

Computer-aided draughting

Computer-aided draughting (CAD) is being used in many firms and time may be saved if the drawing of the component produced with these facilities is used by the part programmer. However, commands used for producing drawings by CAD are different from those used by the part programmer for generating the shape of the component. This is because CAD systems are intended for a wider range of work and are used for design purposes, therefore additional facilities are required such as different line types, dimensioning, cross hatching, insertion of text etc. Figure 13.18 shows a dual-screen computer system where a drawing is being created with the help of a digitizing tablet and, on the left, a plotter on which the drawing of the component could be produced for record purpose if required. The drawings would be stored in numeric format on discs or other media (in the memory of the computer).

The computer used for converting the drawings of the component from CAD to CAM must have software that is able to convert and transfer the details from the CAD file to the CAM file. Initial Graphic Exchange System (IGES) is one such program. Data Exchange File (DXF) is

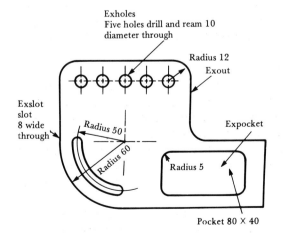

Figure 13.17   Identification of features

Figure 13.18   Computer-aided draughting (courtesy Pathtrace Engineering)

another widely used. Obviously the CAM software must be capable of accepting these files.

The part programmer would access the transferred files (input from the memory) for the initial stage of producing the shape of component on the monitor screen (visual display unit) of the computer. For the purpose of CAM all non-relevant lines would be removed from the drawing on the screen.

It is desirable that all the change or intersecting points of the geometric features (lines, arcs etc.) of the component, are automatically dimensioned from the point selected by the part programmer to be the datum point.

Geometric definition by digitizing

Where the shape of the component is very complex and a three-dimensional model of the component exists, the graphical representation of the component can be generated using a digitizing technique. The surface data of the geometric shape is obtained by actual measurement of the model by moving (tracing) a stylus of a touch trigger probe over the model on a coordinate measuring machine (CMM) or on a three-axis CNC machine tool. The probes used are similar to the spindle mounted probes shown in Figure 10.7. The model can be made out of clay or similar material and created by sculpturing to suit aesthetic conditions or to satisfy experimental tests. Figure 13.19 shows a machine tool being used to digitize a model of a car. The digitizing technique is commenced with the stylus being positioned at a convenient datum point (X, Y and Z) which is recorded in the computer, and the work table is programmed to move a specified distance along the X axis at a low feed rate. The stylus point can be deflected by a very low force and when the point contacts the model the stylus deflects; the movement of the stylus creates a signal that causes the table movement to stop and the head to move on the Z axis until the stylus can move back to the null point. At the end of the programmed travel on the X axis the work table stops, and steps across a small distance on the Y axis; the table movement on the X axis reverses direction and moves back. The sequence of movements on the X and Y axes continues until the model has been traced over. A canned cycle can be used to program the movement of the work table. The movement of the Z axis together with the movement on the X and Y axes is input into the computer which displays the surface generated as a series of lines in a 3D simulation as shown in Figure 13.19.

The values of all the coordinates of the lines and curves that make up the shape are referenced to the datum point established at the start of the digitizing operation. The shape displayed can be reversed by the computer so that a mould can be machined from which components can be produced.

Contact points of the stylus are available in various sizes and shapes. The size and shape of the points has to be selected with consideration to the distance stepped on the Y axis and type and size of the cutting tools that will be used when machining the workpiece. The intermittent stepping on the Y axis results in the workpiece surface being machined as a series of grooves. The smaller the stylus point the smoother the surface but the longer the finish machining time.

When the graphical representation of all the shapes in the component has been defined and seen to be correct, stage one in the creation of the part program is completed.

Stage 2 Machining instructions

The machining operations are specified with reference to the shape displayed, and the path

Figure 13.19 Digitizing (courtesy Anilam Electronics Corp and Renishaw Metrology)

followed by the tool during the positioning and material cutting operations simulates the machining of the component. It is essential that a sequence of operations is planned with all the machining options and tools to be used selected before proceeding with the machining instructions. The following details should also be known before proceeding with the machining instructions.

(a) Metric or imperial units
(b) The coordinates for the tool change
(c) The coordinates of a safe position of the tool when rapid positional movements have to take place. This position may be referred to as the home position on turning centres, or initial, or reference, or clearance plane on machining centres
(d) The coordinates of a safe position for the work or tool when changing the work.

It is recommended that planning sheets and program preparation sheets are completed for record purposes. Included on the planning sheets should be information on details such as:

(a) Cutting speeds and feeds calculated for all the operations
(b) Type of coolant to be used and when to be turned on and off
(c) Tool shapes and sizes with numbers allocated or selected from the tool library. If tools are to be mounted in a turret the turret position in which the tool is to be mounted and the tool register should also be specified.
(d) Tool compensation right or left and when it has to be applied and cancelled.

It is becoming conventional for machining 'technology' to be available within a database in the computer, which can be accessed during the programming. The database should contain technical information on economic speeds and feeds which can be used with different tool materials when cutting different work materials. This facility is not expert system software as the programmers have to select the particular speed and feed based upon their experience. The shape of tools available in the library can be selected and displayed on the screen as shown in Figure 13.20. The shape of the tools can be of assistance in avoiding collisions between tools and work. After inputting details on the type of cutting tool etc., the computer provides options for selecting the most suitable spindle speed and feed rate.

Before any machining instructions are entered it is necessary that all the geometric shapes or features to be machined are displayed on the screen as it would be extremely difficult and virtually impossible to simulate the machining of the component without it. With some systems the geometric shape and size of the work blank can also be displayed on the screen.

Full details have to be input of the positional moves (non-cutting moves) of the tool at rapid feed rate between the machining operations in the sequence that they are to take place. Roughing and finishing operations have to be defined separately.

The positioning moves would be generated as separate blocks with a G00 classification by the post processor. The movements of the tool at a given feed rate during material removal operations would be generated into G01, G02 or G03 classification by the post processor.

Some CAM programs provide the standard fixed cycles and also special cycles for area clearances etc. When a canned cycle or special cycle is selected, prompts normally appear on the screen requesting details. As with manual programming it is necessary to position the cutter at the start point before selecting and activating the cycles.

As an alternative to specifying X, Y and Z coordinates the chosen position of the cutter at the start of cutting, and at subsequent stages throughout the machining sequence, can be indicated by the following two methods:

(a) Where the nodes (corners or positions of the different features) of the component and hole centres etc. are labelled they can be referred to when detailing the movements required; or

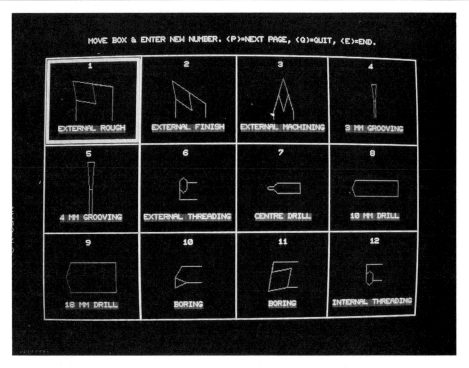

Figure 13.20   Graphic display of tools in library (courtesy Pathtrace Engineering)

(b) Use is made of a special input device such as a mouse which controls the position of a figure which represents the shape of the cutter superimposed on the drawing of the component on the screen. When the figure representing the cutter is at the desired position, an input signal from the keyboard or other device tells the computer to store that position in its memory. It does this with reference to the zero datum at the start of the profile of the workpiece. The cutter representation is moved around the drawing of the component in the sequence required, the position of all the change points being recorded in sequence.

It is extremely time consuming to try and position the mouse at a precise location (to three decimal places) unless it is possible to 'snap on' to a position. It is recommended that the mouse be used only for determining the approximate position and that the desired or actual position is entered at the keyboard.

When all the non-cutting and machining operations have been entered, the resulting initial graphic display is usually in two axes: X and Z for turning operations and X and Y for milling and drilling. It is possible to select alternative displays for milling and drilling on X and Z axes or Z and Y axes. After the movements shown for milling and drilling operations on the monitor in 2D are seen to be satisfactory, it is frequently possible to have a 3D visualization of the movements of the cutter on the screen of the monitor, or a 3D image of the machined component as shown in Figure 13.21. Some systems also have the facility to provide a copy of the display on the screen (screen dump) on a dot matrix printer.

When the graphic display of the tool movements is seen to be satisfactory, all the geometric details of the shape of the component and the required cutter movements etc. are stored in the computer's memory or on magnetic discs. The details are stored in a generalized format fre-

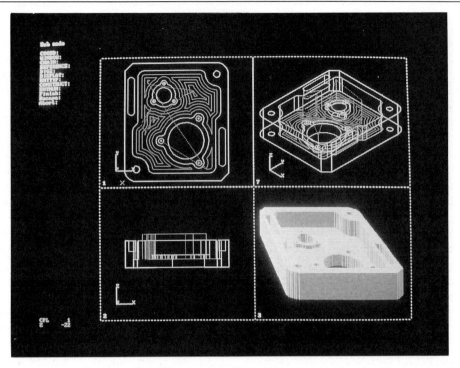

Figure 13.21   Screen display of milling and drilling (courtesy Pathtrace Engineering)

quently referred to as Cutter Location Data (CLDATA) or Cutter Location File (CLFILE).

Stage 3   Post processing

As stated previously, the geometric details of the required cutter movements etc. are converted within the computer's memory into a part program in word addressed or other format that is required by the control system. The conversion of the details is carried out with the use of a computer program known as a post processor. The function of this post processor is similar to that used with the part programming languages. As with the languages, there has to be a different post processor for each machine tool and control system. If another machine with a different control system becomes available and the component has to be produced on that machine, in most cases it is not necessary to repeat the first two stages. It is only necessary to load from the computer's memory the stored program of the geometric details of the required cutter movements etc. Another post processor is then used to convert these geometric and machining de-tails into a suitable part program for the new machine.

Editing of part programs

Most of the software providing CAM facilities has a text editor which can be used to edit any of the files generated during the creation of the part program. The part program in word addressed format may need editing before it is stored or downloaded to the machine tool. The editing may be to provide operators' instructions (within brackets) or to provide an optional program stop (M01) on an operation for some purpose. It may also be necessary to change the spindle speed on some blocks in a turning cycle where the change in diameter is greater than 20 per cent. Additional blocks can be inserted for operations that may have to be carried out manually, such as the changing of clamps. It is also possible to use this editing facility to create part programs for simple components that do not require the full CAM treatment.

Advantages of CAM
There are a number of advantages of computer-aided machining:

(a) No calculation of tangency points for a blending radius is required.

(b) It is comparatively easy to correct errors in tool movements before they take place on the machine.

(c) The large number of errors that can be caused during the actual typing of the part program are prevented.

(d) Once it is seen that the completed movements of the tool or work will correctly define the desired shape of the component to be produced, then a part program can be converted by the post processor within the computer, and either output to a storage facility such as punched tape or magnetic disc, or downloaded to the machine tool without any further work.

(e) Part programs for complicated components can be produced more quickly by this method than by any other.

(f) Since the machining sequence is actually simulated on the screen by showing the path to be followed by the tool around the work, it is possible to reduce the time spent on checking the program without using machine tool time which is expensive. However, it is beneficial to carry out a test of the part program by stepping each block to check that adequate and safe clearances have been provided. As with all other methods of programming the actual size of the component can only be determined when the first component has been machined.

(g) Technology facilities are usually provided, such as the automatic selection of the cutting speeds for various work and cutting tool materials, and calculation of the quickest or shortest cutter path.

Computer-aided machining is extremely useful and is to some extent easier than the other techniques for creating part programs, but requires a different approach. However, to create efficient and economic part programs using CAM, the programmer still requires machining experience to decide how the component should be clamped and to ensure that the best cutting speeds and feeds are used for the particular operation. Figure 13.22 shows a screen display of clamps and work. It has been suggested by various authorities that it will be possible for part programs to be created by drawing office personnel directly after they have created the drawings. For this to be generally possible the training of the personnel concerned will have to be extended and changed, and the computers used will have to have large memories and more extensive technology facilities, and also be capable to make value judgements. It is becoming more normal practice for CAM to be only one of a number of computer programs involved in the integration of computers in the manufacturing industry with skilled personnel responsible for production.

As stated previously, there are a number of firms offering CAM software and the following sections are examples of creating part programs with different software.

CAM Example 1
The following is an example of a CAM program using the turning module of the Pathtrace Manufacturing System (PMS) CAM software version 11.4 supplied by Pathtrace Engineering Systems Ltd. Acknowledgement is given to Pathtrace for the use of the manual and software.

This program was written by the author to whom any errors in the details of the program are attributable and should not be regarded as having been originated by Pathtrace. The purpose of the program is to provide an outline of one technique of computer-aided part programming and is in no way a substitute for the training programme offered by Pathtrace.

So that a comparison can be made between manual programming and computer-aided programming, the component to be used is the same as that programmed in Chapter 12, and will be given the filename DEMO13.

It is essential that a sequence of operations is planned with all the machining options selected

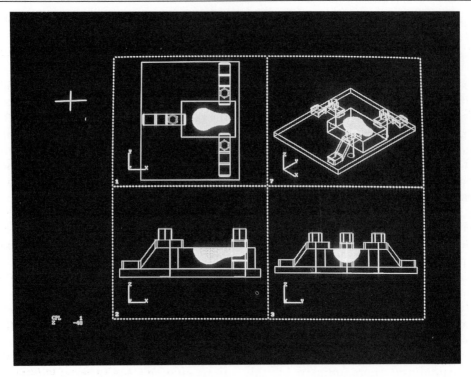

Figure 13.22   Screen display showing work clamping (courtesy Pathtrace Engineering)

before proceeding with the geometry definition or machining instructions. It is recommended that planning sheets and program preparation sheets are completed for record purposes. The speeds and feeds must be calculated for all the operations.

Within the context of this section it is impossible to give full coverage of all the facilities available within the Pathtrace software, and only those required for the writing of the part program will be explained.

In the planning of the operations the following details have been assumed:

1   The machine tool to be used has a six station tool turret and a Fanuc control. The tool registers will be specified but the actual offsets will be individually established on the machine.
2   The sequence of machining operations, tools, speeds and feeds are the same as for the example in Chapter 12, and are also repeated in Table 13.2.

3   The tool numbers quoted in Table 13.2 are assumed numbers from the tool library.
4   The component will be held in a collet chuck.
5   The diameter of the bar is 50 mm and although the bar is 500 mm long, the turned portion is only 65 mm long at one end, which protrudes 75 mm from the face of the collet.
6   A new workpiece is loaded in operation 13 after a machined bar has been removed and the protrusion from the collet adjusted for length in operation 2.

Table 13.2   *Sequence of operations*

1   Load stop for work, tool blank in turret station no. 1; tool offset register no. 1. Tool loaded is no. 363 in tool library.
2   Adjust work to correct length. Operator intervention required.
3   Load facing tool in turret station no. 2; tool offset register no. 2. Facing tool is no. 254 in tool library.

4 Face off end of work at Z zero datum; spindle speed 800 rev/min; feed rate 0.2 mm/rev.

5 Load left-hand turning tool in turret station no. 3; tool offset register no. 3. Left-hand turning tool is no. 1 in tool library.

6 Rough turn outside profile. Spindle speed 800 rev/min; feed rate 0.2 mm/rev.

7 Load 2 mm grooving tool (undercutting) in turret station no. 4; tool offset register no. 4. 2 mm grooving tool is no. 294 in tool library.

8 Form undercut. Spindle speed 2000 rev/min; feed rate 0.05 mm/rev.

9 Load left-hand profiling tool in turret station no. 5; tool offset register no. 5. Left-hand profiling tool is no. 30 in tool library.

10 Finish turn outside profile. Spindle speed 1000 rev/min; feed rate 0.1 mm/rev.

11 Load screw cutting tool in tool turret station register no. 6. Tool no. 325 in tool library.

12 Screw cut M10 thread, 1.5 pitch. Spindle speed 1000 rev/min; 1.5 mm pitch; 0.812 depth of thread.

13 Remove machined component and load new work piece.

The tools can be defined as part of the machining instructions, but one of the facilities (a database) available within Pathtrace is a tool library in which records of 40 tools in each of 50 user defined categories (up to 2000 tools) can be stored. Each record contains the following:

(a) Tool number of up to four digits.
(b) Tool description as an alphanumeric (letters and numbers) sequence of up to 32 characters long. The ISO coding system for designation of indexable insert tool holders can be used very effectively. Letters and numbers are used to designate such features as method of holding the insert; insert shape; hand of tool; width of shank etc. Alternatively a name such as 'left-hand turning tool' can be used.

(c) Type of tool, defined as a number between 0 and 7, which refers to the direction of a line between the tool set point and the centre of the nose radius as shown in Figure 13.23. The type of tool is classified according to the orientation of the line. The line from the set point to the centre of the nose radius for type 0 tool is at the 12 o'clock position and the numbers advance as the orientation of the line changes counterclockwise in 45 degree steps, i.e. the orientation of the line for a type 1 tool and type 7 tool is 45 degrees counterclockwise and 45 degrees clockwise respectively from the 12 o'clock position.

(d) Units imperial or metric.
(e) Turret position (number) – if the tool required special consideration this would be the preferred position; it is possible to change the turret position during programming.
(f) Graphics (number 1 to 400) which fits a graphic shape previously entered in the tool library.
(g) Technology (number 1 to 400) which fits a tool technology table.

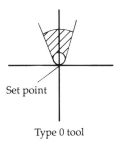

Type 0 tool

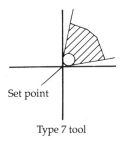

Type 7 tool

Figure 13.23   Tool types

(h) *X* set length/*Z* set length – dimensions which are dependent upon the particular machine tool and post processor configuration.

(i) Nose radius, dimension.

(j) Tool width, dimension of grooving tool or diameter of drill.

(k) Primary angle/secondary angle, angle of sides of tool tip. Measured in an counterclockwise direction with the 3 o'clock position as zero angle.

The development of CAM software is an ongoing task and the content of the software changes to meet developments in control systems and operation of machine tools. In version 11.4 of Pathtrace there are three stages in the creation of a part program:

(a) Stage 1 Construction of geometric shape
(b) Stage 2 Machining instructions
(c) Stage 3 Post processing.

Stage 1 Construction of geometric shapes

Although drawings already produced with CAD software can be used, component shapes can easily be created directly in the CAM software from Pathtrace. It is necessary to have a drawing of the component for which the part program is required, and for convenience the drawing should be dimensioned from a suitable datum point in either absolute and/or incremental dimensions.

The geometric construction option is selected from the menu of the Pathtrace turning module. The commands used for creating the geometric shape which will be used for limiting tool movement are dependent on the fact that the majority of engineering components consist of straight lines and circles. There are seven basic types of commands:

(a) Point to point commands for joining two points with a straight line. The shape is commenced at the starting point which is input as two numbers which represent the *Z* and *X* coordinates of the datum. As successive commands are entered the line will stretch from the current point to the designated point. Values can be entered in either rectangular or polar coordinates; in absolute or incremental units.

(b) Point to circle commands for joining a straight line to a circle in one of four ways: at a tangent; radially; at a specified angle; or at a specified coordinate.

(c) Circle to point commands for joining a circle to a point, leaving the circle under similar conditions as above i.e. tangentially or radially etc.

(d) Circle to circle commands for joining two circles, either by blending them together, or tangentially, or at an intersection.

(e) Go forwards at a specified bearing. The go commands are in two parts, the first specifying the go bearing, the second specifying the destination.

(f) Retrospective commands. These are commands which can be used after the primary construction of the profile. A widely used command enables a blending radius or chamfer to be added to an existing corner. Smooth, another retrospective command can be used to reduce a large number of point-to-point coordinates which describe a shape into a smaller number of circular arcs.

(g) Manipulation commands. These are commands that, as the name suggests, provide a means of manipulating the shape of the profile, such as mirror and rotate etc.

As each set of coordinates are entered the shape being generated is displayed on the screen; the scale of the shape automatically changes so that the largest dimension of the drawing fills the screen. Additionally, as each set of coordinates are entered the change points are automatically labelled with numbers which are known as 'nodes'. The node numbers are automatically upgraded when the retrospective commands are entered.

Stage 2 Machining instructions

When the machining specification option is selected from the menu of the Pathtrace turning module, the screen first displays a list of post processors available within the computer's memory from which it is necessary to choose the one required for the particular control for which the part program is being produced.

The machining instructions consist of two sections: the initial set up; and the designation of the tool path.

Section 1   Initial set-up
In this section the coordinates of positions that will be used throughout the program are established. The various options are selected from a menu which is displayed on the screen.

    U Set machining units
    H Set tool home position
    T Set tool change position
    F Call geometry file
    B Set billet size
    E End initial set up
    M Material technology table

There are other options available in the initial set-up stage which are used for editing purposes. All the options are invoked (activated) by the entry of the respective letters i.e. to set machining units the letter U would be pressed on the keyboard of the computer, and the words machining units would be displayed on the screen and either M for millimetres or I for inches would be entered. Option M (select material technology table) must be activated if it is intended to use these tables to select speeds and feeds in the tool path designation stage.

Section 2   Designation of tool path
The machining instructions contain three types of information:

(a) *Positional (non-cutting)* This is data which can be either node numbers or coordinates, required so that the tool can be moved at rapid feed rate between material removal operations.
(b) *Operational* Information such as specifying the units of feed and speed (mm/rev or mm/min), switching coolant on or off, spindle direction forward or reverse, applying nose radius compensation.
(c) *Material removal* This information consists of feed moves and machining cycles etc. Additionally it is necessary to provide coordinates or node numbers of the start and end of the machining operation.

The post processor uses certain of the information to generate the required G and M codes for the operations to take place. The options available are displayed on two menus. From these menus the following commands are selected; some of the commands are used only once, others are used a number of times.

PA	Profile attributes	For programmers convenience when creating the program
FR/FM	Feed per rev/feed per min	Operational
FP/RV/ST	Spindle FORWARD/REVERSE/STOP	Operational
RA	Rapid move	Positional
TL	Toolchange	Operational
RH	Rapid to home	Positional
CH	Rapid to tool change	Positional
CO/CF	Coolant off/flood	Operational
PS/OS	Program or optional stop	Operational
TU	Simple turning cycle	Material removal and positional
RT	Rough turn area clearance	Material removal and positional
DW	Dwell	Operational
PR	Profiling cycle	Material removal and positional

PN/PY	Pathtrace comp OFF/ON	Operational
NC/CR/CL	Compensation OFF/RIGHT/LEFT	Operational
TH	Threading cycle	Material removal and positional
FE	Feed move	Material removal and operational
EM	End machining	Operational

All the options are invoked (activated) by the entry of the respective letters i.e. to detail a tool change the tool would be returned to the tool-change position by keying in the letters CH. The command 'rapid to tool change' appears on the screen and this has to be accepted by pressing the carriage return key or enter key ⟨CR⟩. The letters TL would be keyed in next and the command 'toolchange' would appear on the screen. This is accepted by ⟨CR⟩ and the screen then displays:

Tool TNR Turr Type

This requires information on the tool number as listed in the tool library, nose radius, turret station and tool type to be supplied (see explanation given previously for type of tool).

As the tool path is programmed, the movements of the tool resulting from the selected options would be shown on the monitor screen. The tool movements are shown in a different colour superimposed on the geometric shape of the profile. Movements at rapid feed are shown as a dotted line and movements at feed rate are shown as a continuous line.

Creation of part program

The commands, with explanations, for the creation of the drawing shown in Figure 13.24 are as follows. Note that at this stage it is not necessary to specify what the units (imperial or metric) of the dimensions are.

Stage 1   Construction of geometric shape
Input
0 0
   Definition of first point at $Z = 0$ and $X = 0$. The Pathtrace turning module always assumes that a Z value is entered before an X value. It is not usually necessary to enter the addresses Z or X.

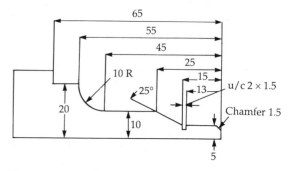

Figure 13.24   Component

LAST 5
   Point-to-point command using rectangular absolute units. A line is drawn from the first coordinate 0 0 to 0 5. LAST denotes that the Z value remains the same as that of the previous (last) coordinate. On a desktop computer with an enhanced 102 key keyboard it is possible to use the key with the name ⟨HOME⟩ printed on it to input the word 'last'. For the X coordinates it is possible to enter diameter values followed by a special symbol i.e. ⟨¦⟩. The computer than automatically halves the entered value for constructing the shape.

I−13
   Point-to-point command using incremental units; defining the end point of line to be at a coordinate with an incremental distance of $Z = -13$ and $X = 0$ from the previous defined node. The previous value of the X coordinate is assumed to remain the same and need not be entered.

I 0 −1.5
   Point-to-point command: similar to above. The incremental Z value of I 0 must be entered, because the software always requires a Z coordinate to be entered first.

I−2
   Point-to-point command: only the incremental Z value entered.

GB90Z−25X10B−25

Go command: constructs a line at 90 degrees from the existing point to intersect with a line constructed from coordinates Z−25 X10 at a bearing of −25 degrees. The GO command is one of the few occasions where the addresses Z and X have to be entered.

The sign of a bearing is taken with reference to the point from which the intersecting line is drawn; zero degree is at the three o'clock position. Anticlockwise angles are positive; clockwise angles are negative.

TC−45 20R−10

Point-to-circle command: constructs a line from the previous point (Z−25 X10) to a tangential point on the circumference of a circle whose centre is at Z−45 X20 with a radius of −10: the negative value indicates that the circle is to be constructed in a clockwise direction.

R−65 20

Circle to point command: constructs a line which leaves the circle radially and advances to the point Z−65 X20

LAST 50⟨⟩

Point-to-point command: constructs a straight line from the previous point to Z−65 X25. The mark ⟨⟩ indicates that the X value is a diameter value.

−68 Last

Point-to-point command.

E   End of primary profile construction.

C2X1.5

Retrospective chamfer command: chamfer node 2 with an incremental X dimension of 1.5. The chamfer will be normal to the bisector of the angle between the lines forming the corner to be chamfered.

BR11 0.25

Retrospective blend radius command: places a blending radius of 0.25 at node 11, this radius will remove a sharp corner as a safety feature. A blending radius can be placed at both external and internal corners.

BR15 0.25

Retrospective blend radius at node 15 similar to BR11 0.25 (safety feature).

E   End of profile definition.

Figure 13.25 shows the screen display at the end of the profile definition. After entering all the

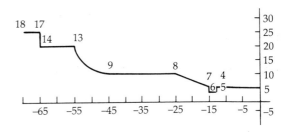

Figure 13.25   Screen display of geometric shape

geometrical details for the external shape of the component they would be saved under file names, one of which has the prefix −G− added by the computer, e.g.

DEMO13 Geometry command file for profile i.e. the entries keyed in by the part programmer.

−G−DEMO13 Geometry tabulation file for profile i.e. the Z, X and radius values generated by the computer.

Files with identifying letters in their name would have to be created if the component had an internal profile (bore) or a cast or formed billet; i.e. for external and internal profiles the filenames could be EDEMO13 and IDEMO13 respectively; a formed billet filename could be BDEMO13.

Stage 2   Machining instructions: section 1 − initial set up

Input command	Keyed entry	Purpose
U	M	Set machining units; (millimetres)
H		Set tool home position;
	Z X	The letters Z and X are displayed on the screen and
	5 30	the value of the coordinates are entered at the keyboard.

The position specified ensures that the tool is clear of the work but will not result in excessive non-cutting time during tool positional moves.

T   Set tool change position.

Z X   This position ensures that
100 50   the tools in the turret are clear of the work when it rotates so that the station required is in position. This position would also be suitable for manual tool changing.

F   Call geometry file.

DEMO13 This is the filename entered previously in the geometry stage.

B   Set billet size.

Z X   If a drawing of a billet had
0.5 25   been created its filename would be entered at this option. The Z value in the billet size is the facing allowance, the X value is the radius of the bar.

E   End initial set up.

After entering all the set-up details the geometric shape would be displayed on the screen in one colour, with the billet outline in another colour.

Stage 2   Machining instructions: Section 2 – designation of tool path

The machining commands for each operation with explanations are as follows:

Input command	Keyed entry	Explanation
PA		Profile attributes; accept with ‹CR›
FR C D(Y/N) L(Y/N)		(Displayed on the screen.)
1 1 Y Y		(Entered at the keyboard.)

This command is only required for convenience, and will not appear in the part program. With this command node numbers will be displayed on the profile which will be referred to in the machining sequences. FR is node number to start from; C is colour of profile; D is display profile; L is label profile with node numbers. The colours available are; 1 green; 2 dark blue; 3 light blue; 4 red; 5 yellow; 6 magenta; 7 white.

Operation no. 1: Load work stop

CH   Rapid to tool change; accept with ‹CR›

The carriage holding the tool turret would move at rapid feed to the tool change positions detailed in the initial set up i.e. Z100 X50 so that the tool change can take place.

TL   Toolchange: ‹CR›

Tool TNR Turr Type   (Displayed on the screen.)

123   1   7   (Entered at the keyboard.)

Call up tool number 123, a blank tool used as work stop. TNR is tool nose radius value and is not required. Turr is turret position. A type 7 tool has the orientation of the line from the set point of the tool and centre of the nose radius at the 45 degrees clockwise from the 12 o'clock position.

RA   Rapid move; ‹CR›

Z coord X coord   (Displayed on the screen.)

0.5   0   (Entered at the keyboard.)

The blank tool will move into position so that the length of the work can be adjusted. The Z value of 0.5 provides a facing allowance.

Operation no. 2: Adjust work to correct length

PS   Program stop; ‹CR›

This is necessary so that the post processor can insert a M00 in the part program so that the operator can carry out the adjustment of the length of the work. The program would be restarted by the operator pressing the start button on the machine control panel.

Operation no. 3: Load facing tool

CH	Rapid to tool change; ‹CR›
TL	Toolchange; ‹CR›
Tool TNR Turr Type	(Displayed on the screen.)
254 0.4 2 7	(Entered at the keyboard.)

Call up tool number 254, a facing tool. TNR is tool nose radius; a value is not necessary since compensation is not being applied but for completeness of entries a value of 0.4 entered. Turr is turret position; type 7 tool.

FR	Feed per rev; ‹CR›

This option selects the units of feed; mm/rev or mm/minute. Actual values of feed will be entered for each operation.

FO	Spindle forward; ‹CR›

This option will cause the spindle to rotate. Actual values of speed will be entered for each operation.

CF	Coolant flood ON; ‹CR›; coolant will stay on until turned off.

Operation no. 4: Face off end of work

RA	Rapid move; ‹CR›
Z coord X coord	(Displayed on the screen.)
1.0 26	(Entered at the keyboard.)

Rapid move to absolute coordinates $Z = 1$ and $X = 26$. This is a suitable start position for the facing cycle. A dotted line is shown on the screen from the tool change position to the coordinates defined. Node numbers can also be referenced by typing the number within single quotes.

TU	Simple turning cycle;

Z coord	X coord	Feed	Speed	Depth	PT	Rap	
							displayed
0	−1.0	0.2	800	1	Y		entered

The Z and X coordinates are the finishing coordinates. The feed and speed are taken from the sequence of operations. Y entered

under PT calls up a Pathtrace cycle. This option can be used either as a facing cycle or a straight turning cycle, the choice of where the tool moves first to the required depth of cut is made by selecting the drive and cut angles which are shown on the screen. For the facing option the tool moves first (cut) from the start position on the Z axis the required depth of cut '1 mm', then feeds (drive) along the X axis to face the work, retracts on the Z axis the depth of cut, and then moves rapid on the X axis to the start position. The cycle repeats until the entered Z and X coordinates have been reached. In this case there will be only one cycle since there is only 0.5 mm facing allowance. The tool moves back to the start position at the end of the cycle.

Operation no. 5: Load left-hand turning tool

CH	Rapid to tool change
TL	Toolchange
Tool TNR Turr Type	(Displayed on the screen.)
384 0.4 3 7	(Entered at the keyboard.)

Call up tool number 384, a left-hand turning tool. TNR is tool nose radius; value of 0.4 entered. Turr is turret position; type 7 tool.

Operation no. 6: Rough turn outside profile

RA	Rapid move
Z coord X coord	(Displayed on the screen.)
1.0 26	(Entered at the keyboard.)

This is to be the start position for the area clearance cycle.

RT	Rough turning area clearance G desc A2 4A7 18

F	SP	D	OZ	OX	PT(Y/N)	CB(Y/N)	RF(Y/N)	
								(Displayed on the screen.)
0.2	800	3	0.5	0.5	Y	Y	Y	(Entered at the keyboard.)

The G description (A2 4A7 18,) is the geometry details which define the start point and end point of the area of the profile to be cleared, the letter designates the profile and the numbers are the nodes i.e. from the start node A2 to 4 link to 7 and on to the end node 18. Nodes 5 and 6 are within the groove at the end of the threaded section. F is feed rate; SP is spindle speed; D is depth of cut; OZ and OX are the finishing allowances (offsets) in the Z and X axes respectively; PT(Y/N) use Pathtrace cycle yes or no; RF(Y/N) rapid moves linking cuts yes or no; CB(Y/N) consider billet form in tool path calculations yes or no. The work would be rotating at the stated speed (SP). The first move made by the tool is at rapid feed rate on the X axis to the stated depth of cut (D). The tool would then be fed along the Z axis at the feed rate (F) until a value equal to the finished profile less the finishing allowance (OZ) in the Z. The tool retracts at feed rate a distance equal to the depth of cut, and then returns on the Z axis at rapid feed to the start position of the cut. The cutting cycles continue until a full depth of cut cannot be taken, and then a profiling pass is made over the complete workpiece taking into account the finishing allowances in both the X and Z axes. The tool returns to the start position at the end of the cycle.

Operation no 7: Load grooving tool

CH				Rapid to tool change
TL				Toolchange
Tool	TNR	Turr	Type	(Displayed on the screen.)
510		4	0	(Entered at the keyboard.)

Call up tool number 510, a 2 mm wide grooving (undercutting) tool. No tool nose radius value required. Turr is turret position; type 0 tool.

Operation no. 8: Form undercut

RA		Rapid move
Z coord	X coord	(Displayed on the screen.)
−15	6.5	(Entered at the keyboard.)

This positions the undercutting tool ready for the feed move.

FE				Feed move
Z coord	X coord	Feed	Speed	
				displayed
−15	3.5	0.05	2000	
				entered

Tool moves to specified Z and X coordinates at feed rate to produce the undercut.

RA		Rapid move
Z coord	X coord	(Displayed on the screen)
−15	6.5	(Entered at the keyboard.)

This positions the undercutting tool clear of the work.

Operation no. 9: Load left-hand profiling tool

CH				Rapid to tool change
TL				Toolchange
Tool	TNR	Turr	Type	(Displayed on the screen.)
390	0.4	5	7	(Entered at the keyboard.)

Call up tool number 390, the finishing tool. The tool nose radius is essential as it will be required when tool nose radius compensation is applied during the finish profiling cycle. Turr is turret position; type 7 tool.

Operation no. 10: Finish turn outside profile

RA		Rapid move
Z coord	X coord	(Displayed on the screen.)
2	0	(Entered at the keyboard.)

This positions the profiling tool clear of the work ready for the finish profiling operation.

	Pathtrace compensa-
PY	tion ON

This activates compensation for the tool nose radius.

FE				Feed move
Z coord	X coord	Feed	Speed	
				displayed
0	0	0.1	1000	
				entered

Tool moves at feed rate to position the tool

at the start of the profiling cycle. This move also enables any compensation to be applied.

PR   Profiling cycle G desc

        A1 4A7 18+2

F SP OZ OX PT(Y/N)

           displayed

0.1 1000 0 0 Y

           entered

Profiling cycles are used for finishing the component and obtaining the required lengths and diameters. The group description which has to be entered are the geometry details of the start and end of the cycle and are specified by the profile letter and nodes e.g. A1 4A7 18+2, from A1 to 4 link to 7 and on to 18+2. The tool will feed 2 mm beyond node 18 to ensure that the full length is machined since compensation is being applied. At the end of the cycle the tool stops at the last node point (18+2) specified.

FE   Feed move

Z coord X coord Feed Speed

          displayed

−65   28 0.1 1000

           entered

Tool retracts at feed rate to a position clear of the work to enable compensation to be removed.

PN   Pathtrace compensation OFF

Operation no. 11: Load screw cutting tool

CH   Rapid to tool change

TL   Toolchange

 Tool TNR Turr Type (Displayed on the screen.)

 615   6 0 (Entered at the keyboard.)

Call up tool number 615 (the screw cutting tool). The tool nose radius is not entered. Turr is turret position; type 0 tool

Operation no. 12: Screw cut M10 × 1.5 mm pitch thread

RA   Rapid move

 Z coord X coord (Displayed on the screen.)

 2   5   (Entered at the keyboard.)

This positions the screw cutting tool ready for the threading operation.

TH   Threading cycle G Desc

        A3−3.5 4+1

P SP TD D PT(Y/N)

           displayed

1.5 1000 0.812 0.2 Y

           entered

The standard Pathtrace threading implementation is a plunge threading cycle capable of turning a single start cylindrical and tapered threads. The group description of the start and end nodes of the length to be threaded defined (A3−3.5 4+1,) starts the threading 3.5 mm away from node 3 and overruns node 4 by 1 mm: TD = total depth of threads: D = depth of first cut. The computer calculates the depth of cut for successive passes so that the volume of material removed per cut is approximately constant.

Operation no. 13: Remove finished component

CH   Rapid to tool change

ST   Spindle stop

CO   Coolant OFF

EM   End of machining

Figure 13.26 shows the screen display when all the details have been entered. The above details are saved under a filename prefixed by −T− i.e. −T−DEMO13. A part program has to be created for the particular machine tool and control.

Stage 3 Post processing

In order that the post processor can create the part program it may be necessary to provide additional information such as the type of chuck (jaw or collet) that is going to be used and distance from the collet face to the end of the component. The word addressed program is saved with prefix −$− i.e. −$−DEMO13.

Part programs generated by the post processor can be edited using the text editor option to provide any instructions required for operator intervention, or changing spindle speeds for operations which are part of a cycle. The text editor can also be used to create part programs manually.

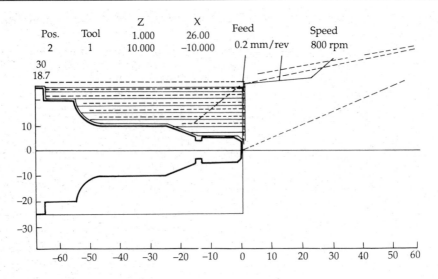

Pos.	Tool	Z 1.000	X 26.00	Feed	Speed
2	1	10.000	−10.000	0.2 mm/rev	800 rpm

Figure 13.26   Screen display showing machining simulation

CAM Example 2

The following section provides an example of a part program created with the Anicam CAM software Version 6.11 supplied by Anilam Electronics Corp. Acknowledgement is given to Anilam for the use of the manual and the software.

The software available from Anilam has a number of modules which can be selected from a menu, but the following example will only be concerned with the Interactive Lathe module.

The example program was written by the author and it is not intended to be taken as being recommended by Anilam. The purpose of this example is only to show an alternative technique of computer-aided part programming. Also there is no pretence that this one example will provide a complete explanation or coverage of all the options provided within the Interactive Lathe module in the Anilam software or is in any way a substitute for the training programme offered by Anilam. As with other software, to be proficient in creating part programs a knowledge of machining practice is essential together with practice and experience in the commands available. As explained previously, until expert system software is available the computer will obey the commands without evaluating the

logicality of the instruction. The importance of time spent planning the machining sequence, calculating speeds and feeds, selecting and identifying the tools to be used before beginning to use the CAM software on the computer cannot be stressed too strongly.

When the Interactive Program Entry – Lathe mode is entered the screen display is in two parts: the upper section – the graphics area with the two axes, where the shape of the component and machining operations are displayed; and the lower section – the text area where a number of lines of text are displayed. Figure 13.27 is a screen display showing the shape of the component after a program has been entered. The tool path movements can be seen as dotted lines using a tool mounted in a rear turret; a drilling operation is also indicated. The continuous line shown is the tool path movement for a finishing operation using a tool mounted in a front turret.

In the text area the first line of text listed details the main functions available for selection such as GENeral, FILE, WINDow, GEOMetry etc. As a function is selected a second line of text is displayed which details the various options available e.g. if GEOM is selected the elements P, L, C etc. are displayed; as the particular function is selected, it is displayed on

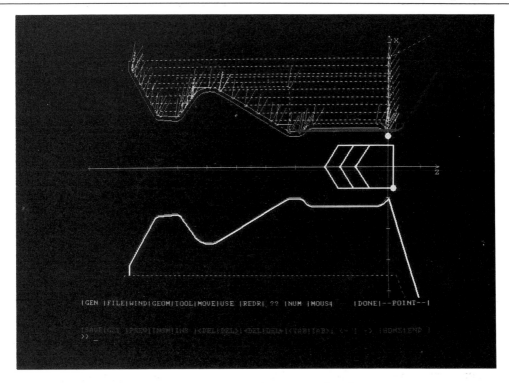

Figure 13.27   Screen display (courtesy Anilam Electronics Corp)

the bottom line. When all the items in the individual command are complete i.e. P1 = XA3.5 ZA0, it is entered by the return key and the point is displayed on the screen. The commands can be entered either by typing the letters and numbers at the keyboard or selecting the function required by moving the cursor using the mouse and by clicking a button on the mouse. The entry of the commands is mainly a keyboard skill or mouse manipulation skill which can be acquired with practice.

There are five steps in creating a part program using the Anilam software.

1   Designate each feature of an existing drawing as either a point, line or circle.
2   Set the graphics drawing area displayed on the monitor to a suitable scale so that component can be displayed.
3   Enter the values of the points, lines or circles into the computer.
4   Specify the machining instructions so that the tool movement can be simulated.

5   Convert the above into a part program for the CNC machine tool to be used (post process).

If an error is made in the context or syntax of a command during the entry of the command it will be detected by the software – a message will be displayed on the screen indicating the type of error and the command can be corrected. Errors in numerical values will not be detected but the position of the point or resulting movement can be seen on the screen.

Step 1   Designate the features of the finished shape of the component

The shape of an existing drawing is defined using points, lines and circles. There can be up to 400 points, 200 lines and 100 circles. Although it is not essential, it is convenient if all the points on each portion of the component to be machined are numbered consecutively. It is necessary to decide on the home position and/or the tool change position together with the start positions for the individual machining operations or cycles that will be used.

287

Step 2 Set the window on the screen

The units, inch or metric input and output should be specified before any other details are entered.

The scale of the graphics drawing area (window) should be set so that the component can be seen clearly at optimum size and the machining area can be displayed on the screen, without excessive wasted space.

Step 3 Define (enter) the numerical values of the points, lines and circles

As the values are entered the points (Pn), lines (Ln) and circles (Cn) are displayed on the screen.

It is possible to import drawings created with the use of CAD software that have been saved in a suitable format i.e. such as data exchange file (DXF) or initial graphics exchange system (IGES). However, it is not difficult, and can be more convenient, to originate the shape of the component on the screen by defining the points lines and circles.

Points can be defined in 16 different ways, ranging from Cartesian or polar absolute or incremental coordinates to the intersection of lines or circles.

A line can be defined in nine different ways, ranging from being parallel with an axis or at a tangent to a circle, to passing through a point at an angle or through two points. Consideration must be given to the direction of alignment of a line, as the path followed by a tool in the machining commands will be in the direction that the line is aligned.

A circle can be defined in 14 different ways, ranging from a circle of specified radius and its centre being at specified coordinates, to a circle of unknown radius tangential to three existing lines. A circle has to be designated as either clockwise or counter-clockwise, because the path followed by a tool around the circle in the machining commands will be in the direction that the circle is designated.

As with other CAM systems a graphical representation of the shape of the component has to be displayed on the screen to which the machining instructions can be related and the tool path simulated.

Step 4 Machining instructions

Before entering any cutting commands it is advisable that all the information on the tools to be used is specified. The tools required can be defined as part of the machining instructions and a tool table created. Alternatively, a tool table can be formed from tools selected from the tool library and material library which are provided as a facility within the Anilam software.

For some applications, only the tool number (#n) nose radius (R:n) and feed rate (F:n) need be defined; these details are essential. A full definition of a tool comprises the following:

TOOL #n R:n F:n S:n L:n Z:n X:n POSI:n DREG:n LREG:n TPA:n LEAD:n TRAI:n

where #n tool number:

R:n nose radius

F:n feed rate in mm/rev

S:n spindle speed in RPM

L:n tool length; in this case the length of the indexable insert in mm

Z:n and X:n are any tool offsets as determined during the presetting of the tools

POSI:n tool position number (from 0 to 8)

DREG:n and LREG:n are registers for diameters (X axis) and length (Z axis) offsets respectively

TPA:n is width of tool, only required for the grooving tool

LEAD:n and TRAI:n are tool leading and trailing angles respectively

If the machine tool control requires the units of feed rate to be a feed number or mm/min the feed rate will be changed from mm/rev by the post processor. Constant surface speed can be defined separately as a spindle speed command i.e. SPEE CSS:n.

It is necessary to designate the tool position (POSI:n) so that nose radius compensation can be applied. The tool position is defined as the position of the centre of the nose radius in relation to the set point. There are nine locations in which a tool can be positioned. Figure 13.28

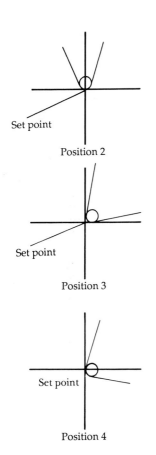

Set point

Position 2

Set point

Position 3

Set point

Position 4

Figure 13.28   Tool positions

shows positions 2, 3 and 4. Tool nose radius, length, width (TPA) and leading and trailing angles are required so that the tool shapes can be drawn on the screen to enable any possible tool interference to be checked.

The measurement of the leading angle is taken to a line extended from the tool face through the set point. Measurement of trail angles are taken to the trailing edge of the tool. Leading and trail angles are measured counter-clockwise from zero degree which is at the 3 o'clock position. Angles measured counter-clockwise are designated positive. Figure 13.29 shows the tools used in the program. A tool is loaded by the command 'MOUNT' followed by the tool number; the tool change (mount) should be made at

the HOME or tool change position.

The tool should be moved at rapid feed rate to the start position for the machining operation. The tool movement for machining can be made by the command FEED and specifying the points, lines or circles that the feed path has to follow. The interactive lathe option of the Anilam software provides a number of canned cycles such as turning, facing, screw threading, grooving and drilling.

Step 5   Post processing
The post processor for the particular control is selected after all the information entered is seen to produce the desired shape.

Creation of part program
The component is the same as that programmed in Chapter 12 and also CAM example 1 together with the same sequence of operations but different tool numbers; it must be noted that the particular sequence of operations planned is not the only sequence and there are a number of other ways that this component could be machined.

Step 1   Labelling the drawing
The drawing of the component is labelled with points, lines and circles as shown in Figure 13.30.

Step 2   Set the scale of the screen display
Input commands
DIM MM
The input and output units are set to millimetres.
WINDow XA‹0:40› ZA‹−70:40›
The size of the window created (0 to 40) on the X axis and (−70 to 40) on the Z axis will provide sufficient space for the tool change position (HOME) (X27 Z20) to be seen clear of the largest radius (25 mm) of the component and 20 mm from the end face of the component on the Z axis. The size of the window will also enable the length (65 mm) to be machined and the billet to be seen.

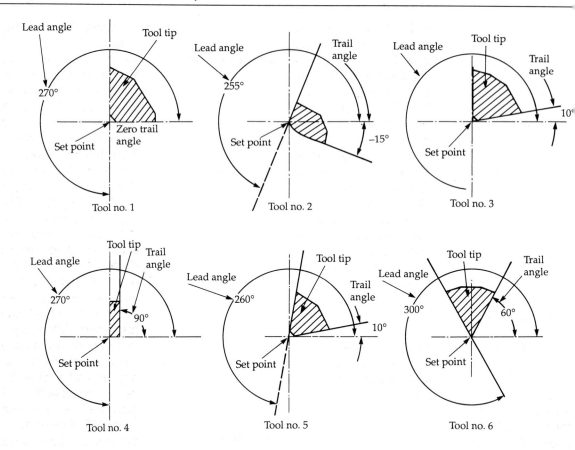

Figure 13.29   Tool shapes

Figure 13.30   Labelling of component

Step 3   Defining the geometry (points Pn, lines Ln and circles Cn)
Input commands
AXES

The axes are displayed.

All the points have been defined in absolute coordinates.

P1 = X 0 Z 0.5	P1 will be required when feeding the work to a stop. The Z value of 0.5 is a facing allowance.
P2 = X 0 Z 0	P2 is a point to which the tool will move when facing off the end of the component.
P3 = X 3.5 Z 3	P3 is the start point of the profile of the component to which the

tool can travel from the home position at rapid feed. The $Z3$ coordinate will ensure that the tool will not collide with the face of the billet at rapid rate and also provides a space for the cutting feed rate to be established at a steady rate before cutting metal.

$P4 = X\,3.5\;Z\,0$ — P4 to P12 are points along the profile to be machined omitting the undercut at the end of the section to be threaded.

$P5 = X\,5\;Z\,-1.5$

$P6 = X\,5\;Z\,-13$ — P6 is the end of the threaded section

$P7 = X\,5\;Z\,-15$ — P7 is a point on line L1 to which the tool will move when turning the profile.

$P8 = X\,10\;Z\,-25$

$P9 = X\,20\;Z\,-55$

$P10 = X\,20\;Z\,-65$

$P11 = X\,25\;Z\,-65$

$P12 = X\,25\;Z\,-67$ — End of the profile.

$P13 = X\,27\;Z\,-67$ — P13 is defined to meet the requirement that the end of a shape must terminate with an X axis movement and also so that the tool can be moved clear of the work.

$P14 = X\,26\;Z\,0$ — Start position clear of the billet for facing off of the profile.

$P15 = X\,7\;Z\,-15$ — Start position for grooving (undercutting)

$P16 = X\,3.5\;Z\,-15$ — End position (depth) for grooving, the groove is 2 mm wide and 1.5 mm deep; since the tool is only 2 mm wide there is no benefit to be gained

$L1 = P16/A\,90$

$L2 = P8/A\,155$

$L3 = P8\,/A\,180$

$L4 = P9\,/A\,90$

$L5 = P10\,/A\,180$

$L6 = P11\,/A\,90$

$L7 = P12\,/A\,180$

$C1 = CW\;L3\;L4\;R\,10$

in using the grooving cycle.

L1 is defined as a line passing through P16 at an angle of 90 degrees. Zero degree is at the 3 o'clock position and clockwise angles are in a positive direction.

L2 is a line passing through P8 at an angle of 155 degrees. The angle is specified as 155 $(180-25)$ so that the direction of alignment of the line is in the direction that the tool will be moving when the component is being machined. Alternatively, the angle could have been stipulated as 25 degrees and during the definition of the tool path the line L2 would be specified as $-L2$.

L3 and L4 have been defined so that the circle C1 can be defined.

L5, L6 and L7 have been defined so that a blending radius can be machined to remove sharp edges (safety considerations) at external corners.

The circle is designated as of clockwise orientation touching lines L3 and L4 and of 10 mm radius.

Figure 13.31 shows the display on the screen when the above commands have been entered.

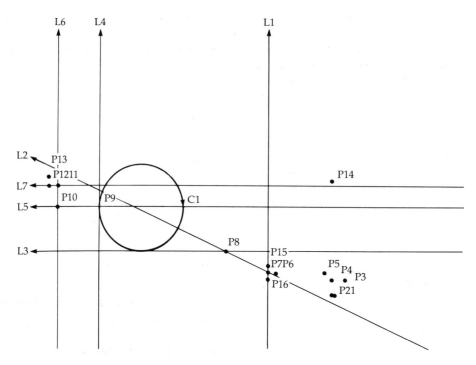

Figure 13.31   Screen display of geometry definition

Step 4   Machining instructions
Input commands

Initialization and tool definition

HOME X27 Z20

The location of the home position has been chosen so that the turret can index without any of the tools colliding with the work. For some machines the position of tool change has to be at a particular location, and for these machines the home position would have to suit the conditions required. A symbol (a cross within a circle) is displayed on the screen at the home position.

The following are tool definition statements, see above for explanation of terms.

TOOL #1 R 0.001 F 1.0 S 100 L12 Z 0 X 0 POSI 3
DREG 1 LREG 2 LEAD 270 TRAI 0

Work stop

TOOL #2 R 0.4 F 0.2 S 800 L 12 Z 0 X 0 POSI 4
DREG 5 LREG 4 LEAD 255 TRAI −15

Facing tool

TOOL #3 R 0.4 F 0.2 S 800 L 12 Z 0 X 0 POSI 3
DREG 5 LREG 6 LEAD 270 TRAI 10

Left-hand turning tool

TOOL #4 R 0.001 F 0.05 S 2000 Z 0 X 0 POSI 2
DREG 7 LREG 8 TPA 2 LEAD 270 TRAI 90

Grooving tool 2 mm wide

TOOL #5 R 0.4 F 0.1 S 1000 L 12 Z 0 X 0 POSI 0
DREG 9 LREG 10 LEAD 260 TRAI 10

Finish turn profiling tool

TOOL #6 R 0.188 F 1.5 S 1000 POSI 2 DREG 11
LREG 12 LEAD 300 TRAI 60

Screw cutting tool

ERASE

Clears the screen. This is required to remove excessive detail being shown on the screen.

AXES

Only the axes are redisplayed.

USE BILLET 75.5 50 75

Billet is displayed on the screen. The billet size has to be designated in relation to the machine datum which is at the face of the chuck. 75.5 is the length from the machine datum to the end of the billet; 50 is the outside diameter of the billet; 75 is the distance from the machine datum to the work datum.

Operation no. 1: Load work stop
MOUNt #1

Turret indexes and presents tool no. 1; the tool blank to be used as a work-stop to the working position. Note: only the letters MOUN need be entered, the tool will be positioned at the home location.

Operation no. 2: Adjust work to correct length
RAPid P1 [Z4]

Rapid approach of work-stop to 4 mm (Z4) away from P1 in a positive Z direction. The movement at rapid feed is shown on the screen as a dotted line. Only the letters RAPI need be entered.

FEED P1

Work-stop feeds to P1 (*X*0 *Z*0.5); feed movement is shown on the screen as a continuous line.

NOTE

Adjust work length to stop. Note required for operator intervention.

Operation no. 3: Load facing tool
HOME

Tool moves back to the home position at rapid feed.

MOUNt #2

Turret indexes and presents tool no. 2 (the facing tool) to the machining position.

COOLANT ON

Coolant is turned on.

Operation no. 4: Face off end of work
RAPId P14

Rapid approach of the facing tool to P14.

FEED P2 [X−1]

Tool moves to 1 mm beyond P2 (*X*0 *Z*0) to ensure that a pip is not left after facing off the billet.

Operation no. 5: Load left-hand turning tool
HOME

Tool returns to home position at rapid feed rate.

FIG S1 P3

The term FIG indicates that a shape is to be defined, and that it is to be called S1 with the starting point at P3

FEED P3 P4 P5 P7 L1 L2 L3 C1 L4 P10 P11 P12 P13

Designation of path to be followed by tool to

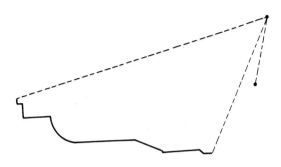

Figure 13.32   Shape S1

produce the desired profile. Figure 13.32 shows the shape S1.

ENDFIG

End of shape definition.

MOUNt #3

Turret indexes and presents tool n. 3 (the left-hand turning tool) to the machining position.

Operation no. 6: Rough turn outside profile
RAPId XA25 ZA1

Rapid approach to start position for rough turning the profile

TURN S1 @3 [X.5 Z.1] STEP 3

Activates the turning cycle to machine the shape S1 taking a depth of cut of 3 mm. The actual cutting will commence at the starting position *X*25 *Z*1 taking a 3 mm depth of cut until the profile defined in FIG S1 is reached but leaving a finishing allowance of 0.1 on the Z axis and 0.5 on the X axis. Tool nose radius compensation is automatically applied in the turning cycle.

Operation no. 7: Load 2 mm wide grooving tool
HOME

Tool returns to home position at rapid feed rate.

MOUNt #4

Turret indexes and presents tool no. 4 (the grooving tool) to the machining position.

Operation no. 8: Form groove (under-cut)
RAPId P15
   Rapid approach to P15; start position for grooving.
FEED P16
   Form groove at feed rate.
FEED XA8 ZA−15
   Retract tool at feed rate clear of groove.

Operation no. 9: Load finish turn profiling tool
HOME
   Tool returns to home position at rapid feed rate.
MOUNt #5
   Turret indexes and presents tool no. 5 (the profiling tool) to the machining position.

Operation no. 10: Finish turn profile (compensation to be applied)
RAPId P3 [X2]
   Rapid approach of profiling tool to 2 mm on the *X* axis from P3.
COMP CAM RIGHT
   Activate nose radius compensation.
FEED P3
   Feed to P3 to enable compensation to be applied.
FEED P3 P4 P5 P7 L1 L2 L3 C1 CCW 0.25 L5 L6 CCW 0.25 L7 L12 P13
   Finish turn (FEED) profile starting at P3. Between C1 and L5 and also between L6 and L7 a counter-clockwise 0.25 mm blending radius is machined to remove a sharp corner.
RAPId XA 28 ZA−67
   Rapid retract of tool to a position (*X*28 *Z*−67) clear of the work to enable compensation to be removed.
COMP CAM OFF
   Remove nose radius compensation.

Operation no. 11: Load screw cutting tool
HOME
   Tool returns to home position at rapid feed rate.
MOUNt #6
   Turret indexes and presents tool no. 6 (the screw-threading tool) to the machining position.

Operation no. 12: Screw cut M10 × 1.5 pitch
RAPId P5 [Z7.5 X2]
   Rapid approach to a position of 7.5 mm on the *Z* axis and 2 mm on the *X* axis away from P5.
USE PTHREAD 1.5 5 6 −0.812 0.257 0.8 4
   Activates screw cutting operation, thread has a pitch 1.5 starting at P5 ending at P6. Thread is 0.812 deep. The first depth of cut is calculated from: total depth of thread/square root of number of passes − 0.812/square root 10 (3.162) = 0.257. It is practical to take 10 passes for a 1.5 pitch thread. Depth of first cut is 0.257; 0.8 mm is pull out distance; and 4 finishing cuts.

Operation no. 13: Remove finished component and load new work piece
HOME
   Tool returns to home position at rapid feed rate.
COOLANT OFF
   Coolant turned off.
DONE
   End of machining instructions.

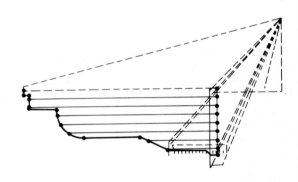

Figure 13.33   *Screen display of machining simulation*

As each command is entered the simulation is displayed on the screen, Figure 13.33 shows the screen display before the command DONE is entered. The program created in the interactive mode is saved under a file name such as DEMO13.IN and two other files are automatically created by the software when the interactive mode is exited DEMO13.GEO and DEMO13.PST. The file with the .GEO extension is used by the plotter program to plot just the AniCam geometry; the file with the .PST extension is to be post processed. The file name of the part program created by the post processor will have an extension to indicate the control for which it has been produced i.e. for a Fanuc control the file will be called DEMO13.FAN

Step 5 Post processing
The post processor is selected after all the information entered is seen to produce the desired shape. The post processor will also arrange the feed rate in the desired units i.e. as feed numbers or feed in mm per min. For an Easiturn CNC lathe the part program contains 148 blocks and for a Fanuc control a part program will contain 155 blocks. Tables 13.3 and 13.4 lists the blocks for the part programs for both of the controls. The tables list the blocks up to the first rough turning cut in Operation no. 6.

Table 13.3 *Word addressed part program for Easiturn CNC lathe*

N1G71L
N2G27X27.0Z20.0L
N3G27X50.4Z25.9L
N4M06T01L
N5M03S0100L
N6M08L
N7G00X0.0Z4.5L
N8G01Z0.5FL
N9G27X50.4Z25.9L
N10M06T02L
N11M03S0800L
N12M08L
N13G00X26.0Z0L
N14L
N15G01X–1.0FL
N16G27X50.4Z25.9L
N17M06T03L

N18M03S0800L
N19M08L
N20G00X25.0L
N21G00Z1.0L
N22G01X27.1Z2.7FL
N23G01X24.1FL
N24G01Z–64.9FL
N25G01X25.5FL
N26G01Z–66.9FL
N27G01X27.1FL
N28G00Z2.7L

Table 13.4 *Word addressed part program for Fanuc Control*

N1 G21
N2 G95
N3 G00 X54. Z20. T00
N4 G00 X100.8 Z25.9 T00
N5 T01
N6 G97 S100
N7 X0 Z4.5 M03
N8 G01 Z0.5 F1.
N9 G00 X100.8 Z25.9 T00
N10 T02
N11 G97 S800
N12 X52. Z0
N13 G01 X–2. F0.2 M08
N14 G00 X100.8 Z25.9 T00
N15 T03
N16 G97
N17 X50.
N18 Z1.
N19 G01 X53.2 Z2.6 F0.2
N20 X47.2
N21 Z–65.
N22 X50.
N23 Z–67.
N24 X53.2
N25 G00 Z2.6

The Anicam editor can be used to insert the program stop (M00) at the correct block in the part program to enable the operator to adjust the length of the work to the work stop.

Questions

13.1 Discuss the different ways that computers can be used in the production of part programs, detailing the advantages and limitations.

13.2 What is a part programming language?

13.3 Detail the three types of information which are normally required for a part programming language.

13.4 What is a post processor, and why is it required?

13.5 Detail the differences between conversational programming and computer-aided machining.

13.6 Specify the different types of information that are required for computer-aided machining.

13.7 Detail eight advantages of using computer-aided machining.

13.8 Explain how an RS232C interface can be used to connect a computer to a control unit of a CNC machine tool.

13.9 In connection with the linking of computers and control units, explain what is meant by baud rate and why it is important.

Preparatory functions: G codes

As explained in Section 8.6, preparatory functions are used as the means of inputting information into the control unit to perform an operation that is essentially non-dimensional in nature.

The preparatory function command consists of a letter and two digits. The letter used is upper case G (capital G). The two digits used can range from 00 to 99, and are a code or a signal which will cause different actions or operations to occur. For convenience it is common practice to refer to preparatory functions as G codes.

There have been attempts made to standardize the particular digits to be used to indicate specific functions. It will be found that the majority of the codes used are the same for the same function on different machine tools and different control systems. However, it is unlikely that all the G codes available will be used in one system on one machine tool. Reference will have to be made to the manual published by the manufacturer of the control system for the G codes available for a particular machine tool.

Not all the 100 codes available have been used for a standardized purpose. There are a number of codes which are provided for the manufacturer of the control system to designate for its particular use, and therefore it is possible for certain G codes to be used for different functions on different systems. Unfortunately it may be found that if a particular standard G code is not being used in the control system by a manufacturer, that G code is allocated for another purpose.

A number of G codes may be programmed in one line provided they do not conflict in their purpose. It is not necessary to keep repeating the majority of the commonly used G codes on every line of the program, as once entered their effect will remain until they are cancelled. These G codes are termed *modal*. There are however a small number of G codes which have to be entered for every line (operation) that they are required. These G codes are termed *non-modal*. The manual for the control system will provide details of which codes are modal and which non-modal. G codes that are modal are cancelled by programming another G code whose effect conflicts with that of the previous G code, or by programming a G code specifically to cancel the previous G code.

Four main groups or categories of preparatory function are used:

(a) To select the movement system
(b) To select the measurement system to use either metric or inch units
(c) To make compensation for variation in tool sizes
(d) To select a preset sequence of events (canned cycles).

The attempts to specify and standardize preparatory functions were made before 1965, and the machines then did not use computer numerical control. Since the development of computer numerical control (CNC) the need for a number of G codes no longer exists. Some of these G codes have been used on some control systems for other functions.

G00 This preparatory function causes the control system to operate in a positioning mode. The code is frequently programmed as G0 because leading zeros are normally suppressed.

Note This code should be read as G nought nought or G zero zero and not as 'go'. It is not normally necessary to program a feed rate when operating in the positioning mode (G00) on CNC machines, as the control unit automatically selects the maximum feed rate. However, on some of the older systems a feed rate has to be programmed.

G00 cancels G01, G02, G03, G06 and any of the canned cycles.

G01 This can be programmed as G1. It causes the control system to operate in a linear interpolation mode. At one time G01 was used to program 'normal' dimensions, but for computer numerical control systems G1 is now entered for all dimensions. (A normal dimension was specified as one between 0.9999 and 9.9999 inches; see G10 and G11 for 'long' and 'short' dimensions.)

It is essential to program a feed rate with G01 for movement to occur. However, feed rate is modal; once a feed rate has been entered in the control unit, regardless of the axis it was originally programmed for or if it was for linear or circular interpolation, it remains effective until changed.

G01 cancels G00, G02, G03, G06 and any of the canned cycles.

G02 This code (normally entered as G2) has to have a feed rate entered in the control unit for movement to occur. However, if a feed rate has been previously entered then it is not necessary to repeat the feed rate if the previous value is suitable.

When programmed with the required data, G02 will result in a movement in a clockwise direction around a circular arc. G02 can be used with:

(a) Multiquadrant or single-quadrant circular interpolation
(b) Polar coordinates
(c) Parabolic interpolation.

G02 cancels G00, G01, G03, G06 and any of the canned cycles.

G03 This is similar to G02 except that the movement is in a counter-clockwise direction.

G03 cancels G00, G01, G02, G06 and any of the canned cycles.

It is important to note the following:

(a) It is essential that one of the movement selection codes G00, G01, G02, G03 or G06 and a feed rate are programmed, or are in effect, in every operation (line) in order that movement of the work or tool can occur.

(b) The codes G00, G01, G02, G03 or G06 are modal, and once one of the codes has been programmed it remains in effect until cancelled by the entry of another. It is not necessary to keep repeating the code on every line.

(c) Additionally it is obvious that only one of the G codes used for the selection of the movement system can be entered on one line, or is in effect in an operation.

(d) The movement system programmed must be available on the machine tool.

G04 Enables the duration of a dwell (stop) to be programmed at the end of a machining operation to allow the cutting action to finish. The dwell may be part of a canned cycle, and the duration of the dwell has to be programmed before the start of the canned cycle in which the dwell is to occur. The duration of the dwell can be conveniently entered at the start of the program.

With some control systems the dwell is programmed as G04/n, where n is the duration of dwell required in units of 0.1 seconds. With some control systems the duration of the dwell is programmed as a number of spindle revolutions. When a dwell is not part of a canned cycle it has to be programmed at the same time as the operation it is part of. This G code is essentially non-modal as it will only be effective either as part of a canned cycle or when separately programmed.

G05 Creates a delay or hold on the execution of the program at the operation where the code is programmed. The difference between a dwell and a delay is that a dwell lasts for the programmed duration and then the program continues, whereas the delay is cancelled by the operator or by some other interlocking action occurring.

G06 The purpose of this code was not originally defined, but it is now used to specify parabolic interpolation. Generally movement can occur in any plane. The arc required is programmed by three points; the first point is assumed to be the end of the previous block, and the intermediate point and end point are programmed in successive blocks. G02 and G03 may be used in conjunction with G06 to indicate direction of movement.
G06 cancels G00, G01, G02, G03 and any of the canned cycles.

G07 Not used for a standard purpose; may be used by different manufacturers for different functions.

G08 Specifies a controlled acceleration of feed rate to programmed value.

G09 Similar to G08 except that it permits a controlled deceleration.
With the development of more efficient drive systems the need for G08 and G09 is limited.

G10 Linear interpolation 'long dimensions'. When the standard for the preparatory functions was being prepared it was found necessary to provide a G code for long dimensions. A long dimension was defined as a movement of between 9.9999 inches and 99.9990.

G11 Linear interpolation 'short dimensions'. Short dimensions are up to 0.9999 inches. The development of computer numerical control systems has tended to make the use of G10 and G11 unnecessary.

G12 Used when operating simultaneously in three axes, but the development of computer numerical control systems has tended to make its use unnecessary.

G13–G16 These codes are used when designating the particular axes on machine tools which have a number of separate machining heads.

G17 There are many control systems which are limited to operating over two axes only in the circular interpolation mode. G17 is used to designate the X and Y axes. For many control systems this code is automatically selected every time the power is switched on, and it is not necessary to enter the code except after programming G18 or G19.
G17 cancels G18 and G19.

G18 Designates the X and Z axes for circular interpolation control.
G18 cancels G17 and G19.

G19 Designates the Y and Z axes for circular interpolation control.
G19 cancels G17 and G18.

G20, **G21** Circular interpolation clockwise movements for long and short dimensions respectively. The development of computer numerical control systems has tended to make their use unnecessary.

G22–G29 Not used for a standard purpose; may be used by different manufacturers for different functions.

G30, **G31** Circular interpolation counter-clockwise movements for long and short dimensions respectively. The development of computer numerical control systems has tended to make their use unnecessary, and G30 and G31 are now used for other functions in some control systems.

G32 Not used for a standard purpose; may be used by different manufacturers for different functions.

G33–G35 For use when screw cutting on a lathe: G33 for constant lead conditions; G34 for increasing lead conditions; and G35 for decreasing lead conditions.

G36–G39 Not used for a standard purpose; may be used by different manufacturers for different functions.

G40 Cancels any cutter compensation applied: see G41 and G42. Cutter compensation can only be cancelled by G40.

G41, G42 Cutter diameter compensation. This can be used in one of three ways:
(a) It will compensate for variation between actual tool diameter and assumed diameter when programming.
(b) It is also possible to program the actual size of the work and apply tool compensation G codes to cause the centre of the tool to be offset automatically the required amount.
(c) It is possible to use different tool numbers for the same tool and, by applying different values of compensation, to use the tool to take a number of roughing cuts.
Code G41 is used when the centre of the cutter is to the left of the workpiece when looking in the direction that the tool is travelling.
Code G42 applies compensation when the centre of the tool is to the right of the workpiece when looking in the direction that the tool is travelling.

G43, G44 May be used when correction has to

be applied for radius on single-point lathe tools. G43 is positive tool offset, and G44 negative tool offset. The size of the nose radius has to be input similarly as in G41 or G42. The majority of control systems now use G41 and G42 for tool nose radius compensation.

G45–G52 Used for other applications of tool compensation, but may also be defined for specialist functions by the manufacturer of the control system.

G53 Cancels any linear shift values; see G54 to G59.

G54–G59 Creates a datum shift. The uses of these codes include:
(a) Tool length compensation
(b) Tool position compensation during turning
(c) Shifting the datum during applications such as pendulum milling. G92 is frequently used to specify the shift in the position of the datum.

G60–G62 Originally planned to be used during point-to-point operations, but computer numerical control has made their use largely obsolete.

G63 Used during tapping operations, but is now very rarely used.

G64 Changes the feed rate. Not required with computer numerical control systems.

G65–G69 Not used for a standard purpose; may be used by different manufacturers for different functions.

G70, G71 These are used to specify the units of the input dimensions: G70 inch units, G71 metric units.

G72–G79 Not used for a standard purpose; may be used by different manufacturers for different functions.

G80 Canned cycle cancel: see G81 to G89. The canned cycles can also be cancelled by any of G00, G01, G02, G03 or G06.

G81–G89 Activates canned cycles. These are predefined sequences of events (movements) stored in the memory of the control unit. The sequences are automatically repeated at programmed locations until cancelled. The use of canned cycles reduces the programming required to perform certain operations. Turning centres and milling/drilling machines use the same number G code to perform canned cycles which are essentially the same but different in context.

Note The cycles G81 to G89 are standardized canned cycles. Manufacturers of control systems may provide other cycles for their systems on other G codes which have not been assigned standard functions.

G81 This cycle is possibly the most frequently used. The tool is advanced at a cutting feed rate to a programmed position, returned to the start position at a rapid rate, and located at the next machining position at a rapid rate; the cycle then recommences. This cycle can be suitable for either drilling or turning.

G82 This cycle is as G81 but in addition has a dwell when the tool reaches its programmed position. Suitable for drilling a 'blind' hole, spot facing, or on lathes for turning to a shoulder.

G83 In this cycle the tool is advanced intermittently; otherwise it is the same as G81. The cycle is called 'peck drilling'. Suitable for deep hole drilling (a deep hole is one whose depth is more than five times its diameter).

G84 In this cycle the tool is advanced at a feed rate which is synchronized with the rotation of the spindle, and when it reaches its programmed Z position the spindle is reversed and the tool is returned to its starting position at the same controlled feed rate. The tool is repositioned and the cycle recommences. The cycle is used when tapping.

G85 With this cycle the spindle continues rotating, but the tool is fed in and out at the same controlled feed rate. Suitable for boring.

G86 This is also a boring cycle. The spindle is stopped at a specified position when the tool reaches the bottom of the bore, when an automatic rapid withdrawal occurs.

G87 This is a chip breaking cycle, where a small withdrawal of the tool occurs to break the swarf of free cutting material.

G88 The only difference from G87 is that a dwell occurs when the tool reaches its programmed position.

G89 With this cycle the only difference from G85 is that a dwell occurs at the bottom of the bore.

G90, G91 These are used to specify the type of input dimensions:
G90 absolute dimensions;
G91 incremental dimensions.

G92 Preset absolute registers. This enables the zero datum position to be changed as part of the program, during the running of the program.

G93 Not used for a standard purpose; may be used by different manufacturers for different functions.

G94 Specifies the feed rate in units of millimetres per minute.

G95 Specifies the feed rate in millimetres per revolution.

G96 Specifies the spindle speed as a constant cutting speed in metres per minute.

G97 Specifies the spindle speed in revolutions per minute.

G98, G99 Not used for a standard purpose; may be used by different manufacturers for different functions.

Note The units of the feed rate specified by G94 and G95 will be changed to inches per minute or inches per revolution respectively if G70 has been selected. Similarly the cutting speed specified by G96 will be specified in feet per minute if G70 has been selected.

Miscellaneous functions: M codes

As explained previously, miscellaneous functions are used in a similar way as preparatory functions to input information for operations which are essentially non-dimensional in character. The main difference between them is that generally preparatory functions are intended to be operative before any dimension instructions programmed in the same block are executed, whereas the majority of miscellaneous functions are usually operative after the motion statements programmed in the same block have been executed.

The code for a miscellaneous function consists of the capital letter M and up to two digits which can range from 00 to 99. Each pair of digits is a code for a signal to be sent from the control unit for some action or machine function to take place. For convenience, miscellaneous functions are frequently referred to as M codes.

As with preparatory functions, certain M codes have become standardized for specific actions. There are a number of M codes which have not been used for a standard function, and the manufacturers of the control system can allocate functions to these codes.

It will be found that usually only one M code is entered in a block. Certain M codes are modal and others are non-modal, similar to G codes. It is self-evident that both the control system and the machine tool must have facilities for the action being coded if the particular function is to

be effective. Generally there are fewer M codes than G codes provided by control systems.

M00 Program stop. When leading zeros are suppressed this code can be programmed as M0, which should be read as M nought or M zero. When this code is programmed it will result in the machine spindle and table movement stopping for some action to occur which cannot be programmed. The program can be continued when the operator presses a button or operates a switch on the control panel. This code is used when it is necessary for the operator to perform some function such as:
   (a) Removing a loose part such as a cover or a bearing housing from a work-piece being machined to allow further operations to take place on the work
   (b) Inspecting the workpiece
   (c) Checking the tool
   (d) Changing the position of the clamps.

M01 Optional program stop. This code can also cause the program to stop, but the decision whether to stop or not is controlled by the operator, who can activate an on/off switch on the control unit to execute the stop. The code is used for

303

such purposes as when the programmer decides that it may be necessary for the operator to check the work or the tool after a number of cuts have been taken on a workpiece. Code M1 would be programmed in the block that the optional stop is required.

M02 End of program. This code has to be entered at the completion of the workpiece and will result in all movements of the machine tables, spindles etc. coming to a halt. Any coolant flowing would be turned off. On some control systems for milling or drilling machines the Z axis would automatically move to its most retracted position without any Z axis coordinate being programmed. The program will automatically be reset to the start. On some control systems for numerical control using punched tape the tape may be rewound to the beginning.

M03 Clockwise rotation of main spindle. The direction of rotation is taken when looking along the main spindle (Z axis) towards the tool or workpiece. This rotation may be also be termed negative (−ve).

M04 Counter-clockwise rotation of main spindle. This rotation may be termed positive (+ve). Interpretation of direction as above.

M05 Stop rotation of main spindle.

M06 Tool change. This will enable the tools programmed as a T word in the same block to be changed. This code will result in all movements of the machine tables, spindles etc. coming to a halt. Any coolant flowing would be turned off. With manual tool change for milling or drilling machines the Z axis may automatically move to its most retracted position without any Z axis coordinate being programmed. With automatic tool change it may be necessary to move the work tables to a suitable position so that the tool

transfer mechanism has room to operate without the mechanism or tools striking the work table etc. The movement of the work tables has to be completed before the tool transfer occurs. On machines with manual tool change the operator would be responsible for activating the button or switch that would cause the program to continue after changing the tools; on automatic tool change machines the program would automatically continue after the tool change.

M07 Coolant on. This code would cause the coolant to be turned on in a mist form.

M08 Coolant on. The coolant would be turned on as a flood or continuous flow.

M09 Coolant off. Either the mist M07 or flow M08 above, or coolant numbers 3 and 4 entered by M50 or M51.

M10 Clamp on. This code will cause the clamping of the machine slides, spindle and workpiece automatically.

M11 Clamp off. The machine or work features clamped during an M10 function would be released.

M12 Not used for a standard purpose; may be used by different control system manufacturers for different functions.

M13 Clockwise (−ve) rotation of main spindle, and coolant on.

M14 Counter-clockwise (+ve) rotation of main spindle and coolant on.

M15 Rapid traverse of machine slides in positive (+ve) direction.

M16 Rapid traverse: of machine slides in negative (−ve) direction.

M17, M18 Not used for a standard purpose; may be used by different control system manufacturers for different functions.

M19 Oriented spindle stop. The main spindle would be stopped at a predefined angular position.

M20–M29 Not used for a standard purpose; may be used by different control system manufacturers for different functions.

M30 End of tape. The only difference between this code and M02 is that this code generates a signal that will cause the tape to be rewound or transfer input to another tape reader; otherwise the same actions occur as if M02 had been programmed.

M31 Interlock bypass. On some machine tools an interlock is provided to prevent the machine working if the interlock switch has not been activated. The interlock could be on a guard, but it may be necessary to bypass the interlock for some reason, such as when carrying out a proving run of the program.

M32-M35 Constant cutting speed. These are mainly used for facing operations on turning machines (lathes) to cause the spindle speed to reduce as the radius of the work being machined by tool increases or vice versa. Each code 32, 33, 34 and 35 is used for a different cutting speed.

M36, M37 Feed range 1 and 2. These select the feed range to be used. The development of machine controls has caused these codes to be obsolescent.

M38, M39 Spindle speed range 1 and 2. These codes are for selection of speed range, but as with M36 and M37 are now obsolescent.

M40–M45 Gear changes. The actual gear change to be used has to be specified in the instruction to the operator. If the machine does not have gear change facilities, these codes may be used for other functions.

M46–M49 Not used for a standard purpose; may be used by different control system manufacturers for different functions.

M50 Coolant number 3 on. On some machines a number of different coolant facilities are provided; for these machines this code would actuate the appropriate pump motor.

M51 Coolant number 4 on. As above but for another coolant.

M52–M54 Not used for a standard purpose; may be used by different control system manufacturers for different functions.

M55 Linear tool shift position 1. Enables the tool length offset to be changed to the first preset value.

M56 Linear tool shift position 2. As above but to a second preset value.

M57–M59 Not used for a standard purpose; may be used by different control system manufacturers for different functions.

M60 Workpiece change. Workpieces are changed automatically.

M61 Linear workpiece shift position 1. Enables the zero datum point to be changed to the first position detailed in the instructions to the operator. This position would be entered during the setting-up stage.

M62 Linear workpiece shift position 2. As above but to the second position.

M63–M67 Not used for a standard purpose; may be used by different control system manufacturers for different functions.

M68 Clamp workpiece. This is effective on machine tools with automatic clamping facilities. The work table would previously have to have been positioned at a suitable location.

M69 Unclamp workpiece. Enables work to be released and removed from the machine. This code and M68 would have to be used with M60.

M70 Not used for a standard purpose; may be used by different control system manufacturers for different functions.

M71, M72 Angular workpiece shift position 1 and position 2. Similar to M55 and M56 except that the move is angular and not linear.

M73–M77 Not used for a standard purpose; may be used by different control system manufacturers for different functions.

M78 Clamp slide. Only the axes not being activated would be affected.

M79 Unclamp slide. Cancels M78.

M80–M99 Not used for a standard purpose; may be used by different control system manufacturers for different functions.

Computer fundamentals

Development and characteristics of computers

Of necessity the explanation that follows is merely an introduction to the subject of computers. There are many textbooks that will give a more complete explanation of their construction and operation.

An electronic digital computer is a device that will obey instructions automatically at speeds ranging from 0.01 to 10 000 MIPS (MIPS is an acronym for 'millions of instructions per second). An electronic digital computer can be made to perform many tasks provided that the instructions and data can be defined and coded in a suitable form. The set of instructions for a computer is called a *program*.

The first electronic digital computer is reported to have been ENIAC (Electronic Numerical Integrator and Calculator), which was completed in 1946. This computer used thermionic valves, and all calculations and storage of numbers were carried out by means of electronic circuitry. The machine was very large and prone to break down frequently. ENIAC was capable of storing only 20 numbers, and consumed 150 kW of power.

There has been a tremendous improvement in reliability and an enormous reduction in size and cost with the development of electronic computers. The stages of computer development were as follows:

(a) In the late 1950s computers were built using separate transistors.
(b) Through the 1960s integrated circuits (ICs) were first developed, i.e. several electronic components were combined (integrated) on a small silicon wafer or chip. The process was known as small-scale integration, and later as large-scale integration (SSI and LSI).
(c) By the late 1980s hundreds of thousands of components were being combined (integrated) on a chip, in very large-scale integration (VLSI).
(d) It is forecast that in the 1990s computers will be capable of intelligent processing of knowledge as well as the present data processing.

At one time it was possible to classify computers as mainframe computers, minicomputers and microcomputers. However, because of the developments in computer technology it is now extremely difficult to subdivide computers into these classifications.

There are four characteristics of electronic computers that make them extremely useful for a wide range of applications:

Automatic operation Once a computer has been provided with the program and the data it will proceed to follow the instructions without any intervention or supervision.

Speed The speed of operation has been indicated above in the number of MIPS that can be performed by the computers.

Memory There are computers now with memory sizes ranging from 16K to 1000M. A kilobyte (K) is actually 1024 bytes but is usually taken as 1000 bytes, and a megabyte (M) is a million bytes. A byte is 8 bits, and a bit is one binary digit. All instructions and operations carried out in a computer are binary coded.

Accuracy It is impossible to state a value of the accuracy capability of computers; they will give answers either as exact as required, or meaningless. In the majority of cases the meaningless answer is due to incorrect data being entered. GIGO is a word used with computers which indicates this; GIGO means garbage in, garbage out.

These factors make computers a valuable and essential addition to engineering from design applications to sales control.

Computer components

There are six main sections in a computer system, as shown in Figure C.1:

Input
Memory
Arithmetic
Control
Output
Bus

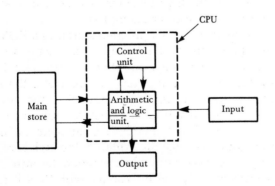

Figure C.1 Main parts of a computer

Input

The input unit receives information in binary coded format. The code widely used is ASCII (American Standard Code for Information Interchange). The most common format used for numerical control is in an eight-bit structure; there are computers with other bit structures. Information can be presented to the input unit from a variety of sources, such as direct from a keyboard, or from signals generated by process control equipment linked (interfaced) to the computer. Information can also be input from storage media such as punched tape, magnetic tape and magnetic discs.

Memory

Internal memory can be divided into two main types: read only memory (ROM) and random access memory (RAM).

Read only memory is where all the instructions of what to do with the input data are usually stored. In a ROM the information is stored by the actual pattern of connections between the transistors on a chip; the information is permanent and cannot be changed. The information stored on a ROM is non-volatile, and remains when the power is switched off. Where the computer has to perform only one task such as control of a machine, there is only one program in ROM and the computer is said to be dedicated. When the computer is for general purpose use and has to be capable of a variety of applications, there have to be a number of ROMs each with a program, or the information has to be fed into RAM from a backing store.

Random access memory can be both added (written) to and read from. The information stored in RAM is volatile, so the information will be lost when the power is switched off unless there is battery backup power supply. The part program data is stored in RAM. Most computers in control units have battery backup; this allows a RAM to hold a number of programs at a time, but each program will have to be identified. The RAM on computers in control units usually has a memory capacity of 16K to 64K. It is possible to delete (remove) and change programs stored in RAM.

There are other forms of memory, such as programmable read only memory (PROM) and erasable programmable read only memory (EPROM). All the ROMs etc. are contained within the computer, and are referred to as immediate access stores because they can be accessed directly and very quickly.

Arithmetic

The *arithmetic logic unit* (ALU) consists mainly of a series of logic gates, and is where all the calculations are carried out. The units are bistable, i.e. can be either on or off. This makes binary format particularly suitable as a language for the operation of computers, because binary has two symbols which can represent an on and an off state. An on state is represented by a high voltage and an off state by a low voltage.

Control

This unit controls the operation of the computer and causes data to be sent to and from all the sections of the computer. The control unit and arithmetic unit form the *central processing unit* (CPU). The control incorporates a *clock* whose speed determines the speed of the computer. All the instruction steps in pulses of high or low voltages are synchronized to the pulses from the clock. The CPU is a microprocessor in a microcomputer. Microprocessors are also used in the larger computers, but not as the CPU.

Output

This unit converts the information into the form required for the particular equipment to be driven by the computer.

Bus

All the units are connected by a number of parallel wires which are known as a bus. There are various buses in a computer. The *address* bus is used by the CPU to indicate from which part of memory the instructions are to be taken. Information is transmitted along the *data* bus to and from the CPU. The *control* bus is used for conveying instructions between the CPU and the various input and output devices.

There are usually 16 lines for the address bus, 8 lines for the data bus, 2 lines for the power supply and other lines for the control bus.

Operation of computers

For a computer to be able to carry out any task it has to be programmed. When a computer is switched on, the flow of the current will energize the CPU, and a series of checks will occur to ensure that all the circuitry etc. is working correctly. If everything is functioning correctly, an instruction stored in a ROM will be sent to the output unit to display a message on the screen. This series of checks and instructions is part of the operating system program. There are a number of operating system instructions, such as what to do when a program is to be loaded.

The programs stored within ROM and RAM are in a language known as *machine code*, which is different for each type of computer. Machine code is referred to as a low-level language because it is in binary format and close to the operational level. Machine code is fast in operation but difficult and tedious to write. To improve communication between human beings and computers, various machine independent languages known as *high-level languages* have been developed. There are many different high-level languages but, in order for the computer to be able to carry out a task programmed in such a language, the program has to be converted to a format which the CPU can cope with. There are two techniques of translating the high-level language: one uses an interpreter, the other a compiler.

An *interpreter* is a program stored in a ROM that picks up the English-type instructions of the high-level language when the program is loaded into the computer, and translates them line by line into machine code when the program is actually run. The operation of a computer is so fast that the CPU appears to be executing the program written in the high-level language. It is easy to edit a program, because it is translated by the interpreter each time it is run.

A *compiler* is a program that takes the program written in the high-level language and produces a new version of the program translated into machine code. The program written in the high-level language is known as the *source code*; the machine code program is known as the *object code*. It is the object code program which is loaded and run when required. A compiled program operates faster than a program that is interpreted. However, it is extremely difficult to edit a program in object code, and so if a program has to be amended it is necessary to go back to the program written in the source code and to create another object code program. When the computer is used for process control applications, the speed of operation becomes important.

There is another language between the high-level languages and machine code, known as an *assembly language*. This is a mnemonic code using symbols which are converted to machine code by a program known commonly as an *assembler*.

There are a number of terms used in conjunction with computers:

Hardware	The actual wires, chips, switches etc. that form the solid parts of the computer. The reliability of individual integrated circuit chips is extremely high; countless millions of operations can be performed accurately in the lifetime of a chip, which is over twenty years.
Software	The programs that are loaded into the memory to enable the computer to perform a particular application such as part programs.
Firmware	The programs that are stored on ROMs and are non-volatile.
Liveware	The people who staff and use computers.

Index